Aspects of the Life Record

Major Events	Dominant Forms[b]		Systems
	Homo		Quaternary[c]
			Tertiary[c]
Grasses become abundant	Mammals	Flowering plants	
Horses first appear			
Extinction of dinosaurs			Cretaceous
Birds first appear			Jurassic
Dinosaurs first appear	Reptiles	Conifer and cycad plants	Triassic
			Permian
	Amphibia		
Coal-forming swamps			Pennsylvanian
			Mississippian
		Spore-bearing land plants	Devonian
	Fish		Silurian
Vertebrates first appear (fish)			Ordovician
	Marine invertebrates	Marine plants	
First abundant fossil record (marine invertebrates)			Cambrian
Primitive marine plants and invertebrates			Precambrian
One-celled organisms			

Source: L. D. Leet, S. Judson, and M. E. Kauffman, *Physical Geology*, 6th ed., copyright © 1982 by Prentice-Hall, Inc., Englewood Cliffs, N.J. Reprinted by permission of Prentice-Hall, Inc.

[c]Some geologists prefer to use the term Cenozoic for the Quaternary and the Tertiary.

[d]In most European and some American literature Pennsylvanian and Mississipian are combined in a period called the Carboniferous.

[e]Subdivisions not firmly established.

GENERAL GEOLOGY FOR ENGINEERS

GENERAL GEOLOGY FOR ENGINEERS

Alan E. Kehew
Western Michigan University

Prentice Hall, Englewood Cliffs, New Jersey 07632

Library of Congress Cataloging-in-Publication Data

KEHEW, ALAN E.
  General geology for engineers / Alan E. Kehew.
    p.   cm.
  Includes bibliographies and index.
  ISBN 0-13-350406-9
  1. Engineering geology. I. Title.
TA705.K38 1988   87-30882
550—dc19   CIP

Editorial/production supervision: Kathleen M. Lafferty
Cover and chapter opening design: Meryl Poweski
Manufacturing buyer: Paula Benevento
Cover photograph: Courtesy of Chevron U.S.A., Inc.
Credits for chapter-opening photographs are listed at
  the end of the preface.

 © 1988 by Prentice-Hall, Inc.
A Paramount Communications Company
Englewood Cliffs, New Jersey 07632

Printed in the United States of America
10 9 8 7 6 5 4

ISBN 0-13-350406-9

Prentice-Hall International (UK) Limited, *London*
Prentice-Hall of Australia Pty. Limited, *Sydney*
Prentice-Hall Canada Inc., *Toronto*
Prentice-Hall Hispanoamericana, S.A., *Mexico*
Prentice-Hall of India Private Limited, *New Delhi*
Prentice-Hall of Japan, Inc., *Tokyo*
Simon & Schuster Asia Pte. Ltd., *Singapore*
Editora Prentice-Hall do Brasil, Ltda., *Rio de Janeiro*

To Dick and Betty Kehew, who set a good example in life to follow;
to Kay, for her constant love and support;
and to Melissa, Michelle, and Elizabeth, who tolerate my well-intentioned
but sometimes misguided performance as a father.

CONTENTS

Preface *xiii*

Chapter 1 Introduction *1*

    Geology as a science *2*

    Geologic time *8*

        Relative time *9*, Absolute time *11*

    Application of geology to engineering *14*

        Geology and the construction site *14*, Development of natural resources *16*,

        Water resources and the environment *18*

    The earth: a dynamic planet *19*

        The solar system *19*, Differentiation of the earth *19*

    Plate tectonics *22*

    Major features of the earth's surface *23*

    Summary and conclusions *23*

    References and suggestions for further reading *24*

    Problems *24*

PART I EARTH MATERIALS

Chapter 2 Minerals *25*

    The nature of minerals *26*

        Internal structure *26*, Crystals *30*

    Physical properties *32*

        Cleavage and fracture *32*, Hardness *34*, Color and streak *35*, Luster *35*,

        Specific gravity *36*, Other properties *36*

    Mineral groups *36*

        Silicates *37*, Oxides *45*, Halides and sulfides *45*, Sulfates and

        carbonates *46*, Native elements *46*

    Summary and conclusions *47*

    References and suggestions for further reading *47*

    Problems *47*

Chapter 3   Igneous Rocks and Processes   *49*
            Properties of igneous rocks   *50*
                Texture   *50*,   Color and composition   *52*
            Volcanism   *55*
                Volcanoes and plate tectonics   *55*,   Eruptive products   *56*,   Types of eruptions   *59*,
                Remnant volcanic landforms   *63*
            Intrusive processes   *65*
                Types of plutons   *65*,   Crystallization of magmas   *68*
            Engineering and igneous rocks   *69*
            Case study 3-1   Volcanic hazards: Mount St. Helens   *70*
            Summary and conclusions   *73*
            References and suggestions for further reading   *74*
            Problems   *74*

Chapter 4   Sedimentary Rocks and Processes   *75*
            Origin of sedimentary rocks   *76*
            Characteristics of sedimentary rocks   *79*
                Texture   *79*,   Sorting   *85*,   Sedimentary structures   *85*,   Color   *90*,
                Fossils   *91*,   Stratigraphy   *91*
            Engineering in sedimentary rocks   *94*
            Case study 4-1   Radioactive waste disposal in sedimentary rocks   *94*
            Summary and conclusions   *98*
            References and suggestions for further reading   *99*
            Problems   *99*

Chapter 5   Metamorphic Rocks and Processes   *101*
            Metamorphic processes   *102*
                Types of metamorphism   *103*
            Characteristics of metamorphic rocks   *107*
            Engineering in metamorphic rock terrains   *112*
            Case study 5-1   Failure of the St. Francis Dam   *112*
            Case study 5-2   Tunneling problems in metamorphic rocks   *115*
            Summary and conclusions   *116*
            References and suggestions for further reading   *117*
            Problems   *117*

Chapter 6   Mechanics of Earth Materials   *119*
            General types of earth materials   *120*
                Rocks, soils, and fluids   *120*,   Phase relationships   *120*
            Stress, strain, and deformational characteristics   *123*
                Stress   *123*,   Deformation—response to stress   *124*
            Engineering classification of intact rock   *130*
                Igneous rocks   *132*,   Sedimentary rocks   *132*,   Metamorphic rocks   *134*
            Rock-mass properties   *135*
            Engineering properties of soils   *136*
                Index properties and classification   *136*,   Shear strength   *142*,   Settlement and
                consolidation   *143*,   Clay minerals   *147*
            Case study 6-1   Bearing capacity failure in weak soil   *148*
            Case study 6-2   The Gros Ventre slide: Role of discontinuities in rock-mass
                stability   *149*
            Summary and conclusions   *149*
            References and suggestions for further reading   *151*
            Problems   *151*

PART II   INTERNAL GEOLOGIC PROCESSES

Chapter 7   Earthquakes   *153*
Occurrence   *155*
Elastic rebound theory   *155*,   Detection and recording   *157*,   Magnitude and intensity   *159*
Seismic waves   *160*
Wave types   *160*,   Time-distance relations   *162*
Earthquake hazards   *164*
Seismic risks and land-use planning   *168*
Location of earthquakes   *169*,   Ground motion   *170*
Earthquake engineering   *174*
Earthquake prediction   *178*
Case study 7-1   The San Fernando earthquake   *181*
Summary and conclusions   *185*
References and suggestions for further reading   *186*
Problems   *187*

Chapter 8   Heat Flow, Gravity, and Magnetism   *189*
The earth's internal heat   *190*
Heat flow   *190*,   Variations in heat flow   *193*
Gravity   *194*
Measurement   *195*,   Isostasy   *197*
Magnetism   *198*
The earth's magnetic field   *198*,   Paleomagnetism   *199*
Summary and conclusions   *201*
References and suggestions for further reading   *202*
Problems   *202*

Chapter 9   Crustal Deformation and Plate Tectonics   *203*
Mechanics of deformation   *204*
Measurement and interpretation of structure   *206*
Strike and dip   *207*,   Geologic maps   *208*
Folds   *209*
Parts of a fold   *210*,   Types of folds   *213*
Fractures   *213*
Joints   *214*,   Faults   *215*
Plate tectonics   *220*
Continental drift   *220*,   The theory of sea-floor spreading   *222*,   Plates and plate margins   *223*
Case study 9-1   Effects of rock structure on dam construction   *225*
Summary and conclusions   *227*
References and suggestions for further reading   *228*
Problems   *228*

PART III   EXTERNAL GEOLOGIC PROCESSES

Chapter 10   Weathering and Erosion   *229*
Mechanical weathering   *231*
Chemical weathering   *234*
Stability   *236*
Rock weathering and engineering   *238*

Erosion *238*
   Erosion by water *239*,   Erosion by wind *245*
Case study 10-1   Erosion, sedimentation, and reservoir capacity *246*
Summary and conclusions *247*
References and suggestions for further reading *248*
Problems *248*

Chapter 11   Soils, Soil Hazards, and Land Subsidence *249*
The soil profile *250*
Soil-forming factors *252*
Soil classification *254*
Soil hazards *255*
   Expansive soils *256*,   Hydrocompaction *259*,   Liquefaction *261*
Land Subsidence *261*
Summary and conclusions *265*
References and suggestions for further reading *266*
Problems *266*

Chapter 12   Groundwater *267*
Groundwater flow *268*
   Darcy's Law *268*,   Darcy's Law under field conditions *270*,
   The water table *270*,   Groundwater flow systems
   and flow nets *272*,   Recharge and discharge processes *274*
Groundwater resources *277*
   Aquifers *277*,   Production of water from aquifers *279*,
   Geologic setting of aquifers *282*
Groundwater contamination *286*
   Natural groundwater quality *286*,   Generation and movement of contaminants *287*,
   Attenuation mechanisms *290*,   Waste disposal *291*,   Other sources of
   groundwater contamination *294*
Groundwater and construction *294*
Case study 12-1   Groundwater mining and the High Plains Aquifer *295*
Case study 12-2   Industrial waste disposal: A threat to groundwater
   quality *296*
Summary and conclusions *298*
References and suggestions for further reading *298*
Problems *299*

Chapter 13   Mass Movement and Slope Stability *301*
Types of slope movement *302*
   Falls and topples *303*,   Slides *303*,   Lateral spreads *306*,   Flows *309*,
   Complex slope movements *312*
Causes of slope movements *316*
   Geological setting of slopes *316*,   Influence of water *318*,   Processes that reduce
   the safety factor *320*
Slope stability analysis and design *323*
   Recognition of unstable slopes *323*,   Stability analysis *325*,   Preventive and
   remedial measures *326*
Case study 13-1   Landslide remediation *328*
Summary and conclusions *329*
References and suggestions for further reading *330*
Problems *330*

Chapter 14   Rivers   *333*

   River basin hydrology and morphology   *334*
      The drainage basin   *334*,   Stream patterns   *334*,   Stream order
      and drainage density   *336*,   The hydrologic budget   *337*,   Development of
      drainage networks   *338*

   Stream hydraulics   *340*
      Flow types   *340*,   Discharge and velocity   *342*

   Stream sediment   *344*
      Entrainment   *344*,   Transportation   *346*

   Depositional processes   *347*
      Meandering streams   *348*,   Braided streams   *352*,   Alluvial fans   *355*,
      Deltas   *356*

   Equilibrium in river systems   *358*
      The graded stream   *358*,   Causes and effects of disequilibrium   *359*

   Flooding   *363*
      Flood magnitude   *364*,   Flood frequency   *366*,   Effect of land-use changes   *367*,
      Flood control   *367*

   Case study 14-1   Future shift of the Mississippi River?   *369*
   Case study 14-2   Flooding and land use   *370*
   Summary and conclusions   *372*
   References and suggestions for further reading   *372*
   Problems   *373*

Chapter 15   Oceans and Coasts   *375*

   The ocean basins   *376*
      Oceanic circulation   *376*,   Topography   *376*,   Oceanic sedimentation   *378*

   Continental margins   *379*
      Topography   *379*,   Turbidity currents   *380*

   Shorelines   *381*
      Waves   *381*,   Tides   *385*,   Longshore currents   *388*,   Coastal morphology   *389*,
      Beaches   *392*,   Types of coasts   *394*,
      Sea-level changes   *396*

   Coastal management   *398*
      Coastal hazards   *398*,   Coastal engineering structures   *402*,   Implications for
      coastal development   *404*

   Case study 15-1   Human interference with coastal processes   *405*
   Case study 15-2   Where not to build a swimming pool   *406*
   Summary and conclusions   *407*
   References and suggestions for further reading   *408*
   Problems   *408*

Chapter 16   Glacial Processes and Permafrost   *409*

   Glaciers   *411*
      Mass balance   *411*,   Movement   *412*,   Erosion   *414*,   Deposition   *415*

   Valley glaciers   *420*
   Continental glaciers   *424*
   Engineering in glaciated regions   *427*
   The Pleistocene Epoch   *429*
   Permafrost   *431*
      Extent and subsurface conditions   *431*,   Freeze and thaw   *431*,   Mass
      movement   *434*,   Engineering problems in permafrost regions   *436*

   Case study 16-1   Subsurface complexity in glacial terrain   *437*
   Summary and conclusions   *438*
   References and suggestions for further reading   *439*
   Problems   *439*

Index   *441*

PREFACE

The content and technical level of a geology text for engineers is a matter of broadly differing opinion. The type of geology course offered to engineering students at universities around the United States varies from introductory physical geology to engineering geology. This situation results from differences in philosophy, engineering curricula, and geology faculty available to teach such a course. This text is a compromise, based on my own experience, between these extremes.

The concept for *General Geology for Engineers* evolved from a course of the same name that I taught at the University of North Dakota. The enrollment included civil and geological engineering students who were required to take the course and mechanical and miscellaneous other engineering majors who took the course as a science elective. This course was the only geology course in the civil engineering curriculum, but, for the geological engineering majors, it had to serve as a foundation for numerous future geology courses. The academic level of the students varied; however, the majority of them completed the course in their sophomore year. As a result, most had taken little, if any, engineering mechanics, soil mechanics, or hydrology.

These conditions made structuring the course very difficult. My basic approach was to cover the essentials of physical geology and simultaneously introduce the students to the engineering and environmental applications of geology. In order to do this, many of the traditional topics of physical geology were omitted or only briefly discussed. On the other hand, it was apparent that students could not gain a full understanding of the applied aspects of geology without a good foundation in rocks, minerals, and structural geology. Several physical geology texts were tried, and in addition an engineering geology text was used one year. All these proved to be unsatisfactory because they failed to cover part of the desired course content. Students requested a reference other than the lecture presentations for either the basic geology or the engineering aspects, depending on which type of text was used.

I eventually concluded that the only solution for this course was to write a text combining the two areas at an introductory level. *General Geology for Engineers* is the result. Its objective is to provide a reasonably comprehensive introduction to physical geology and also to demonstrate the importance of geology to engineers by including introductory mechanics, hydraulics, and case studies that illustrate interactions between geology and engineering. The approach is mostly nonquantitative, although appropriate basic formulas are introduced where necessary. Enough material is present so that individual instructors can use the book to achieve their own personal compromise between physical geology and engineering geology.

It would have been impossible to write this text without the support of numerous individuals and organizations. My colleagues and supervisors in the department of Geology and Geological Engineering at the University of North Dakota and in the North Dakota Geological Survey provided help and encouragement in many ways. Don Halvorson, former Geology Department chairperson and state geologist, maintained a stimulating environment for research and creative activity in the department. Others to whom I am particularly grateful include Debby Kirby, for word processing assistance, and Ken Dorsher, for drafting and artistic contributions. My new colleagues at Western Michigan University have been equally supportive of this project.

The editors and staff at Prentice Hall have been remarkably tolerant of my publishing inexperience. These include Dave Gordon, who provided initial guidance, and Holly Hodder, who very capably brought the project to completion. Special thanks are due to Kathleen Lafferty, production editor, who made the endless production process seem almost bearable.

I feel a great debt of gratitude to individuals and organizations who provided photographs, permissions, and other materials for the book and to those who reviewed all or part of the manuscript. Among the former, numerous employees of the U.S. Geological Survey and the USDA Soil Conservation Service were exceptionally prompt, courteous, and helpful. As to the latter, the manuscript was improved considerably by suggestions and comments of those who critically reviewed successive drafts. These astute colleagues include Lawrence Herber, California State Polytechnic University; John Lewis, George Washington University; John Lemish, Iowa State University; John Williams, San Jose State University; Gary Robbins, Texas A&M University; Daniel P. Spangler, University of Florida, Gainesville; Charles Baskerville, U.S. Geological Survey; Robert B. Johnson, Colorado State University; Stan Miller, University of Idaho; Brian Stimpson, The University of Manitoba; and John M. Sharp, Jr., The University of Texas at Austin.

CHAPTER OPENING PHOTOGRAPHS

Chapter 1   The Grand Canyon. (J. R. Balsley; U.S. Geological Survey.)
Chapter 2   Quartz crystals. (A. E. Kehew.)
Chapter 3   Devil's Postpile, California. (University of Colorado.)
Chapter 4   Bryce Canyon, Utah. (R. A. Kehew.)
Chapter 5   Migmatite. (C. C. Hawley; U.S. Geological Survey.)
Chapter 6   The leaning tower of Pisa. (Italian Government Tourist Office, E.N.I.T.)
Chapter 7   Fault scarp formed during the 1959 Hebgen Lake earthquake, Montana. (I. J. Witkind; U.S. Geological Survey.)
Chapter 8   The Geysers geothermal area, California. (U.S. Geological Survey.)

Chapter 9   The San Andreas fault, California. (R. E. Wallace; U.S. Geological Survey.)

Chapter 10   Exfoliation in granitic rocks. (N. K. Huber; U.S. Geological Survey.)

Chapter 11   Damage to an apartment building by expansive soil, Texas. (USDA Soil Conservation Service.)

Chapter 12   Comal springs, Texas. (A. E. Kehew.)

Chapter 13   Failure of quick clay, Nicolet, Québec. (H. Laliberté; Gouvernement du Québec.)

Chapter 14   River terraces, Montana. (A. E. Kehew.)

Chapter 15   Development on a barrier island. (U.S. Geological Survey.)

Chapter 16   A terminal moraine deposited by a mountain glacier, Montana. (A. E. Kehew.)

CHAPTER
1
INTRODUCTION

People study geology—the science of the earth—for different reasons. Even among those who become professional geologists, there is a great diversity in purpose and objective. For engineers, an understanding of geology is a tool that is utilized in the design and construction of any structure that is in contact with the earth or that interacts with its natural materials in any way. Unfortunately, the application of geology to engineering is not a simple task. One cannot refer to a manual for the solution to a geological problem. Instead, the solution must come from experience and judgment—qualities that are based on a sound knowledge of the fundamental principles of geology. For this reason, the engineer must approach geology as he or she approaches chemistry or physics—as a science that must be assimilated by thorough and systematic study. We will begin our study of geology accordingly, by examining its origins and foundations as a science.

GEOLOGY AS A SCIENCE

The growth of the science of geology has changed the way that we regard ourselves and our environment on the planet Earth. Modern western civilization evolved with a strong reliance on literal interpretations of the opening chapters of the Bible. Early scientists applied these concepts to their observations of the natural environment. Accordingly, the earth was thought to be quite young, and its formation was attributed to a series of events that were rapid and cataclysmic in intensity. This school of thought, which became known as *catastrophism*, held that the earth's surface was generally a stable, nonchanging place that had remained quiescent except for violent upheavals during which mountains were thrust upward from the plains and canyons were formed as the earth's surface was ripped apart. The related theory of *neptunism* involved the presence of a primordial ocean from which rocks at the earth's surface precipitated. There was considered to be a specific sequence of precipitation as the ocean level dropped in order to account for observations of rocks that were made in western Europe.

An important early contribution to geology was made by Nicolaus Steno (1638–1687), a Danish physician working in Italy. Steno observed the rocks exposed in cliffs and stream banks and also studied the *sediment*, or particles of rock and soil being deposited in streams and in the Mediterranean Sea. He realized that the exposed rocks had many similarities to the modern sediments and had, in fact, originated as sediments at some time in the distant past. To formulate these ideas, Steno used true scientific thought because he recognized the order in nature.

Steno also stated three principles that are fundamental to modern geology. First, the *Principle of Original Horizontality* states that layers of sediment are always deposited in horizontal sheets as they can be observed to be accumulating on the bed of a stream or ocean (Figure 1-1). Thus when we see layers, or *beds*, of rock which appear to be tilted or folded (Figure 1-2), we must realize that subsequent events have deformed the originally horizontal beds. The *Principle of Original Continuity* states that accumulations of sediment are deposited in continuous sheets up to the point where they terminate against a solid surface (Figure 1-1). There is no better illustration of these two principles than in the Grand Canyon (Figure 1-3), where individual horizontal rock layers can be traced for miles along the side of the canyon. Steno's third principle is known as the *Principle of Superposition*, which states that in a vertical sequence of sedimentary rock layers, the oldest layer lies at the base and each successive layer above is younger than the bed it overlies. Although

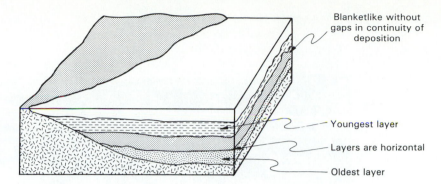

Blanketlike without gaps in continuity of deposition

Youngest layer

Layers are horizontal

Oldest layer

Figure 1-1 Principles proposed by Nicolaus Steno. These principles are essential to an understanding of sedimentary rocks. (From E. A. Hay and A. L. McAlester, *Physical Geology: Principles and Perspectives*, 2d ed., copyright © 1984 by Prentice-Hall, Inc., Englewood Cliffs, N.J.)

Figure 1-2 Beds of sedimentary rocks that have been bent into a fold from their original horizontal orientation. (A. Keith; photo courtesy of U.S. Geological Survey.)

this principle may seem intuitively obvious to us, in Steno's era it required acceptance of the radical idea that many rocks are formed by gradual deposition of sediment, the same processes that can be observed to be occurring in a stream, lake, or ocean.

The final great leap into the modern era of geology can be attributed to the ideas of James Hutton, an eighteenth-century Scottish farmer. Hutton challenged the ideas of catastrophism and neptunism by proposing that the earth's surface is not a stable, unchanging landscape. Instead, changes are occurring continuously, although mostly at very slow rates. Therefore, valleys are formed by the slow erosion of streams that flow within them and mountains are uplifted gradually to their elevated positions. Hutton also suggested that

Figure 1-3   In the Grand Canyon, horizontal beds of sedimentary rock can be traced for great distances. (J. R. Balsley; photo courtesy of U.S. Geological Survey.)

the rocks exposed on the landscape had formed by the same gradual processes operating in the past.

These ideas are incorporated into the *Principle of Uniformitarianism*, the principle that the physical and chemical laws governing natural processes have been constant throughout the history of the earth, and its evolution has been continuous rather than catastrophic. This principle does not mean that the rates of natural processes are always constant; we know that erosion of a river valley takes place at a much faster rate during a flood than during a normal flow period. Thus, although the processes may vary in intensity and rate in time and space, they are still controlled by the unchanging natural laws of physics and chemistry. The Principle of Uniformitarianism is often summed up by the phrase "the present is the key to the past," because it gives geologists the ability to interpret the origin of ancient rocks in terms of natural processes that can be observed in progress at the present.

Hutton's idea of uniformitarianism required an acceptance of the great antiquity of the earth. A single rock exposure along the coast of Scotland emphatically drove this point home. The exposure consisted of horizontal beds of rocks resting upon tilted rocks of a different type (Figure 1-4). Using Steno's principles, Hutton realized that this exposure must represent a complex sequence of events that occurred over a great length of time. First, the lower sequence of rock beds was deposited in horizontal layers of sediment (Figure 1-5). Next, these sediments hardened into rocks and became deformed and tilted into their present orientation. Deformation of this type is associated with major movements of the earth's crust, the type that may result in lateral compression and vertical uplift of rocks to form a mountain range. It is interesting to imagine the awe that Hutton must have felt as he began to realize that the next event in the history of that outcrop must have been a period of erosion representing an enormous amount of geologic time. A mountain range

Figure 1-4   The exposure of rock at Siccar Point, Scotland, studied by James Hutton. The unconformity (plane marked by lens cover) represents a long period of geologic time between the formation of the vertical rock beds in the lower part of the outcrop and the nearly horizontal rock beds overlying them. The geologic events involved in the history of these rock units are shown in Figure 1-5. (Photo courtesy of John Bluemle.)

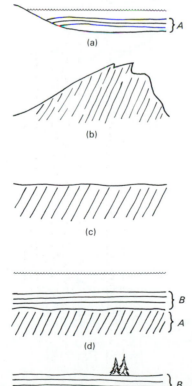

(a)

(b)

(c)

(d)

(e)

Figure 1-5   Generalized sequence of events comprising the geologic history of the rock exposure shown in Figure 1-4. (a) Deposition of the rocks of sequence *A*. (b) Deformation and uplift. (c) Erosion of the rocks of sequence *A*. (d) Deposition of the rocks of sequence *B* forming an unconformity between sequences *A* and *B*. (e) Uplift and erosion of both rock sequences as a unit.

must have been gradually worn down by erosion to a low-lying plain on which the sediments of rock unit B were deposited, perhaps by a sea gradually spreading over the plain. The surface separating the two rock sequences is called an *unconformity*. The tremendous amount of time that it represents must have convinced Hutton that he was right about the great age of the earth.

By studying rock exposures in many areas, Hutton became convinced that rocks in the crust follow a pattern that has been repeated over and over again throughout geologic time. Hutton's chain of events is known as the *geologic cycle* (Figure 1-6). Although the cycle has no definite beginning or end, it is convenient to think first of sediments being deposited in continuous horizontal beds according to the principles of Steno. As successive layers are deposited, the beds near the base of the sequence become compacted and begin the transformation from sediments to *sedimentary rocks*, a process known as *lithification*.

With continued deposition, the sedimentary rocks are deeply buried and subjected to higher temperatures and pressures. Under these conditions, the rocks are deformed; layers may be bent, broken, or heated to the melting point. The melting or partial melting of a rock produces *magma*, a molten liquid which can later cool to form an *igneous* rock. If the rocks are highly altered by heat and pressure but not to the point of melting, the original rock is changed to such a great degree that it is now called a *metamorphic* rock. The close association of deformation, metamorphism, igneous activity, and mountain belts suggests that these activities are all related. It was obvious to Hutton and most other early geologists that thick sequences of sedimentary rocks are intensely deformed and altered in the process of mountain building (Figure 1-7).

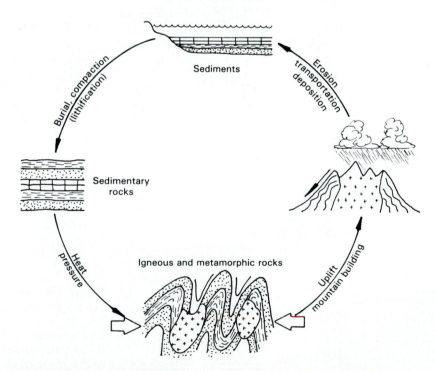

Figure 1-6 The geologic cycle. Rocks in the earth's crust follow a cycle that includes deposition of sediments, formation of sedimentary rocks, conversion to igneous or metamorphic rocks, and erosion to form sediments.

Figure 1-7 An exposure of a metamorphic rock, produced by alteration under high temperature and pressure. (C. C. Hawley; photo courtesy of U.S. Geological Survey.)

The final step in the cycle occurs after the uplift of a mountain belt. Exposed to the destructive agents of gravity, wind, rain, and ice, mountains are gradually lowered by erosion. The sediments produced by erosion are transported by rivers to the ocean where the cycle can continue in a new location. Given enough time, mountain ranges can be worn down to gentle, low-lying plains. The interior of Canada contains regions of igneous and metamorphic rocks that once must have been at the core of towering mountains but now are reduced to an elevation barely above sea level. In the United States, both the Appalachian and Rocky mountains are following the path of the geologic cycle (Figure 1-8). The Appalachians are older and thus erosion has reduced them to a lower elevation than the Rockies.

The controversy set off by Hutton's ideas raged in scientific circles for many years. It was not until the mid-nineteenth century that uniformitarianism was accepted by the majority of scientists.

Figure 1-8 The Grand Tetons, a mountain range consisting of igneous and metamorphic rocks. Uplift and erosion have produced the striking, rugged topography.

The work of Steno, Hutton, and many others paved the way for the recognition of geology as a modern science. The practice of geology requires the observation of rock or sediments and interpretation of the origin and sequence of events that has led to the present condition. The only change in this approach since the time of Hutton lies in the ever-increasing sophistication and complexity of the tools and instruments that we have available to study the evidence we find in the natural landscape. In this respect, geology differs from other sciences because the conclusions and interpretations of past geologic events cannot be verified by experimentation. Geologic study is not unlike the procedure followed by a detective who collects clues and evidence in order to solve a crime.

GEOLOGIC TIME

Geologists are, in part, historians. Our goal is to decipher the physical, chemical, and biological evolution of our planet from its origin to the present time. Like historians who study the history of the human race, geologists must have a method of determining the time at which various events in the earth's history took place. Imagine how difficult it would be to study the history of the United States if there were no concept of time; how could we comprehend the differences between the exploration of the West and the space program of the twentieth century if we did not understand the amount of time and technological progress that separates the two events?

Geologists in some ways face a much more formidable task in the investigation of the earth's history. There are no written records available. The only direct evidence that exists from the geologic past is the sequence of rocks that we can observe at the earth's surface. In the twentieth century we have learned to use the natural radioactivity in certain types of rocks to directly estimate their age; with this method we can determine the *absolute* time of an event.

Geologic pioneers like Hutton, however, did not have radioactive age dating available to them. Hutton could only speculate about the amount of time represented by an unconformity in a rock exposure. Confronted with this problem, geologists began to work out the history of the rock record using *relative* time. This method is an attempt to construct a time scale based upon the relative ages of rock units by establishing their relationship to other rock units.

Relative Time

Steno's Principle of Superposition is a useful tool in determining the relative age of sedimentary rocks. In an undisturbed layered sequence, the beds are successively younger from bottom to top. Another method that can be used with sedimentary rocks is *correlation*, the tracing of rock units from one area to another. In addition, the Principle of Original Continuity can be used to extend the correlation of rock units with similar characteristics across areas in which the rocks are not present or are not exposed (Figure 1-9).

More complicated rock exposures require additional techniques to establish relative time. The *Principle of Cross-cutting Relationships*, for example, is particularly useful in some places. It states that any rock that cuts across or penetrates a second rock body is younger than the rock it penetrates. This principle is often called upon for interpretation of sequences containing igneous rocks which were forced (intruded) into existing rocks in a liquid state and then cooled to a solid state. Because the igneous rocks often cut across the layering of sedimentary or metamorphic rocks, they can be identified as younger than the rocks which enclose them (Figure 1-10). In some places a layer of igneous rocks lies between two beds of sedimentary rocks. In these instances, it is sometimes possible to determine the relative age by examining the boundaries or *contacts* of adjacent rock units. If the heat of the molten rock altered the rocks both above and below the igneous rock at the time it was intruded, then the igneous rock penetrated the sequence and is younger than both the overlying and underlying units. If only the upper contact of the sedimentary rock below the igneous rock was altered, then the igneous rock was

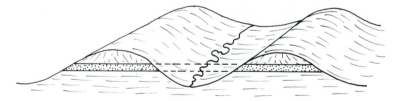

Figure 1-9 Beds of sedimentary rock can be traced across areas in which they are not present using correlation based on the Principles of Superposition and Original Continuity.

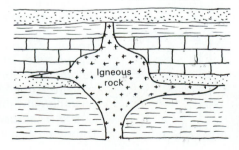

Figure 1-10 A body of igneous rock penetrates and cuts across the bedding of a series of sedimentary rocks. It is therefore younger than the sedimentary rocks.

probably formed as a lava flow on the ground surface and thus is younger than the rock beneath it but older than the rock above it (Figure 1-11).

It is often possible to use several of the principles described above to determine the sequence of events of a rock exposure (Figure 1-12). We therefore would be using relative time to compare the ages of rock units but still have no idea of the absolute age.

However useful the above methods may be, the foundation of relative age dating is based upon *fossils*, the visible remains of organisms preserved in rocks. William Smith (1769-1839), an English civil engineer, was one of the first people to recognize the usefulness of fossils as tools for determining relative age and correlation of sedimentary rocks. While supervising the construction of canals and other engineering works over a period of many years, Smith discovered that certain rock units always contained the same group, or *assemblage*, of fossils. The shells of marine organisms were most useful in this work. By using superposition and other principles, Smith was able to group these rock units in order of relative age on the basis of correlation of fossil assemblages. The method was later applied throughout Europe, and a *geologic col-*

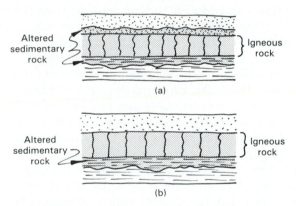

(a)

(b)

Figure 1-11   (a) Sedimentary rocks are altered both above and below their contacts with the igneous rock. Therefore, the igneous rock body is younger than both sedimentary rock units. (b) Only the lower sedimentary rock unit is altered near the contact with the igneous rock. In this case, the igneous rock is a lava flow that is younger than the lower sedimentary rock unit but older than the upper sedimentary rock unit.

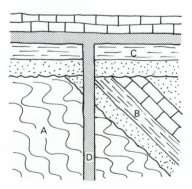

Figure 1-12   The Principles of Original Horizontality, Original Continuity, Cross-cutting Relationships, and Superposition can be used to unravel the sequence of events in this rock exposure. They include (1) deposition of rock sequence A, (2) metamorphism of A, (3) uplift and erosion of A to produce an unconformity, (4) deposition of sequence B, (5) deformation, uplift, and erosion of sequences A and B to produce an unconformity, (6) deposition of sequence C, and (7) intrusion of igneous rock body D through sequences A, B, and C.

umn was constructed for all sedimentary rocks known in that region (Figure 1-13). The rocks of each *system* contained a distinct and unique assemblage of fossils. Although certain fossils were similar to those in other systems, consistent differences between systems were evident. Rocks beneath the Cambrian system were generally thought to be nonfossiliferous and therefore could not easily be divided. They were given the name Precambrian.

	Systems	Where First Described
	Quaternary	Europe
	Tertiary	Europe
	Cretaceous	England
	Jurassic	Jura Mts. Europe
Youngest	Triassic	Germany
	Permian	Russia
	Carboniferous	Great Britain
	Devonian	Great Britain (Devonshire)
	Silurian	Great Britain (Wales)
	Ordovician	Great Britain (Wales)
	Cambrian	Great Britain (Wales)
Oldest	Precambrian Era	Nonfossiliferous, therefore not well subdivided

Figure 1-13 The geologic column, showing the names of the rock systems and the regions from which they were named.

The reasons for the changes in fossil characteristics were not known in the time of William Smith. Charles Darwin later provided an explanation for this phenomenon with his theory of evolution. The idea of systematic change of organisms over long periods of time was consistent with the fossil record.

Absolute Time

The construction of the geologic column was the culmination of relative age dating of rocks. Throughout the nineteenth century, geologists and physicists began to speculate about the age of the earth and the ages of the systems of the geologic column. The estimates became older and older. In the early twentieth century, one of the triumphs of modern geology was achieved with the application of radioactivity to the dating of rocks. Finally, a method for determining the absolute ages of certain rocks was available.

Radioactive particles (atoms) are disseminated throughout many types of rocks in trace amounts. These particles decay spontaneously by emitting atomic particles from the nucleus of the atom. Most of the radioactive particles are *isotopes* of elements; that is, they are forms of a certain element that differ slightly in atomic mass from other isotopes of the same element. (Atomic particles are more fully discussed in Chapter 2.) Upon radioactive decay, isotopes are altered to different elements by the loss or gain of atomic particles by the nucleus.

A basic assumption of radioactive dating is that each radioactive isotope (uranium 235, for example) will decay at a constant rate. In the example of uranium 235, 235 is the mass number of the isotope, that is, the number of protons and neutrons in the nucleus. The decay rate is often described using the *half-life* of the isotope. The half-life is the amount of time required for the decay of half of any amount of a particular isotope. For example, the half-life of uranium 235 is 0.7 billion years. After this amount of time, only one-half of the original amount of uranium 235 will remain. After two half-lives, only one-fourth of the original amount will be left.

A radioactive isotope that decays to another isotope is called the *parent*; the isotope produced by the decay is known as the *daughter*. In order to determine the age of a rock, the amount of both parent and daughter isotopes must be measured. This analysis is performed with a *mass spectrometer*, an instrument that can measure minute amounts of matter with great accuracy. The most useful parent-daughter isotope pairs are shown in Figure 1-14. With the exception of carbon 14, all these isotopes can be used for dating very old rocks. The date obtained, however, must be carefully analyzed. It can be considered to be accurate only if the rock has not been substantially altered since its formation. The age of an igneous rock, therefore, would represent the time that it cooled from a magma. Sedimentary and metamorphic rocks present some difficult dating problems. Sedimentary rocks consist of particles eroded from older rocks. The date obtained from an igneous grain incorporated into a sedimentary rock would thus represent the cooling of the igneous rock rather than the deposition of the sediment to form the sedimentary rock. In metamorphism, parent and daughter particles can be separated by heat and partial melting. Therefore, dates obtained from metamorphic rocks may or may not yield the correct age of metamorphism.

Carbon 14 serves a different function in radioactive dating because of its short half-life. The value of carbon 14 lies in its utilization for dating very recent sediments—particularly those from the last part of the ice age. Carbon 14 is a radioactive isotope of carbon produced by the activity of cosmic rays in the earth's atmosphere. It mixes with the normal carbon (carbon 12) in the atmosphere and is taken up by plants and animals, so organic material contains carbon 14 at the same concentration as the atmosphere at the time of death of the organism. Wood fragments and bone buried by the sediment deposited by glaciers can be used for dating the time of glacial advance. Because of the short half-life, 75,000 years is about the practical limit for dating by the carbon 14 method.

Application of radioactive dating to rocks began soon after the discovery of radioactivity. By dating rocks from all over the world and integrating these dates with the geologic column, scientists were able to construct an absolute chronology of geologic time. The geologic time scale (Figure 1-15) is the result. Geologic time is divided into two eons, Precambrian and Phanerozoic. The Phanerozoic eon is divided into eras, periods, and epochs. The periods are named after the rock systems in the geologic column; thus, finally, geologists had a means for relating evolutionary changes in organisms to absolute time.

The ages of rocks determined by radioactive dating made it necessary to greatly increase our conception of the age of the earth. Phanerozoic time alone, which includes most of the fossil record, is only a small percentage of geologic time (Figure 1-15). The oldest rocks dated extend back to about 4 billion years. If the age of the earth were reduced to a 24-hour scale, the human species would not appear until the last few seconds. Thus radioactive dating has had

Radioactive parent nuclide		Stable daughter nuclide		Half-life (years)
Potassium 40	K	Argon 40	Ar	1.3 billion
Rubidium 87	Rb	Strontium 87	Sr	47 billion
Uranium 235	U	Lead 207	Pb	0.7 billion
Uranium 238	U	Lead 206	Pb	4.5 billion
Carbon 14	C	Nitrogen 14	N	5600

Figure 1-14 Radioactive isotopes used for geologic age dating.

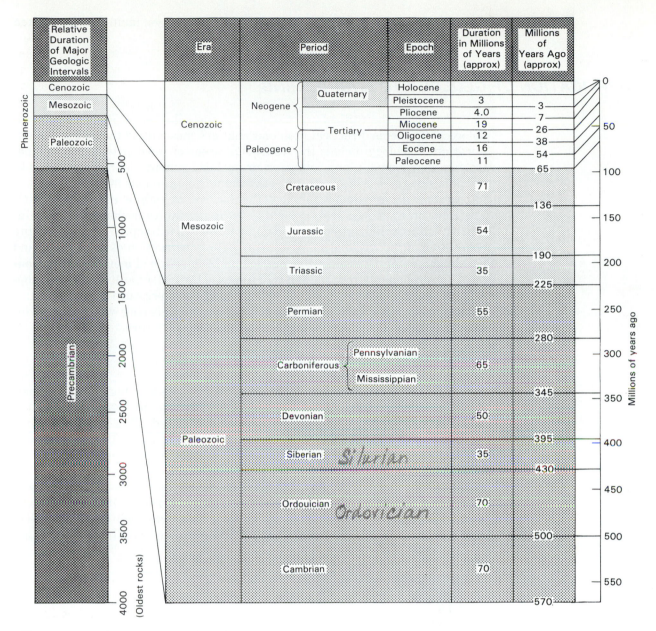

Figure 1-15   The geologic time scale.

a major effect upon our thinking about the age of the earth and the other planets.

Although the oldest rocks on Earth yield radioactive ages of about 4 billion years, Earth is considered to be about 4.7 billion years old. Our estimate for this age comes from dating meteorites that fall to the earth. These rocky fragments consistently yield ages of 4.5 to 4.7 billion years. They are considered to be composed of material formed at the time of origin of the solar system. The oldest rocks dated from the moon have similar ages, suggesting that the moon was formed about the same time as the meteorites. Why are there no rocks on Earth from the period of time between 4.7 and 4 billion years ago? The answer to this question is that Earth, as will be emphasized throughout this text, is an extremely dynamic planet. Rocks of the outermost layer of the earth, the crust, are continually transformed through the geologic cycle. The rocks of that early period of the earth's history have been melted and remelted so many times that no original rocks remain.

APPLICATION OF GEOLOGY TO ENGINEERING

Engineering is the utilization of science in the solution of practical or technological problems. Practical problems that are most likely to require the application of geology include the traditional areas of construction, mining, petroleum development, and water resources development and management. In the past several decades, however, environmental problems have increased in importance. An excellent example is the disposal of high-level nuclear waste. The solution to this critical problem is going to require perhaps the most intensive collaboration of geology and engineering yet attempted. The increasing population of the world is going to require the construction of more roads and buildings; it will necessitate the location and extraction of more metallic, non-metallic, and energy resources; it will increase the risk of loss of life and property in natural disasters; and it will intensify the pressure on our natural environment. For all these reasons, geology and engineering will become more and more interrelated in the immediate and distant future.

Geology and the Construction Site

In the realm of construction of civil engineering works, there are two basic ways in which geology must be considered. First, the geologic setting of the construction site must be described and characterized. This is accomplished by mapping the distribution of geological materials and by collecting and testing samples of rock and soil from the surface as well as from below the surface of the site. The problem with these procedures is that natural materials are rarely homogeneous or continuous. Because of the high costs of drilling, sampling, and testing involved in site investigation (Figure 1-16), prediction of certain engineering properties of the geologic materials and their response to engineering activity often relies on a thorough understanding of the origin

Figure 1-16 Collection of a soil sample with a truck-mounted auger. (Photo courtesy of USDA Soil Conservation Service.)

and geologic history of the rocks and soil present. The amount of geologic investigation that is necessary depends upon the cost, design, purpose, and desired level of safety of the project. A dam across a canyon above a city requires a much more detailed investigation than a typical suburban housing development. Failures of large engineering projects, although very rare, often can be attributed to an inadequate understanding of the geology of the site.

The second major interaction between geology and engineering concerns the threat of hazardous geologic processes. This problem is often more relevant to the location of the structure rather than to its design or construction. The list of hazards includes spectacular geologic events such as volcanic eruptions, earthquakes, landslides, and floods (Figure 1-17). Other less sensational processes such as land subsidence, soil swelling, frost heave, and shoreline erosion result in billions of dollars of property damage in the United States every year.

The critical task in dealing with geologic hazards is how to predict where and when these events will take place. Often the problem of when the hazardous process will occur is more difficult than where. Although geologists have made great strides in understanding the causes and effects of natural hazards, the ability to predict their time and place remains an elusive goal.

An important characteristic of natural hazards is that there is usually a statistical relationship between the magnitude and frequency of the event. Magnitude is an indication of the size or intensity of an event; the amount of energy released by an earthquake, the size of a landslide, or the amount of flow in a flood are examples of magnitude. The frequency, on the other hand, is a measure of how often, on the average, an event of a particular size will occur. Frequency can be expressed as the probability of occurrence within a specific period, or the *recurrence interval* of the event. A flood with a 1% chance of taking place within a period of any one year has a recurrence interval of 100 years.

Figure 1-17 Damage from the Big Thompson flood of 1976 in Colorado, an event of high magnitude and low frequency. (W. R. Hansen; photo courtesy of U.S. Geological Survey.)

Careful recording of floods and other types of hydrologic data indicates that there is an inverse relationship between magnitude and frequency (Figure 1-18). These data can be fitted with the equations of theoretical statistical distributions with the result that the probability of events of high magnitude and low frequency can be predicted. Other geologic processes, including earthquakes and landslides, seem to follow a similar pattern. Engineers should understand this relationship and remember not to overlook the possibility of a destructive event of large magnitude. Critical structures such as nuclear power plants, schools, and hospitals must be designed to withstand large, infrequent events.

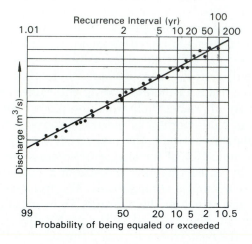

Figure 1-18 The relationship between magnitude and recurrence interval for floods from a hypothetical river. Each point represents the magnitude and calculated recurrence interval of a particular flood.

When certain or extensive loss of life is not a prime consideration in an engineering project, the economics of the project must also be considered in terms of the magnitude-frequency relationship. For example, a bridge that is built to withstand a flood of very high magnitude and low frequency would have high initial costs but low maintenance costs because the smaller, more frequent floods would cause little damage. Alternatively, the bridge could be built to withstand only small floods. In this case, the initial construction costs would be lower, but the maintenance costs would be higher because floods that might damage the bridge are more frequent than large floods. In order to minimize the long-term costs of the structure, both initial costs and maintenance must be considered (Figure 1-19). The optimal total cost would be realized by designing the bridge for a flood of moderate recurrence interval.

Development of Natural Resources

The search for mineral deposits and petroleum is carried out by geologic exploration. When a resource is located, it must be delineated and assessed in order to determine if it is economically recoverable. If it is determined to be so, the extraction of the resource is conducted by mining and petroleum engineers. These engineers must have a sound knowledge of geology because of the geologic control of the location and characteristics of the deposit.

Mining engineers must design the type of mine to be utilized, the equipment necessary, the mining procedures, and the reclamation of the site (Figure 1-20). The geologic setting of the deposit influences many decisions that must be made in this process.

The production of petroleum involves drilling through great thicknesses of rock. The drilling process requires the petroleum engineer to be familiar

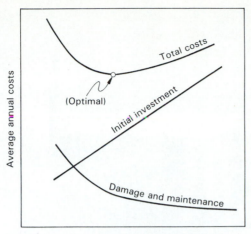

Average annual costs

Total costs

(Optimal)

Initial investment

Damage and maintenance

Average replacement period
(recurrence interval)

Figure 1-19 Economics of the design of hydraulic structures. The optimal cost of the structure is associated with an event of moderate recurrence interval. (From J. E. Costa and V. R. Baker, *Surficial Geology: Building with the Earth*, copyright © 1981. Reprinted by permission of John Wiley & Sons Inc., New York.)

Figure 1-20 Extraction of coal from an open-pit mine. About 20 m of overburden has been removed above the coal bed.

with all relevant physical and chemical conditions that can be encountered at depth (Figure 1-21). This encompasses a broad range of problems associated with rocks, liquids, and gases that are present in the subsurface. If a petroleum reservoir is located, petroleum engineers must develop it in the most efficient and economic manner possible. Extensive information must be gathered concerning the size, shape, and rock characteristics of the reservoir. New technologies are constantly being tested to produce the maximum amount of oil possible from the reservoir.

Figure 1-21 A drilling rig exploring for oil in the Williston Basin,
North Dakota.

Water Resources and the Environment

The development and protection of water resources is becoming one of society's
most important concerns. Future increases in water supply are going to rely
more heavily upon groundwater than surface water. The occurrence, movement, and quality of groundwater is therefore a matter of great significance.
Engineers and geologists together are developing techniques to manage the
utilization of groundwater supplies in order to ensure their availability for the
future. Overuse of groundwater for irrigation or other needs can lead to depletion and exhaustion of the supply (Figure 1-22).

Figure 1-22 Crop irrigation places a heavy demand upon water
resources. (Photo courtesy of USDA Soil Conservation Service.)

Along with the need to wisely develop groundwater resources is the increasing necessity of protecting the chemical quality of groundwater. Past, and perhaps present, waste disposal practices have contaminated groundwater supplies in all parts of the United States. Development of new methods to rehabilitate groundwater reservoirs will require the talents of a wide variety of scientists and engineers. Geology will be critical to the effort because an understanding of the subsurface geological environment is necessary for dealing with any aspect of groundwater occurrence or movement.

THE EARTH: A DYNAMIC PLANET

The Solar System

The planet on which we live is only one of nine planetary bodies that revolve around the sun in the solar system (Figure 1-23). In order to put Earth in its proper perspective, we should briefly examine the characteristics of our planetary neighbors. One of the most basic observations is that there are two groups of planets: four smaller planets orbiting close to the sun and four larger planets occupying the outer reaches of the solar system. Pluto is somewhat of an exception to this grouping. The inner, or *terrestrial*, planets are dense bodies mainly composed of iron and silicate (containing the element silicon) rocks. The four *giant* planets are much lighter in density because of their gaseous composition.

Although many ideas have been proposed to explain the origin of the solar system, this question is still a matter of debate. A possible hypothesis begins with a large diffuse mass of gas and dust slowly rotating in space. For some reason, this *solar nebula* began to contract under gravitational forces and to increase its rotational velocity. At some point, a concentration of matter formed at the center of the nebula. Compression of matter raised its temperature to the point at which nuclear fusion was initiated. Thus the sun, a body composed of 99% hydrogen and helium, was born. The matter rotating around the newly formed sun gradually cooled and condensed into the nine planets. The final composition of the planets was controlled by their initial position in the nebula with respect to the sun. Because the temperature was highest in the vicinity of the sun and decreased with distance away from the sun, the terrestrial planets were formed of material with relatively high boiling points. Thus Mercury, Venus, Earth, and Mars are dense bodies of iron and silicate rocks. In fact, four elements—iron, oxygen, silicon, and magnesium—make up about 90% of these planets. Volatile elements were carried away from the inner part of the nebula by matter streaming from the sun and perhaps the inadequate gravitational pull of the terrestrial planets before they attained their total mass. Volatile gases like water, methane, and ammonia were driven to the cooler regions of the giant planets. Jupiter, Saturn, Uranus, and Neptune retained their volatiles because of their greater gravitational attraction and are therefore more similar in composition to the original solar nebula.

Differentiation of the Earth

After initial condensation of the earth, its internal structure was very much different than at present. It probably consisted of a homogenous accumulation of rock material. At this point, a gradual heating of the earth must have taken place. The rise in temperature was probably caused by compression of the newly condensed matter and also by the energy released by the decay of ra-

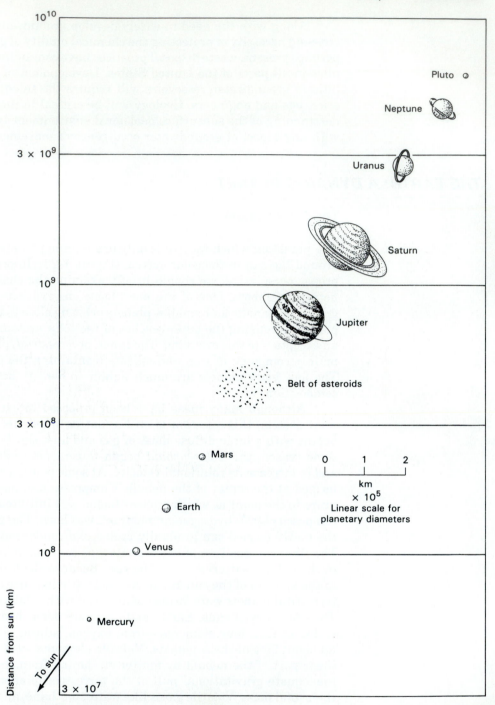

Figure 1-23 Relative sizes and positions of the planets. (From S. Judson, M. E. Kauffman, and L. D. Leet, *Physical Geology*, 7th ed., copyright © 1987 by Prentice-Hall, Inc., Englewood Cliffs, N.J.)

dioactive particles disseminated throughout the earth. The consequence of the temperature rise was that the melting point of iron was exceeded in some parts of the earth. The dense molten iron then began to migrate toward the center of the earth (Figure 1-24). Gradually a molten iron core began to develop at the earth's center. This process released gravitational energy in the form of heating, leading to more melting and partial melting. The end result was a

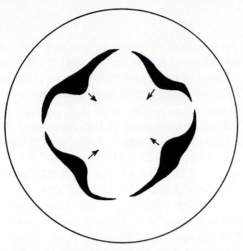

Figure 1-24 Sinking of molten iron toward the center of the earth to form its core.

density-stratified planet (Figure 1-25). The dense core was overlain by a mantle composed of iron and magnesium silicate rocks. The lightest materials accumulated in a very thin layer near the surface to form the crust. Earlier we mentioned that no rocks older than 4.0 billion years remain at the earth's surface. The period of time between the formation of the earth at about 4.7 billion years and 4.0 billion years may be accounted for by this zonation or *differentiation* of the earth. The original rocks may have been melted and their components segregated by density during this interval.

After differentiation the earth consisted of a molten core, which is only partially molten now, and a cool exterior. Heat therefore flows from the core to the surface. This flow of heat is extremely important because it indirectly or directly causes many of the processes that occur in the earth's crust.

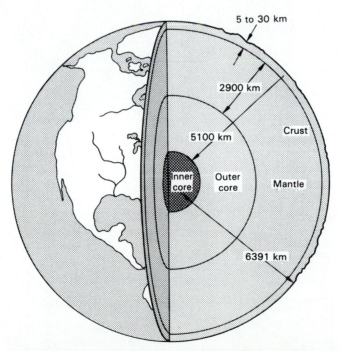

Figure 1-25 Internal structure of the earth. (From S. Judson, M. E. Kauffman, and L. D. Leet, *Physical Geology*, 7th ed., copyright © 1987 by Prentice-Hall, Inc., Englewood Cliffs, N.J.)

PLATE TECTONICS

One of the greatest accomplishments in geology during the past few decades has been the formulation of a new theory called *plate tectonics*. Although this model will be discussed in greater detail later in this text, it is introduced now to reinforce the concept of the earth as a dynamic, ever-changing planet.

According to this theory, the earth is composed of an upper, rigid layer, the *lithosphere*, which is composed of the crust and upper mantle. The lithosphere is broken into about 12 major *plates*, which slowly move laterally over the earth's surface (Figure 1-26). Movement of these plates may be driven by the flow of material in a layer that directly underlies the lithosphere called the *asthenosphere* (Figure 1-27). Rock in the asthenosphere is soft and plastic because it is near its melting point. Heat flow and plastic flow of rock in the asthenosphere is thought to occur in cells consisting of rising segments, areas of lateral flow beneath the lithosphere, and descending limbs. The rising material (just below its melting point) spreads out laterally at the base of the lithosphere, carrying the rigid lithospheric plates along its path. Volcanoes erupt to form new lithosphere in the cracks formed as the plates split apart. The plates move until they reach a zone of downward flow in the asthenosphere.

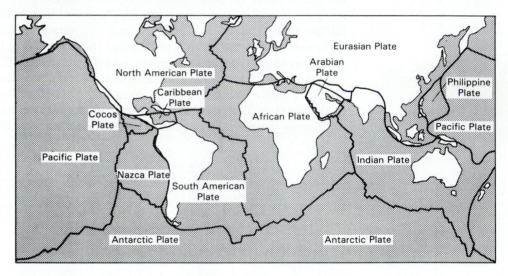

Figure 1-26   Major plates of the lithosphere.

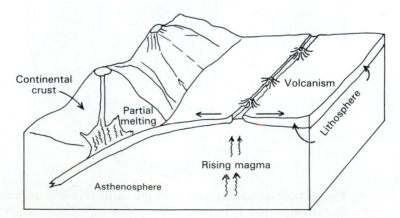

Figure 1-27   Configuration of the lithosphere and asthenosphere.

Here, the lithospheric plates are dragged down into the asthenosphere and may become partially melted (Figure 1-27).

The movement and interaction of lithospheric plates has probably been continuous since differentiation. The interaction of plates can explain the location of many geologic phenomena, including the distribution of volcanoes, earthquakes, and most mountain ranges. In fact, the movement of lithospheric plates can be considered to be a driving force in Hutton's geologic cycle.

MAJOR FEATURES OF THE EARTH'S SURFACE

The earth's surface can be divided into two dominant topographic features: the continents and the ocean basins (Figure 1-28). The rocks that lie beneath the continents differ greatly from the rocks of the ocean basins. Continental rocks are low in density and include the oldest rocks on earth. These are the lowest density materials concentrated by differentiation and subsequent plate interactions. The ocean basins are composed of denser, younger rocks. Volcanic eruptions continually produce new oceanic lithosphere at places where plates crack and split apart. In later sections of this book, the rocks of the continents and ocean basins will be described in more detail.

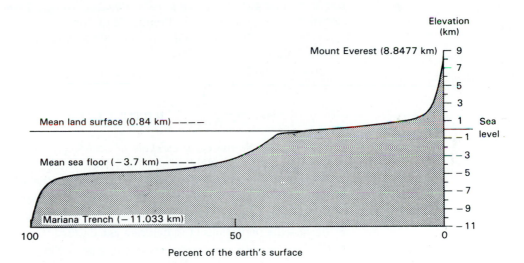

Figure 1-28 The two predominant levels of the earth's surface: the continents and ocean basins. (From W. K. Hamblin, *The Earth's Dynamic Systems*, 4th ed., copyright © 1985 by Macmillan Publishing Co., Inc., New York.)

SUMMARY AND CONCLUSIONS

The development of geology over the past four centuries has been a remarkable scientific journey. We have seen our conception of the earth's history change greatly. Hutton realized the role of continuing change in the evolution of the earth's surface. He recognized geologic processes in action as well as the evidence of past geologic processes in the rocks that he studied.

Although Hutton and others had begun to suspect that the earth was very old, its true age was not determined until the method of radioactive age dating was discovered. Dating provided a method of quantifying the divisions of the geologic column. These divisions were incorporated into the geologic time scale.

The early history of the planet Earth includes a period of density stratification. Since formation of the core, crust, and mantle, continual circulation and alteration of material has taken place. Hutton's geologic cycle is now integrated into the model of plate tectonics.

The need for integration of geology and engineering is increasing. Detailed geological input is required in construction and development, in recovery of natural resources, and in solving problems of groundwater contamination. The type of geologic information required by engineers includes a comprehensive interpretation of surface and subsurface geologic materials, and also a thorough analysis of hazardous geologic processes that could affect an engineering project.

REFERENCES AND SUGGESTIONS FOR FURTHER READING

COSTA, J. E., and V. R. BAKER. 1981. *Surficial Geology—Building with the Earth*. New York: John Wiley.

HAY, E. A., and A. L. MCALESTER. 1984. *Physical Geology: Principles and Perspectives*, 2d ed. Englewood Cliffs, N.J.: Prentice-Hall, Inc.

JUDSON, S., M. E. KAUFFMAN, and L. D. LEET. 1987. *Physical Geology*, 7th ed. Englewood Cliffs, N.J.: Prentice-Hall, Inc.

PROBLEMS

1. In what way do the early concepts of catastrophism and neptunism differ from modern scientific thought?
2. What were the contributions of Nicolaus Steno to the science of geology?
3. Why is Hutton's Principle of Uniformitarianism so important to modern geology?
4. What forms of evidence are used in relative age dating?
5. Discuss the limitations of radioactive age dating.
6. Choose a recently constructed engineering project in your area (such as a building, highway, or power plant). Describe all the ways in which this project interacts with the geologic environment both during construction and throughout the life of the structure.
7. What role should engineers play in the development and protection of water resources?
8. Summarize the earth's early geologic evolution following its initial formation.
9. What are the major ideas of the theory of plate tectonics?

CHAPTER
2
MINERALS

0   1   2   3 cm

The minerals, rocks, and soils that occur at and beneath the earth's surface are the materials with which the engineer must work. Unlike the materials used in buildings and other structures, which have uniform, homogeneous properties, these natural materials are notoriously variable and nonhomogeneous. The task for engineers designing any structure is to evaluate the distribution and properties of the natural materials present at the site and then base the design upon this assessment. It is impossible to evaluate natural materials at specific sites without a general understanding of the physical and chemical characteristics of earth materials. In addition, the engineering properties must be ascertained. In this part of the text we will concentrate upon the origin and characteristics of the minerals and rocks that make up the earth's crust. In following sections we will turn our attention to geologic processes operating both deep within the earth and at its surface.

An in-depth study of geology usually begins with an introduction to minerals. This particular point of departure should not come as a surprise, considering that the earth's solid surface is composed of rocks and soils that are primarily mineral aggregates. An understanding of mountains, volcanoes, earthquakes, and all other geologic phenomena could never be complete without some knowledge of the type and state of matter that makes up the solid part of the earth.

Knowledge of minerals is essential for the engineer who deals with earth materials. Minerals are partially responsible for the physical and mechanical properties of rock and soil encountered in mines, tunnels, and excavations. Likewise, dams, embankments, and other structures built from earth materials function because engineers have successfully utilized the properties of rock and soil in the project design. In industry, minerals are directly incorporated into chemicals, abrasives, and fertilizers and are processed into thousands of other useful products.

THE NATURE OF MINERALS

Minerals are inorganic, naturally occurring solids. In order to define the term mineral completely, however, we must further state that minerals are crystalline substances that have characteristic internal structures and chemical compositions that are fixed or that vary within fixed limits. The distinct internal structure of each mineral results in diagnostic external properties that can be used in mineral identification. In subsequent sections, we will examine each component of the definition of a mineral.

Internal Structure

At the heart of the definition of a mineral is the concept of a regular internal structure. It is this characteristic that distinguishes minerals from other types of solid, inorganic matter. A mineral, therefore, is a chemical compound in which elements react and combine to form a regular arrangement of particles within the solid.

The fundamental unit of each chemical element in the periodic table is the atom, the smallest amount of the element that retains its characteristic properties. An atom, however, is composed of even smaller particles. The most common model of atomic structure consists of a central concentration of mass, the nucleus, surrounded by concentric shells inhabited by minute charged subatomic particles called electrons (Figure 2-1). Electrons are held in the vicinity of a nucleus because of the attraction of opposite electrical charges. By con-

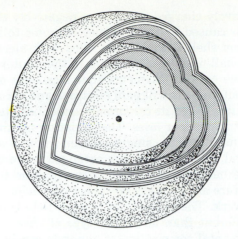

Figure 2-1 Electron cells around the nucleus of an atom. (From L. D. Leet, S. Judson, and M. E. Kauffman, *Physical Geology*, 5th ed., copyright © 1978 by Prentice-Hall, Inc., Englewood Cliffs, N.J.)

vention, electrons are considered to be negatively charged and the nucleus is considered to be positively charged.

The nucleus of the atom can also be subdivided. Positive charges are concentrated in particles called *protons*. The number of protons in an atomic nucleus defines the *atomic number* of the element. Thus an atom of sodium has 11 protons in the nucleus balanced electrically by 11 electrons orbiting around the nucleus in three shells (Figure 2-2). Particles that are neither positively nor negatively charged also reside in the nucleus. These *neutrons* have about the same mass as a proton, so the *atomic mass* of an atom is basically the mass of the protons plus the neutrons. The mass of an electron is about three orders of magnitude less than the mass of a proton or neutron and can be neglected in determining the atomic mass. Unlike the atomic number, the atomic mass of an element is not fixed.

Nuclei of an element may vary slightly in the number of neutrons that they contain. *Isotope* is the name given to an atom that has more neutrons than the most common configuration of the element. For example, hydrogen, with one proton in the nucleus, has an atomic mass of approximately one. With the addition of one neutron to the nucleus, an isotope of hydrogen called deuterium is formed. Another hydrogen isotope, tritium, contains two neutrons in the nucleus. Even though the properties of these three forms of hydrogen vary slightly, they are all unmistakably hydrogen.

A final type of atomic particle with which we must be familiar is the *ion*. Ions form when an atom gains or loses electrons, so the number of electrons is then no longer equal to the number of protons. The atomic particle then

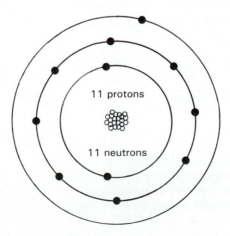

11 protons

11 neutrons

Figure 2-2 Atomic structure of sodium.

becomes a positively charged *cation* or a negatively charged *anion*. The formation of ions is caused by the tendency of atoms to attain a state in which the outer electron shell is completely filled with electrons. Thus atoms like sodium with a small number of electrons in their outer shells will readily lose them and become cations. Elements that lack only one or two electrons to complete their outer shells will gain electrons and become anions. The charge of a cation or anion, called its *oxidation number*, is determined by the number of electrons that it gains or loses. The oxidation number of the sodium ion is $+1$ because sodium loses the single electron from its outer shell.

In a mineral, atoms or ions are arranged in an orderly fashion to form the regular internal structure. Among the reasons that particles assume a particular structure are *chemical bonds* between adjacent atoms or ions in the internal framework. The presence of interparticle bonds indicates that minerals are no different than any other type of *chemical compound*. As compounds, the elemental composition of minerals can be described by chemical formulas that specify the exact proportions of elements in the mineral. *Quartz*, for example, has the formula SiO_2, meaning that there are two oxygen ions for every silicon ion in the structure.

Two types of bonds are of particular importance for geologic applications (Figure 2-3). An *ionic bond* is formed when one atom donates one or more electrons to an atom of another element. The first atom becomes a cation and the second, an anion. As illustrated in Figure 2-3, when an electron is transferred from an atom of sodium to an atom of chlorine, both ions then have achieved a stable state because their outer shells are filled with electrons. The resulting compound is called sodium chloride (common table salt), has the formula NaCl, and is given the mineral name *halite*. Elements that do not have as strong a tendency to lose or gain electrons may form *covalent bonds*, in which electrons are shared between atoms to complete the outer shell. Examples of covalent bonding are methane (Figure 2-3) and water. The type and number of bonds that particles in a mineral form with adjacent particles have a strong influence on the properties of the mineral. In fact, two entirely different minerals can have the same chemical compostion. *Diamond* and *graph-*

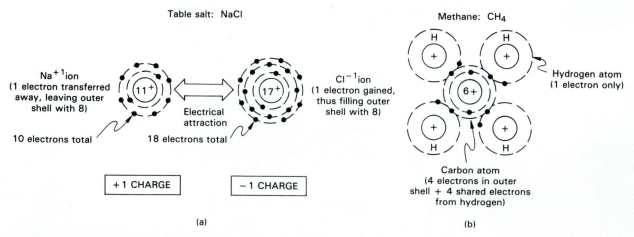

Figure 2-3 Common types of atomic bonding in minerals. (a) Ionic bonding (electron transfer). (b) Covalent bonding (electron sharing). (From E. A. Hay and A. L. McAlester, *Physical Geology: Principles and Perspectives*, 2d ed., copyright © 1984 by Prentice-Hall, Inc., Englewood Cliffs, N.J.)

ite, for example, are both composed only of carbon. In diamond, the carbon atoms form a three-dimensional framework with each atom bonded strongly to four adjacent atoms (Figure 2-4). The graphite structure consists of strong bonds between each carbon atom and three neighbors, all within a single plane. This results in sheets or layers of graphite with much weaker bonds between layers than within layers. Because of these differences in internal structure, diamond is the hardest mineral known and graphite is one of the softest.

The regular internal structure of minerals that we have alluded to justifies their description as *crystalline.* A crystalline solid is one in which a regular arrangement of atoms is repeated throughout the entire substance. The smallest unit of this structure is called the *unit cell.* The relative size of the ions plays an important role in the manner in which ions are packed in a particular structure. The unit cell of halite is composed of six chloride ions surrounding each sodium ion. The repetition of the unit cell in three directions produces a cubic structure as shown in Figure 2-5. The presence of a regular, unique internal structure for each mineral was suspected by mineralogists for several centuries before proof was obtained in the early part of the twentieth century. In the process called *x-ray diffraction,* x-rays are passed through min-

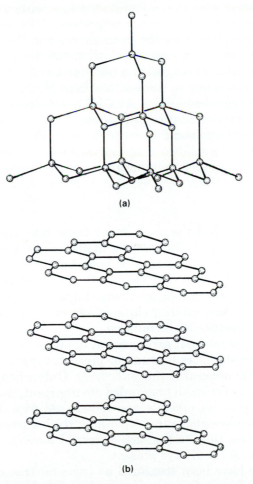

(a)

(b)

Figure 2-4 The internal structure of (a) diamond and (b) graphite, both composed entirely of carbon. (After Linus Pauling, *General Chemistry,* copyright © 1953 by W.H. Freeman and Co., San Francisco.)

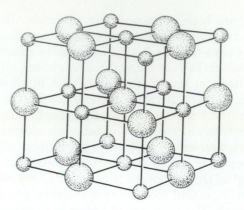

Figure 2-5 Structure of halite, showing arrangement of chloride ions (large spheres) and sodium ions (small spheres). (From S. Judson, M. E. Kauffman, and L. D. Leet, *Physical Geology*, 7th ed., copyright © 1987 by Prentice-Hall, Inc., Englewood Cliffs, N.J.)

eral samples. Photographic images produced by the x-rays as they emerge from the sample show the regular pattern of atomic particles in the structure.

As we have seen in the cases of quartz and halite, the chemical composition of a mineral can be expressed by a chemical formula. For some minerals, the chemical formula is fixed. There are minerals, however, that have a variable composition. If two ions have similar size and charge, they may occupy the same types of sites in a mineral structure without altering the structure and without greatly affecting the properties of the mineral. Magnesium and iron form cations that have the same charge and the same approximate size. In the silicate mineral *olivine*, magnesium and iron can substitute freely. The varying proportions of magnesium and iron constitute a *solid-solution series*. Thus olivine can range in composition from Mg_2SiO_4 to Fe_2SiO_4. These two end members represent olivine with all magnesium or all iron. In addition, any intermediate composition is possible. A sample of olivine with 50% iron and 50% magnesium would have the formula $(Mg_{0.5}, Fe_{0.5})_2 SiO_4$. Although the composition of a solid-solution–series mineral is variable, it varies within fixed limits. The limits for olivine are the magnesium and iron end members.

Crystals

We have now reached the point in our discussion where we can consider the exterior appearance of minerals. Normally, minerals will be present as irregular grains in a rock. Occasionally, however, we will be lucky enough to find a mineral specimen bounded by smooth, plane surfaces arranged in a symmetrical fashion around the specimen (Figure 2-6). This sample is said to be a *crystal*. The plane, exterior surfaces, or *crystal faces*, are manifestations of the internal structure of the mineral.

The unique internal arrangement of atoms in each mineral determines the possible crystal faces that can be present. Crystals form slowly by adding unit cells of the mineral to the outer faces. Only when crystals are allowed to grow in this manner in an uncrowded environment, that is, unhindered by the growth of neighboring crystals, are well-shaped crystal faces developed. These conditions are somewhat rare, explaining the occurrence of most minerals as irregular grains. Despite the lack of crystal development, the characteristic internal structure is always present.

Crystals have been treasured as gems for thousands of years. The scientific study of crystals took great strides when Nicolaus Steno recognized in 1669 that the angles between corresponding crystal faces of the same mineral are always the same. This holds true no matter what size crystals are mea-

0   1   2   3 cm

Figure 2-6   A crystal of garnet. The exterior surface of the specimen is composed of smooth, planar crystal faces arranged in a regular geometrical pattern.

sured. Steno's observation is now known as the *Law of Constancy of Interfacial Angles* (Figure 2-7).

The location of crystal faces on a crystal is quantified by their orientation with respect to the *crystallographic axes*, which are imaginary lines drawn through the crystal connecting the centers of corresponding crystal faces on opposite sides of the crystal. Despite the large number of possible crystal-face orientations that have been observed in natural crystals, all crystals have been grouped into six *crystal systems* based upon their geometrical properties. The definition of each system is based upon the number of crystallographic axes and the angles between the axes. Illustrations and examples of the crystal systems are presented in Figure 2-8.

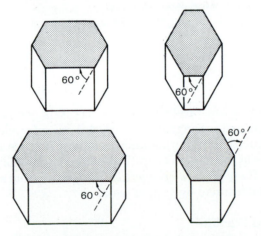

Figure 2-7   Crystals of the same mineral, no matter what size, have the same angle between corresponding crystal faces. (From E. A. Hay and A. L. McAlester, *Physical Geology: Principles and Perspectives*, 2d ed., copyright © 1984 by Prentice-Hall, Inc., Englewood Cliffs, N.J.)

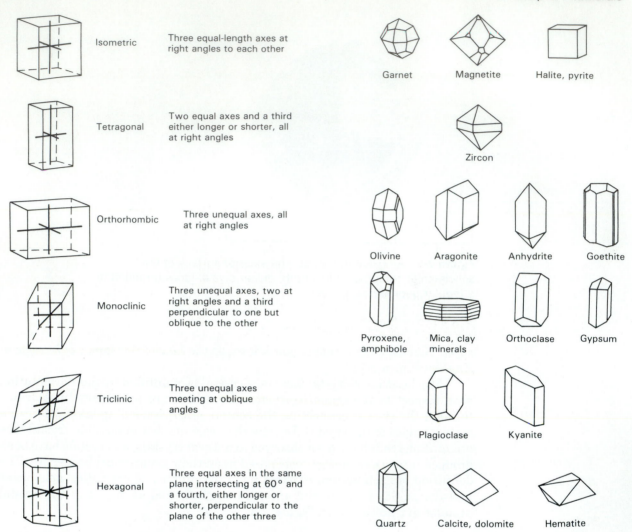

Figure 2-8 The six crystal systems with representative minerals and crystal forms. (From E. A. Hay and A. L. McAlester, *Physical Geology: Principles and Perspectives*, 2d ed., copyright © 1984 by Prentice-Hall, Inc., Englewood Cliffs, N.J.)

PHYSICAL PROPERTIES

Because minerals are rarely present as crystals, we must use criteria other than the shape and arrangement of crystal faces for identification. Although there are about 2000 minerals known, only a few are abundant in the most common rocks of the earth's crust. The common rock-forming minerals can be identified by their *physical properties*, which are characteristics that can be observed or determined by simple tests.

Cleavage and Fracture

The internal structure of a mineral is responsible for an extremely useful group of properties involving its strength. For example, there are two ways in which a mineral can break. First, the breakage can take place along regularly spaced planes of weakness caused by weak bonds in those directions within the internal structure. This type of breakage is called *cleavage* and the planes are called cleavage planes. The second type of breakage occurs in minerals that lack preferred directions of weakness. The surfaces of rupture are more irregular in these minerals and the type of breakage is known as *fracture*.

Cleavage planes can sometimes be confused with crystal faces because both develop in response to the internal arrangement of atoms in a mineral. Crystal faces, however, do not necessarily represent planes of weak bonds. For example, quartz has no cleavage but often exhibits well-developed crystal faces. Cleavage is classified as *perfect* if the cleavage planes are level and smooth. *Biotite* and *muscovite*, the micas, are examples of minerals with excellent cleavage in one direction (Figure 2-9). Cleavage can be developed in as many as six directions, creating a variety of cleavage types. Several common minerals develop cleavage in three directions. When the three planes are mutually perpendicular, the cleavage form is called *cubic* (Figure 2-10). If the three directions of cleavage do not intersect at right angles, *rhombohedral cleavage* can be developed (Figure 2-11).

Most fractures are relatively rough and irregular. Fracture is the form of breakage displayed by materials composed of atoms that are evenly spaced

Figure 2-9 Muscovite, showing excellent cleavage in one direction. Samples split readily parallel to the cleavage planes into very thin plates.

Figure 2-10 Halite, a mineral with well-developed cubic cleavage.

0 | 2 3 cm

Figure 2-11 Rhombohedral cleavage in the mineral calcite. The three directions of cleavage intersect at angles of 75° and 105°.

and equally attracted to adjacent particles in all directions. Occasionally, when materials of this type break, the surface is marked with smooth, concentric depressions (Figure 2-12). This type of fracture is called *conchoidal fracture*. Quartz sometimes displays this surface feature.

Hardness

The strength of atomic bonds, along with the size and density of packing of atoms or ions in a mineral, determines its *hardness*. The physical test used to determine the hardness of a mineral involves scratching it with various materials. A mineral with a high hardness value cannot easily be scratched by most substances. A soft mineral, on the other hand, can be scratched by any harder material. Tests of this type have shown that minerals vary quite widely in hardness. A relative scale used to compare minerals is called the *Mohs Hardness Scale* (Table 2-1). The scale is made up of 10 levels arranged in order

0 | 2 3 cm

Figure 2-12 Obsidian, a type of volcanic glass that produces conchoidal fracture upon breakage.

TABLE 2-1
The Mohs Hardness Scale

Mineral	Common object
1. Talc	
2. Gypsum	Fingernail ($2\frac{1}{2}$)
3. Calcite	Penny (3)
4. Fluorite	
5. Apatite	Knife blade ($5\frac{1}{2}$)
6. Orthoclase	File (6), window glass (6)
7. Quartz	Unglazed porcelain streak plate (7)
8. Topaz	
9. Corundum	
10. Diamond	

of increasing hardness, each represented by a common mineral. It is important to remember, however, that the scale is relative and the hardness increments between the levels are not necessarily equal.

In addition to the common objects listed in Table 2-1, the minerals on the scale can be used for hardness tests if available. Thus if a sample of quartz will scratch another mineral, the hardness of the unknown mineral must be less than seven.

Color and Streak

It would simplify mineral identification greatly if each mineral had a unique color that never varied from sample to sample. Unfortunately, this is not the case; many minerals occur in a wide variety of colors.

The property of color is caused by the absorption of selective wavelengths of white light. The color of a mineral is associated with the wavelengths that are not absorbed. These portions of the spectrum are reflected from the surface of the mineral. The absorption of light energy is related to the configuration of electrons around the nucleus in certain elements. In the transition elements, the outer electron shell contains one or two electrons before the adjacent inner shell is filled. Upon stimulation by light energy, the loosely held electrons in the outer shell vibrate easily and in doing so, readily absorb energy. Even a minute percentage of such elements present in a mineral as an impurity can supply color to the mineral. For example, pure quartz is colorless, but only a small amount of manganese or titanium can impart a pink hue to the mineral. Quartz of this color is known as rose quartz.

The color of finely powdered mineral particles produced by scraping the specimen across a porcelain (*streak*) plate is often more diagnostic than the bulk color of the mineral. The mineral *hematite*, for example, produces a reddish brown streak, even though the sample may have a metallic gray appearance. The limitation of a streak plate is that it can only be used on minerals with a hardness less than seven.

Luster

Luster is a property that results from the manner in which light is reflected from a mineral. All minerals can be classified with respect to luster as either *metallic* or *nonmetallic*. Metallic luster is caused by high surface reflectivity of light by the opaque minerals, minerals that strongly absorb light. The native metals such as gold and silver, as well as numerous other minerals (Figure 2-

13), have the appearance that we normally associate with metals because of their metallic luster.

The remaining minerals have various types of nonmetallic luster. These include the brilliant, reflective luster of diamond and other gems known as *adamantine luster*. One of the most common varieties of nonmetallic luster is *vitreous luster*, which is best illustrated by common glass. Quartz is a mineral with vitreous luster. Other minerals have luster that can be described as *greasy, waxy, silky, pearly,* or *dull (earthy)*.

0   1   2   3 cm

Figure 2-13 The mineral galena has metallic luster. Cubic cleavage is also evident.

Specific Gravity

Density provides a simple way of identifying some minerals that look similar simply by lifting the specimen. When the density is expressed as the ratio of the weight of a mineral to the weight of an equal volume of water, it is called the *specific gravity*. It is most useful for minerals composed of heavy elements such as iron, nickel, and lead. The mineral *galena* (Figure 2-13), which contains lead and sulfur, has a specific gravity of 7.57. A sample of galena is much heavier than a specimen of quartz of about the same size because quartz has a specific gravity of 2.65.

Other Properties

A few other physical properties may be useful for identifying specific minerals. Examples are magnetism, radioactivity, and taste, smell, or touch. *Magnetite*, an iron-bearing mineral, is the best example of a magnetic mineral. Highly magnetic varieties of magnetite called lodestone were used as the first compasses in the early days of navigation.

MINERAL GROUPS

Minerals are classified according to chemical composition and structure. The composition of the most common rock-forming minerals is limited by the abundance of elements in the crust. In fact, only eight elements constitute about 98% of the weight of the earth's crust (Table 2-2). Most of the minerals are

TABLE 2-2
Elemental Abundance in the Earth's Crust

Element	Symbol	Weight percent
Oxygen	O	46.60
Silicon	Si	27.72
Aluminum	Al	8.13
Iron	Fe	5.00
Calcium	Ca	3.63
Sodium	Na	2.83
Potassium	K	2.59
Magnesium	Mg	2.09

SOURCE: From Brian Mason, *Principles of Geochemistry*, © 1958 by John Wiley & Sons, New York.

members of a group characterized by combinations of the two most abundant elements, oxygen and silicon. This group is called the *silicate group* because all its members contain a specific structural combination of silicon and oxygen, even though most silicate minerals also contain other elements. Similarly, the other major mineral groups are composed of specific compositional units, usually anions or ion groups. We will begin our description of these groups with the silicates because of their abundance in rock-forming minerals.

Silicates

Every silicate mineral contains a basic structural unit called the *silica tetrahedron*. As shown in Figure 2-14, its structure involves a central silicon cation with an oxidation number of $+4$ surrounded by four oxygen anions, each with an oxidation number of -2. The four oxygen anions form the corners of a four-sided geometric form called a tetrahedron.

In its isolated state, the silica tetrahedron has a charge of -4. In order to form a mineral with an overall neutral charge, the negative charge must be balanced by adding cations or by forming linked tetrahedra in which the oxygen ions are shared by adjacent tetrahedra. The vast number of silicate minerals are formed by utilizing both of these mechanisms.

The silicate group is subdivided by the way in which silica tetrahedra interact within the structure. The most basic structural form consists of isolated tetrahedra with no sharing of oxygens. Olivine (Figure 2-15) is an example of a mineral with isolated tetrahedra. Magnesium and iron are added in various proportions to balance the charge, forming the solid-solution series

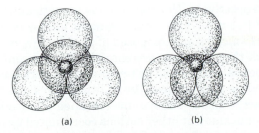

(a)   (b)

Figure 2-14   The silica tetrahedron shown from (a) top and (b) side views. A central silicon ion is surrounded by four oxygen ions. (From S. Judson, M. E. Kauffman, and L. D. Leet, *Physical Geology*, 7th ed., copyright © 1987 by Prentice-Hall, Inc., Englewood Cliffs, N.J.)

Figure 2-15 (a) Photo of the mineral olivine. (b) Structure of
olivine is characterized by isolated silica tetrahedra and regularly
spaced magnesium and/or iron ions. (From E. A. Hay and A. L.
McAlester, *Physical Geology: Principles and Perspectives,* 2d ed.,
copyright © 1984 by Prentice-Hall, Inc., Englewood Cliffs, N.J.)

that we previously considered. The physical properties of olivine are listed
along with other minerals in Table 2-3.

In other subdivisions of the silicate group, the linking pattern of silica
tetrahedra becomes successively more complex. One common type of linkage
produces chains of tetrahedra. Two types of chains are formed; the *pyroxene*
group consists of silicate minerals with single chains (Figure 2-16), and the
amphibole group contains double chains of silica tetrahedra (Figure 2-17). *Augite* and *hornblende* are common representatives of the pyroxene and amphibole groups, respectively. Hornblende has a particularly recognizable form of
cleavage with two directions intersecting at angles of 56° and 124°.

The *sheet silicates* are composed of thin layers, or sheets, of silica tetrahedra in which three oxygens are shared with adjacent tetrahedra (Figure 2-
18). These sheets are described as tetrahedral sheets because the silicon atoms
are surrounded by four oxygens. In addition, the sheet silicates contain other
structural types of sheets. For example, octahedral sheets (Figure 2-18) contain
aluminum, magnesium, or other cations coordinated by six oxygen or hydroxyl
atoms. The geometric figure defined by the arrangement of oxygen atoms is
an octahedron.

Important minerals within the sheet silicate group include the micas and
the clay minerals. Different minerals are formed by different ways of combining sheets and by variations in charge caused by ionic substitutions within

TABLE 2-3
Physical Properties of Common Rock-Forming Minerals

Mineral	Chemical formula	Color	Cleavage directions	Hardness	Specific gravity	Other properties
Silicates						
Augite (pyroxene)	$Ca(Mg,Fe,Al)(Al,Si_2O_6)$	Dark green to black	2 at 90°	5–6	3.2–3.6	
Biotite (mica)	$K(Mg,Fe)_3AlSi_3O_{10}(OH)_2$	Black	1	$2\frac{1}{2}$–3	2.8–3.2	
Garnet	$(Ca,Mg,Fe,Mn)_3(Al,Fe,Cr)_2(SiO_4)_3$	Dark red, brown, green	0	$6\frac{1}{2}$–$7\frac{1}{2}$	3.5–4.3	
Hornblende (amphibole)	$(Na,Ca)_2(Mg,Fe,Al)_5Si_6(Si,Al)_2O_{22}(OH)_2$	Dark green to black	2 at 56° and 124°	5–6	2.9–3.2	
Muscovite (mica)	$KAl_3(AlSi_3O_{10})(OH)_2$	Colorless to pale green	1	2–$2\frac{1}{2}$	2.8–2.9	
Olivine	$(Mg,Fe)_2SiO_4$	Pale green to black	None	$6\frac{1}{2}$–7	3.3–4.4	
Orthoclase (feldspar)	$KAlSi_3O_8$	White, gray, or pink	2 at 90°	6	2.6	
Plagioclase (feldspar)	$(Ca,Na)(Al_2Si)AlSi_2O_8$	White to gray	2 at 90°	6	2.6–2.7	Striations
Quartz	SiO_2	Colorless to white but often tinted	None	7	2.6	
Oxides						
Hematite	Fe_2O_3	Reddish brown to black	None	$5\frac{1}{2}$–$6\frac{1}{2}$	5.26	
Goethite (limonite)	$FeO \cdot OH$	Yellowish brown to dark brown	None	5–$5\frac{1}{2}$	4.37	Limonite is noncrystalline
Magnetite	Fe_3O_4	Black	None	6	5.18	Strongly magnetic
Halides and sulfides						
Halite	$NaCl$	Colorless or white	3 at 90°	$2\frac{1}{2}$	2.16	
Pyrite	FeS_2	Pale brassy yellow	3 at 90°	6–$6\frac{1}{2}$	5.02	Sometimes called "fool's gold"
Chalcopyrite	$CuFeS_2$	Brassy yellow	2 at 90°	$3\frac{1}{2}$–4	4.1–4.3	Ore of copper
Sphalerite	ZnS	Brown to yellow	6 at 120°	$3\frac{1}{2}$–4	3.9	Ore of zinc
Galena	PbS	Lead gray	3 at 90°	$2\frac{1}{2}$	7.54	Ore of lead
Sulfates and carbonates						
Gypsum	$CaSO_4 \cdot 2H_2O$	Colorless to white	1	2	2.32	
Calcite	$CaCO_3$	White to colorless	3 at 75°	3	2.72	Forms limestone
Dolomite	$CaMg(CO_3)_2$	Pink, white, or gray	3 at 74°	3.5–4	2.85	

(a)

(b)

Figure 2-16 (a) Photo of pyroxene. (b) In pyroxene, single chains of silica tetrahedra are bound together by calcium and magnesium ions. (From E. A. Hay and A. L. McAlester, *Physical Geology: Principles and Perspectives*, 2d ed., copyright © 1984 by Prentice-Hall, Inc., Englewood Cliffs, N.J.)

Figure 2-17 Hornblende, a member of the amphibole group of silicate minerals. The distinctive cleavage of hornblende consists of two planes intersecting at angles of 56° and 124°.

the structures. The clay minerals are especially significant to engineering because of the tendency of some clays to absorb water and swell. The effects of swelling soils on foundations can be very destructive.

One of the most basic structural configurations is the combination of one tetrahedral sheet and one octahedral sheet to form the clay mineral *kaolinite* (Figure 2-19). Aluminum ions fill the octahedral sites. Hydroxyls substitute for some of the oxygens to balance the charge of the structural unit. Adjacent

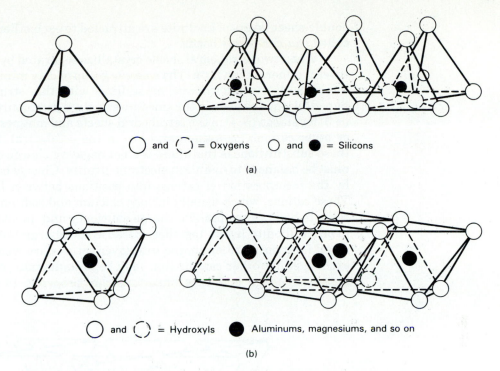

○ and ◌ = Oxygens   ○ and ● = Silicons

(a)

○ and ◌ = Hydroxyls   ● Aluminums, magnesiums, and so on

(b)

Figure 2-18 The sheet silicates are composed of two types of sheets: (a) A tetrahedral sheet of linked silica tetrahedra (single tetrahedron shown at left) and (b) an octahedral sheet (single octahedron shown at left) in which cations are surrounded by oxygens in octahedral coordination. (From R. E. Grim, *Clay Mineralogy*, 2d ed., copyright © 1968 by McGraw-Hill, Inc., New York.)

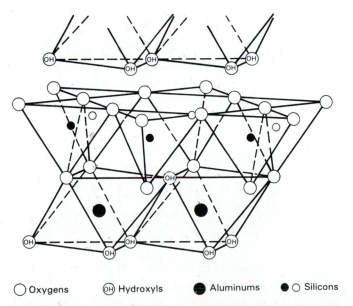

○ Oxygens   ⊕ Hydroxyls   ● Aluminums   ● ○ Silicons

Figure 2-19 The kaolinite structure forms layers composed of one octahedral sheet and one tetrahedral sheet. Adjacent layers are held together by van der Waals bonds. (From R. E. Grim, *Clay Mineralogy*, 2d ed., copyright © 1968 by McGraw-Hill, Inc., New York.)

double-sheet layers of kaolinite are attracted to each other by very weak forces called *van der Waals* bonds.

When two tetrahedral sheets crystallize separated by an octahedral sheet, several minerals can form. The *smectite* group of clay minerals, which includes the mineral *montmorillonite*, crystallizes with this structure (Figure 2-20). Substitution of cations in the smectite structure commonly includes aluminum (+3) for silicon (+4) in the tetrahedral sheets and magnesium (+2), iron (+2), or other cations for aluminum (+3) in the octahedral sheets. The result of these substitutions is that there is a net negative charge on the structure that must be balanced to maintain electroneutrality. Charge balance is maintained by the incorporation of cations into positions between layers (Figure 2-20). These cations, which usually include calcium and sodium in montmorillonite, serve to balance the positive charge deficiency and, in addition, bond adjacent montmorillonite layers together. Because the charge imbalance is relatively small for montmorillonite, the interlayer cations are weakly held and can be exchanged for other cations that may be brought into contact with the clay. *Cation exchange* is a very important clay property. In addition to cations,

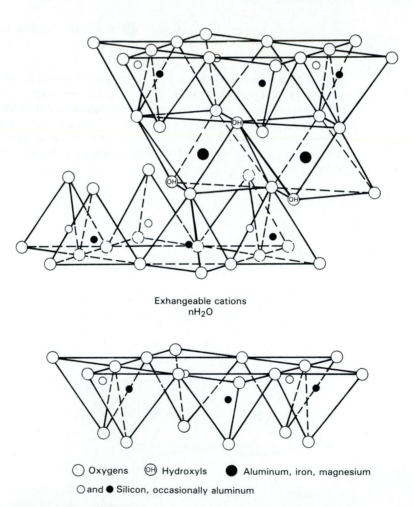

Exhangeable cations
nH₂O

○ Oxygens   ⊙ Hydroxyls   ● Aluminum, iron, magnesium

○ and ● Silicon, occasionally aluminum

Figure 2-20 Two tetrahedral sheets and one octahedral sheet form the smectite structure. Layers are separated by exchangeable cations and water molecules. (From R. E. Grim, *Clay Mineralogy*, 2d ed., copyright © 1968 by McGraw-Hill, Inc., New York.)

interlayer sites contain variable amounts of water. If water comes into contact with the clay, water molecules are drawn into the interlayer spaces and the structure expands with a force sufficient to lift or crack the foundation of fully loaded buildings.

The structure of the micas is very similar to the smectites. The excellent cleavage of muscovite (light mica) and biotite (dark mica) is developed along the planes of the individual layers (Figures 2-9, 2-21). The micas differ from the smectites in that aluminum ions always substitute for silicon in the tetrahedral layers in the micas, resulting in a larger charge imbalance. Potassium ions fill the interlayer positions in micas and, unlike the smectites, are tightly bound and not exchangeable. *Illite is a clay mineral similar in struc*ture *to muscovite* but in which exchange of interlayer potassium does occur because of variations in ion substitutions.

The final group of silicate minerals, the *framework silicates* contains the *feldspars* and quartz, which are the most abundant minerals in the earth's crust. The framework refers to the three-dimensional linking of silica tetrahedra, forming strong bonds in all directions. There are two important types of feldspar. *Plagioclase feldspar* includes a number of minerals in a solid-solution series with sodium and calcium as end members. Plagioclase can sometimes be recognized by fine, closely spaced lines, or *striations* on cleavage faces of the specimen. Striations are caused by defects in the crystal structure that formed as the mineral was crystallizing. *Orthoclase* is the other main type of

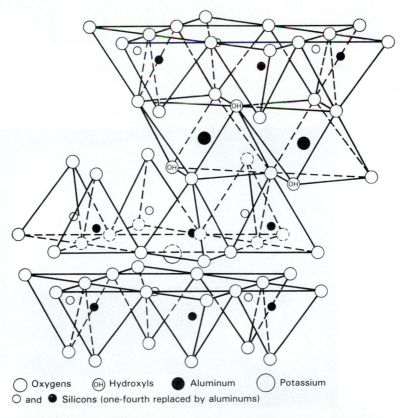

○ Oxygens   ⊕ Hydroxyls   ● Aluminum   ○ Potassium
○ and ● Silicons (one-fourth replaced by aluminums)

Figure 2-21 The muscovite structure is similar to the smectites except that layers are more strongly bonded by nonexchangeable potassium cations. (From R. E. Grim, *Clay Mineralogy*, 2d ed., copyright © 1968 by McGraw-Hill, Inc., New York.)

feldspar. It differs from plagioclase in its lack of striations and its high potassium content (Figure 2-22).

Quartz (Figure 2-23) is a framework silicate composed entirely of silica tetrahedra in which all four oxygen ions are shared. The development of strong bonds in all directions and the lack of additional elements in the structure results in the absence of cleavage. The color of quartz is extremely variable due to minute amounts of impurities within the structure.

Figure 2-22   Orthoclase, one of the two major types of feldspar, is a framework silicate mineral. Note two directions of cleavage at right angles.

Figure 2-23   Quartz, one of the most common silicate minerals, occurs as well-developed crystals (lower left), but more commonly in massive form (upper right).

The silicate minerals are used for many purposes in our society. Most gem stones, including emerald, topaz, jade, and turquoise are silicate minerals. In addition, silicate minerals are mined as ores for various metals and used in a variety of industrial processes. The clay minerals are particularly valuable in industry and manufacturing, with uses in brick, tile, plastics, rubber, paint, ceramics, and paper.

Oxides

Minerals that form by combination of various cations with oxygen are called *oxides*. Among the oxide minerals are important ore minerals of iron, aluminum, chromium, and other metals. The iron oxide minerals are particularly common. Iron and oxygen can form several different minerals depending on the oxidation number of the iron ions. In hematite (Figure 2-24) and *geothite*, iron is present in the oxidized ($+3$) state, while in magnetite, iron is present in both the oxidized and reduced ($+2$) states.

Halides and Sulfides

Minerals of the halide and sulfide group contain anions of fluorine, chlorine, bromine, iodine, and sulfur as the framework anion. Minerals composed of the first four elements of the preceding list constitute the halide minerals. An example of a halide mineral is halite, or common salt (Figure 2-10). The sulfide mineral group constitutes a very important source of metallic ores of iron, copper, zinc, and lead. *Pyrite* (Figure 2-25) is a common mineral often mistaken for gold because of its metallic luster. For this reason, it is sometimes called fool's gold.

0 1 2 3 cm

Figure 2-24 Hematite, a common iron oxide mineral.

Figure 2-25 Pyrite, an important iron sulfide mineral, can be recognized by its brassy yellow color, cubic cleavage, and metallic luster. The specimen on the left contains thin parallel lines, or striations, on its exterior surfaces.

Sulfates and Carbonates

Sulfates and carbonates consist of framework radicals similar to the silica tetrahedra in that an anion group is the basis of the structure. The sulfate anion group consists of sulfur and four oxygen ions (SO_4^{2-}). An important mineral in this group is *gypsum,* the main ingredient in building materials such as sheet rock.

In carbonate minerals the basic building block is the carbonate ion (CO_3^{2-}). The most important carbonate minerals are *calcite* (Figure 2-11), which combines calcium with the carbonate ion, and *dolomite*, which contains calcium and magnesium in its structure. Calcite is often found in hot-spring deposits and caves. Mammoth Hot Springs in Yellowstone National Park (Figure 4-9) has deposited large amounts of *travertine* (a variety of calcite) in a beautiful series of terraces and pools. Among the many uses of calcite and dolomite are building stones and the production of lime and Portland cement. Lime is valuable for its acid neutralizing properties with applications in agriculture and industry, and Portland cement is an essential ingredient of concrete.

Native Elements

Certain minerals consist of a single element. Examples are gold, silver, copper, sulfur, and carbon. Elemental gold, silver, and copper are relatively rare in nature; they more commonly occur as components of other minerals. Sulfur is utilized in the production of sulfuric acid and other chemicals. Graphite and diamond (Figure 2-4) are two minerals composed of carbon. The difference in structure between graphite and diamond is largely due to pressure at the time of formation. Diamond forms in high-pressure geologic environments deep within the earth and therefore develops a dense structure. Its great hardness is attributed to a three-dimensional network of strong covalent bonds. Graphite is a soft, greasy mineral with perfect cleavage in one direction.

SUMMARY AND CONCLUSIONS

The rocks of the earth's crust are aggregates of minerals, which are inorganic crystalline solids with a chemical composition that is fixed or varies within fixed limits. Within each mineral, atoms and ions are arranged in a specific structural pattern that determines the properties we can observe on the exterior surfaces of the sample.

When a mineral slowly crystallizes in unobstructed surroundings, crystals form. The smooth, planar crystal faces reflect the internal structure of the mineral and, for a specific mineral, maintain uniform angles between corresponding faces on crystals of any size. The conditions necessary for the formation of crystals are rare, so, instead, minerals usually occur as irregular masses in rocks because of competition for space with other minerals during crystallization.

Identification of most common rock-forming minerals is possible by using observation and simple tests to determine the physical properties of the specimen. The most useful physical properties include color, streak, luster, cleavage, hardness, and specific gravity.

Composition is the main criterion for dividing minerals into classes of chemically similar minerals. The basic structural unit used in classification is the major anion or anion group in the mineral. Silicates, comprising the most abundant group, are defined by the silica tetrahedron, a structural unit of four oxygen anions and one silicon cation. Subgroups of silicates are formed by different ways of arranging and linking silica tetrahedra in the structure. Oxides, carbonates and sulfates, halides and sulfides, and native elements are the classes of nonsilicate minerals.

REFERENCES AND SUGGESTIONS FOR FURTHER READING

HAY, A. E., and A. L. McALESTER. 1984. *Physical Geology: Principles and Perspectives,* 2d ed. Englewood Cliffs, N.J.: Prentice-Hall, Inc.

GRIM, R. E. *Clay Mineralogy,* 2d ed. New York: McGraw-Hill.

JUDSON, S., M. E. KAUFFMAN, and L. D. LEET. 1987. *Physical Geology,* 7th ed. Englewood Cliffs, N.J.: Prentice-Hall, Inc.

MASON, B. 1958. *Principles of Geochemistry:* New York: John Wiley & Sons.

TENNISSEN, A. C. 1983. *Nature of Earth Materials.* Englewood Cliffs, N.J.: Prentice-Hall, Inc.

PROBLEMS

1. List the characteristics of each major atomic particle.
2. How do ionic and covalent bonds differ?
3. Why are only crystalline solids included in the definition of a mineral?
4. How does the internal structure of a mineral relate to its external appearance?
5. What is mineral cleavage?
6. For which common minerals is hardness a very useful diagnostic physical property?
7. Why are the silicate minerals so important?

8. Briefly describe the structure of clay minerals. What is it about these minerals that makes their presence in soil so important to engineering?

9. What is cation exchange?

10. If you allow a bucket of seawater to totally evaporate, various minerals will precipitate. Into which mineral groups do you think most of these minerals will fit?

CHAPTER
3
IGNEOUS ROCKS AND PROCESSES

Let us assume that we are in charge of constructing a highway tunnel through a mountain. Obviously, we will need to know as much as possible about the properties and characteristics of the rocks along the route of the tunnel. This information will influence the design and construction of the tunnel as well as the safety of the project. The tunneling conditions would be much different in a sequence of volcanic lava flows than they would be in alternating beds of sandstone and shale. With an understanding of how rocks form, it is possible to make general predictions about their spatial distribution and their physical properties.

Preliminary information typically available for such a tunneling project consists of rock samples taken from holes drilled at intervals along the proposed path of the tunnel. Our task is to predict the rock types and rock conditions between the test hole locations. The only way to do this is to use geological reasoning and interpretation based on sound knowledge of how different rock types are formed. Thus although it may seem academic to study the processes of rock formation, an understanding of these processes must be applied in every major rock-engineering project.

PROPERTIES OF IGNEOUS ROCKS

When a rock sample is examined, much can be inferred about its origin by observing its color, texture, and mineral content. The term *texture* refers to the size, shape, and arrangement of the mineral grains in a rock. In this chapter we are concerned with igneous rocks, rocks that formed by subsurface cooling and solidification from a molten liquid called *magma*, or *lava*, if cooling takes place at the earth's surface. In subsequent chapters we will examine the other two fundamental rock types, sedimentary and metamorphic.

Texture

The texture of igneous rocks (Table 3-1) is variable but usually consists of an interlocking network of mineral crystals that grew from the magma or lava

TABLE 3-1

The Texture of the Igneous Rocks

Rock type	Description and interpretation
Extrusive rocks	
Glassy	Noncrystalline, very fine grained; very rapid cooling
Aphanitic	Uniformly fine grained; rapid cooling
Porphyritic	Large phenocrysts within fine-grained groundmass; two-stage cooling process
Vesicular	Numerous small holes on surface; gas escape during cooling
Intrusive rocks	
Phaneritic	Uniformly coarse grained; slow, gradual cooling in subsurface

Figure 3-1 Rhyolite, an igneous rock with aphanitic texture. Individual crystals are not visible with the unaided eye.

during cooling. If the crystals are too small for observation by the naked eye, the texture is called *aphanitic* (Figure 3-1). These extremely fine-grained rocks are associated with volcanic processes because of the rapid cooling that occurs when a magma reaches the earth's surface in a volcanic eruption and becomes lava. Crystal size is one of the major differences between volcanic, or *extrusive*, rocks and rocks that formed beneath the earth's surface, or *intrusive* rocks. The growth of large crystals in a rock takes long periods of time and conditions of gradual, undisturbed cooling. The intrusive rocks that formed under these conditions have a texture called *phaneritic* (Figure 3-2). Phaneritic rocks are composed of an interlocking network of large crystals easily visible in hand specimens.

When a volcano erupts, the magma brought to the surface is quickly cooled by contact with the atmosphere, which is much cooler than the sub-

Figure 3-2 Phaneritic texture is characteristic of coarse-grained rocks. This specimen of granite contains large grains of orthoclase, quartz, and biotite.

surface *magma chamber* that contained the magma prior to eruption. At times, this cooling is so rapid that crystallization is prevented. The silica tetrahedra are frozen in place before they can attain an orderly arrangement necessary for mineral growth. Cooling at these rates leads to the formation of *glassy* texture, the glass in this case being a natural, noncrystalline material. *Obsidian* (Figure 2-12) is a good example of a rock composed primarily of volcanic glass.

Several other types of texture are associated with volcanic rocks. Rocks that are composed of crystals of two sizes have *porphyritic texture* (Figure 3-3). The interpretation of this texture is that crystals began to separate from the magma in the magma chamber or within the conduit leading from the magma chamber to the surface. These early-formed crystals grew to a substantial size by the time they reached the surface, enclosed in the still-liquid magma. Once exposed to the air, the remainder of the magma crystallized rapidly as much smaller crystals. The resulting rock therefore consists of large *phenocrysts*, the early crystals, surrounded by a fine-grained crystalline *groundmass*.

In addition to liquid, magmas contain significant amounts of dissolved gases that are kept in solution by the great pressure exerted on the magma chamber by the overlying rocks. Water vapor is the most abundant gas, although carbon dioxide (CO_2), sulfur dioxide (SO_2), and other gases are present. When a magma reaches the earth's surface, the pressure decreases and the dissolved gases subsequently migrate to the surface of the lava and escape to the atmosphere. The upper surfaces of solidified lava flows are given an appearance much like Swiss cheese by the small holes, or *vesicles*, through which gases escaped (Figure 3-4). These rocks are said to have a *vesicular texture*. The types of igneous rock texture are summarized in Table 3-1.

Color and Composition

The chemical and mineralogical composition of igneous rocks is a reflection of the composition of the magma from which the rocks crystallized. Magmas are

Figure 3-3 Porphyritic texture is composed of large phenocrysts embedded in an aphanitic groundmass.

0 1 2 3 cm

Figure 3-4   A sample of vesicular basalt. As this lava rapidly cooled, gases escaped to the air through the holes (vesicles) in the rock.

variable in composition, most importantly in the amount of silica (SiO_2) that they contain. The silica content ranges from less than 45% to more than 66%. Rocks that are rich in silica are called *silicic*, or *felsic*, rocks and those that are low in silica are called *mafic* rocks. Fortunately, color provides a valuable clue for identification of igneous rocks because the silicic rocks are mainly composed of light-colored minerals like quartz and feldspar, whereas the mafic rocks are dark colored because of the abundance of *ferromagnesian* minerals. The dark-colored ferromagnesian minerals, which, as their name implies, are rich in iron and magnesium, include olivine, pyroxene, and hornblende. The major igneous rock types fall into categories of high, intermediate, and low silica content. Figure 3-5 shows the principal minerals contained in each of these rock types.

Texture and silica content provide a basis for classifying the igneous rocks. Aphanitic and phaneritic rocks of the same silica and mineral content have different rock names. For example, *granite* and *rhyolite* are both light-colored silicic rocks containing mostly quartz and orthoclase. They receive different names, however, because granite is a phaneritic intrusive rock and rhyolite is an aphanitic extrusive rock. In the intermediate range of silica content (52–66%), *diorite* and *andesite* are the coarse-grained and fine-grained equivalents, respectively. The common intrusive rock *granodiorite* falls between granite and diorite in silica content. As Figure 3-5 indicates, granodiorite contains more plagioclase feldspar than orthoclase.

On the mafic side of the chart, *gabbro* and *basalt* (Figure 3-6) are important rock types. Basalt is the most common type of volcanic rock present on the earth's surface. The equivalent intrusive rock is gabbro. Rocks that are even lower in silica than gabbro have been discovered in the crust and are inferred to be much more common in the mantle. These rocks, which are largely composed of ferromagnesian minerals are known as *ultramafic rocks*.

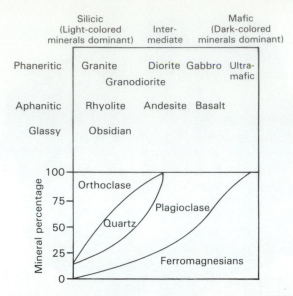

Figure 3-5 Classification of the igneous rocks. (Modified from L. Pirsson and A. Knopf, *Rocks and Rock Minerals*, John Wiley & Sons, New York, 1926.)

Figure 3-6 Basalt is an aphanitic rock of mafic composition. The majority of lavas on the earth are basaltic.

The most common rock with a glassy texture is obsidian (Figure 2-12). Despite its high silica content, obsidian is black in color. This exception to the usual light color of silica-rich rocks is due to the presence of impurities.

Figure 3-5 is a handy guide to the igneous rocks. By observing the color, it is possible to estimate the silica content of an igneous rock and thereby to predict its major minerals. The texture of the rock indicates whether it is intrusive or extrusive. Together, the color and texture can be used to assign the proper rock name.

VOLCANISM

Volcanism is the most obvious igneous process. Major eruptions generate worldwide interest because of their spectacular power and beauty and, as Mount St. Helens proved so clearly, their danger and destructiveness. In addition to the areas of the earth that have been affected by volcanic activity in historic times, vast regions of the earth, including all the ocean basins, are underlain by volcanic rocks erupted earlier in geologic time.

Volcanoes and Plate Tectonics

One of the most interesting applications of the theory of plate tectonics is its use in explaining the distribution of active volcanoes on the earth's surface. Most of these are concentrated in narrow bands along the edges of lithospheric plates. Divergent plate margins are characterized by voluminous outpourings of basaltic lavas at midoceanic ridges (Figure 3-7). These lavas form the foundation of new oceanic crust that fills the void left as plates move away from each other. Convergent plate margins (Figure 3-7) also generate volcanic activity. Here, the descending plate in a subduction zone is partially melted. The magmas formed at depth migrate to the surface to form chains of volcanic mountain ranges like the Cascades of the northwestern United States. The volcanic rocks of convergent plate margins are usually intermediate in composition. Andesite, named for the Andes mountains of South America, is a common volcanic rock at convergent plate margins.

Volcanism is not restricted to plate margins. A number of volcanic areas, such as the Hawaiian Islands, occur distant from plate margins. The volcanic activity is attributed to localized zones of high heat flow extending downward, deep into the mantle. These *mantle plumes* are assumed to remain in fixed

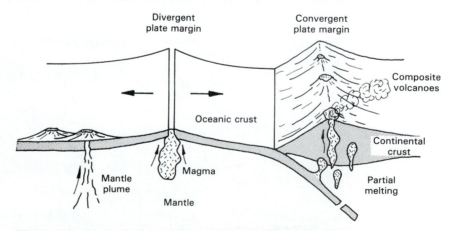

Figure 3-7 Volcanism is common at both divergent and convergent plate margins. Fissure eruptions produce new oceanic crust at divergent margins; composite volcanoes of andesitic composition are common at convergent margins, where partial melting of the crust occurs.

positions as the plates above move along their current paths. As long as the plate movement is in a straight line, a linear chain of volcanoes is produced.

Eruptive Products

Volcanoes erupt quite a wide variety of materials. Aside from the gases mentioned earlier, eruptive products generally can be divided into lava and *pyroclastic material*. Lava, the name for magma after it reaches the surface, flows downslope away from the eruptive vent at a rate dependent on its temperature and silica content. Basaltic lavas, which are low in silica and are erupted at high temperatures (~1100°C), are fluid and may flow great distances. Even so, there are variations in the flow characteristics of basaltic lavas. Highly fluid basaltic lava flows develop a smooth, ropy surface that is given the Hawaiian name *pahoehoe* (Figure 3-8). Lava flows that are more viscous and move at a slower rate, form a rough, jagged upper surface as blocks of partially solidified lava in the upper parts of the flow are broken apart by the slowly advancing mass beneath. The rubbly flows, which can cut a person's boots to shreads within a short hiking distance are known as *aa*, also a name of Hawaiian origin (Figure 3-9). Occasionally, the molten center of a lava flow may flow out from beneath the solidified crust, leaving a natural tunnel known as a *lava tube* (Figure 3-10).

Rhyolite, one of the most silicic types of lava, is much more viscous than basaltic lava. Rhyolitic flows form plugs or domes in the vent area of a volcano (Figure 3-11); they are just too viscous to move very far from their point of extrusion.

The separation of gas from magma as it rises in a volcano often occurs with explosive force. Particles of all sizes are blasted into the air and carried away from the vent. Collectively, all materials of this type are called *pyroclastic*. The individual particles may consist of rock, minerals, or glass. The smaller particles, called *volcanic ash*, may be carried upward thousands of meters into the atmosphere and transported around the earth several times as clouds of dust.

The coarse particles, which range in size from about 6 cm to many meters in diameter, fall close to the volcano or upon the mountain itself, where they

Figure 3-8 Pahoehoe lava has a smooth, sometimes ropy appearance.

Figure 3-9   An aa basalt lava flow with a typical rough, jagged surface.

Figure 3-10   Roof collapse of a lava tube. The tube forms by cooling of the lava-flow surface and outflow of the molten interior.

then may bounce or roll downward to the base of the slope (Figure 3-12). Rocks formed from volcanic ash are given the name *tuff*, whereas *breccias* are rocks formed from coarse-grained pyroclastic material.

One of the most destructive volcanic phenomena is a type of pyroclastic flow called *nuée ardente*, or fiery cloud (Figure 3-13). In these events, pyroclastic material, buoyed up by gases and dust at an extremely high temperature, flows down the sides of a volcano at great speeds. In 1902, a nuée ardente from the volcano Mount Pelée on the Caribbean island of Martinique killed 28,000 people in the town of St. Pierre (Figure 3-14). Flows of pyroclastic material form distinctive rocks called *welded tuffs*. Particles of rock and vol-

Figure 3-11   A lava dome that formed in the crater of Mount St. Helens after the May 18, 1980, eruption. (R. Tilling; photo courtesy of U.S. Geological Survey.)

Figure 3-12   Two large pyroclastic fragments on the flank of a volcano in the Canary Islands. (University of Colorado.)

canic glass contained in these flows are still very hot when they are deposited on the ground surface. Thus as the material cools, the soft, high-temperature particles become "welded" together to form a hard, dense rock unlike other pyroclastic materials.

Another type of flow produced by volcanic eruptions is called a *lahar*. Lahars, which are similar to mudflows, are generated when pyroclastic material becomes saturated with water from melting snow or glacial ice. The resulting mass of debris flows down the mountainside in existing stream valleys, burying roads, bridges, and buildings in its path. The deposition of debris from lahars in stream channels leads to increased flooding because of the reduction in the capacity of the river channels.

Figure 3-13 A nuée ardente flowing down the flank of a volcano.
(University of Colorado.)

Figure 3-14 The remains of St. Pierre, Martinique, West Indies,
after the nuée ardente of May 8, 1902. (Howell Williams.)

Types of Eruptions

Volcanoes erupt in a variety of styles. Although we normally think of volcanoes
in terms of majestic mountains issuing clouds of dust and gas from a vent
located in a depression, or *crater* at the top of the mountain, there are several
other types of eruptions. Frequently, eruption occurs from a linear crack, or
fissure, in the land surface rather than from a central vent. Fissure eruptions
are common along the midoceanic ridges, where lithospheric plates (Chapter
1) are being pulled apart. Basaltic lava formed by partial melting of the asthenosphere moves upward to the surface to form new crustal material as the
plates move away from each other. The entire oceanic crust is generated in
this fashion. Iceland is a portion of midoceanic ridge that projects above sea
level. Here, fissure eruptions can be observed on land (Figure 3-15).

Figure 3-15 The line of small cones
marks the site of a fissure eruption that
produced the surficial lava flows in the
valley.

Voluminous fissure eruptions that produce basaltic lavas also occur away
from plate boundaries. The highly fluid lava spreads out from fissures over
vast areas. The Columbia River and Snake River plateaus in the northwestern
United States (Figure 3-16) are underlain by an immense volume of *flood
basalts* originating from fissure eruptions. Beneath these plateaus, basalt flows
are stacked on top of each other to form a basalt sequence thousands of meters
thick (Figure 3-17).

Eruptions in which the vent is a single point, rather than a fissure, can

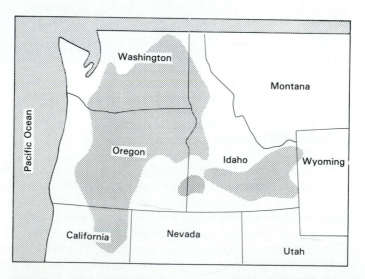

Figure 3-16 Map showing the extent of the Columbia River and
Snake River flood basalts. (From S. Judson, M. E. Kauffman, and
L. D. Leet, *Physical Geology*, 7th ed., copyright © 1987 by
Prentice-Hall, Inc., Englewood Cliffs, N.J.)

Figure 3-17 The Columbia River basalts in eastern Washington consist of a sequence of nearly horizontal lava flows, stacked one upon another.

be divided into several categories. If the eruption produces mainly lava of basaltic composition, the fluid lavas will spread laterally from the vent in all directions, producing a symmetrical cone with gently sloping sides (Figure 3-18). The slope angles of these *shield volcanoes* are low as the basaltic lava flows can travel great distances because of their low viscosity.

Basaltic lava flows erupted from shield volcanoes and fissures frequently exhibit a distinctive surface appearance called *columnar jointing*. During cooling of lava, contraction initiates a series of cracks, or joints, that form a polygonal pattern on the flow surface. Viewed from the side of a lava flow, the pattern of joints looks like a series of columns (Figure 3-19).

Composite volcanoes are formed when a volcano erupts both lava and abundant pyroclastic material. The resulting cone has a layered internal struc-

Figure 3-18 The gentle slopes and symmetrical profile of this small volcano in Iceland are typical of shield volcanoes.

Figure 3-19 Columnar jointing at Devil's Postpile, California.
(University of Colorado.)

ture (Figure 3-20) consisting of alternating layers of lava and pyroclastics. The
lava of composite volcanoes is more viscous than that of shield volcanoes and
falls into the intermediate range of composition. Andesite is a common rock
type associated with composite volcanoes. Because of the high viscosity, andesitic lavas do not flow very far, and the side slopes of composite volcanoes
are quite steep. Volcanoes in the Cascade Range, including Mount St. Helens,
are composite volcanoes (Figure 3-21).

The origin of andesitic lava is somewhat different than basaltic lava.
Composite volcanoes occur above descending plates at convergent plate boundaries (Figure 3-7). As oceanic crust is forced downward into the mantle, where
the temperature is much higher, *partial melting* may occur. In the process of
partial melting, some minerals in a rock melt before others. The minerals that
melt first are richer in silica than the minerals remaining in the rock, so the
magma produced is more silicic in composition than the host rock. By partial
melting, basaltic oceanic crust may produce magma of andesitic composition.

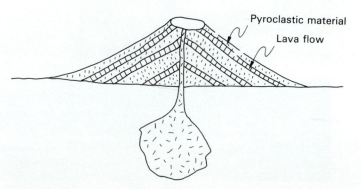

Figure 3-20 Alternating strata of lava and pyroclastics in a
composite volcano.

Figure 3-21 Mount Rainier, a composite volcano in the Cascade Range. (Howell Williams.)

The composition of the magma may also be influenced by partial melting of sediments on the subducted slab or of other crustal rocks.

In some volcanic eruptions, the gas content is so high that the eruption is extremely explosive. This situation produces mainly pyroclastic eruptive products. The larger clasts that fall around the eruptive vent form a *cinder cone* (Figure 3-22). The finer particles may be carried by winds for thousands of kilometers from the vent.

Remnant Volcanic Landforms

We have described a crater as a small depression in the vent area of a volcano. Depressions of much larger size are sometimes present in volcanic regions. The formation of these *calderas* is attributed to the partial collapse of the volcanic mountain into the subsurface void left by a major eruption of magma from the magma chamber. The crust above the magma chamber subsides along

Figure 3-22 A large cinder cone in northern Iceland.

a series of fractures (Figure 3-23). A spectacular example of a caldera is Crater Lake in Oregon (Figure 3-24).

After a volcano becomes dormant, erosion begins to attack the volcanic cone. During long periods of geologic time, entire volcanoes can be eroded away, along with the rock surrounding the volcano. Erosion to this degree may expose

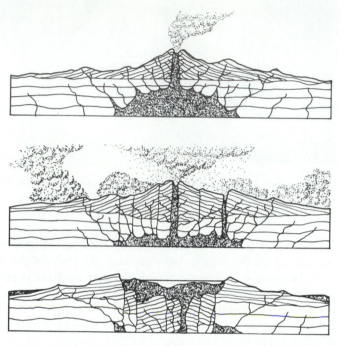

Figure 3-23   Hypothetical sequence of events in the formation of a caldera. (After Howell Williams, "Calderas and Their Origin," Bull. Univ. Calif. Dept. Geol. Sci., vol. 25, pp. 239–346, 1941.)

Figure 3-24   Crater Lake, Oregon, is a caldera. (Washington National Guard; photo courtesy of U.S. Geological Survey.)

Figure 3-25 Shiprock, a diatreme in New Mexico. The thin, dark, linear feature extending outward from the diatreme is a type of intrusion called a dike. (University of Colorado.)

the pipe or conduit that led from the magma chamber to the vent. These pipes are either filled with solidified magma or breccia that was formed by the explosive escape of gases from the magma chamber. In either case, the rock in the volcanic pipe may be more resistant to erosion than the surrounding rock and may therefore project above the landscape as an imposing monument to past volcanism. Shiprock is a breccia-filled pipe, or *diatreme*, that towers above the New Mexico landscape (Figure 3-25).

INTRUSIVE PROCESSES

Intrusion refers to the movement of magma from a magma chamber to a different subsurface location. Bodies of rock formed by the intrusive magma are called *plutons*. Rocks that make up plutons usually have phaneritic texture because the cooling time was sufficient to allow the formation of large crystals. It is only after erosion has removed the overlying rocks that we are able to observe intrusive rocks at the earth's surface.

Types of Plutons

Plutons differ in terms of size, shape, and relationship to the rocks that were intruded by the magma, which are older rocks known as *country rocks*. A major group of plutons is classified as *tabular* because they are thin in one dimension as compared with the other two dimensions (Figure 3-26). If a tabular pluton is roughly parallel to the layering of the country rocks, it is said to be *concordant* and is called a *sill*. *Discordant* plutons cut across the layering of the country rock. A tabular pluton of this type is called a *dike* (Figure 3-27). Dikes may represent the nearly planar cracks through which magma was moved to the surface to produce fissure eruptions. At the close of the eruption, the magma remaining in the crack connecting the magma chamber with the fissure solidifies to become a dike.

Nontabular plutons include *laccoliths, stocks,* and *batholiths*. The shapes of these plutons are thick as well as broad. A laccolith is a concordant pluton

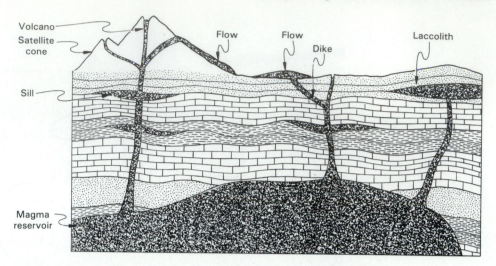

Figure 3-26 Types of plutons. (From S. Judson, M. E. Kauffman, and L. D. Leet, *Physical Geology*, 7th ed., copyright © 1987 by Prentice-Hall, Inc., Englewood Cliffs, N.J.)

Figure 3-27 A basaltic dike cutting metamorphic rocks along the Maine coast. (See also Figure 3-25.)

that causes doming of the overlying country rocks (Figure 3-28). The rocks above are bowed upward during the intrusion of the magma. Nontabular, discordant plutons are called stocks if their surface exposure covers an area of less than 100 km² and batholiths if they are exposed over a larger area. Batholiths are truly immense bodies of rock that form the cores of entire mountain ranges. The Idaho and Sawtooth batholiths in central Idaho are excellent examples (Figure 3-29).

   The origin of batholiths has been a controversial subject in geology for many years. The rock type of most batholiths is granite or granodiorite; this highly silicic composition rules out a mantle source for the magma. Therefore, the magma must have originated in the continental crust, the part of the crust that approximates the composition of granite. The major problem concerning

Figure 3-28 Laccoliths in Texas exposed by erosion of the overlying sediments. (C. C. Albritton, Jr.; photo courtesy of U.S. Geological Survey.)

Figure 3-29 Granitic rocks of the Sawtooth batholith, central Idaho. (T. H. Kiilsgard; photo courtesy of U.S. Geological Survey.)

batholiths is to account for the huge volume of rock that previously occupied the space now filled by the batholith. A possible explanation for some batholiths is that existing rock was converted to granite, essentially in place. Ideas to explain the mechanism for this conversion range from actual melting to a solid-state process in which existing rocks react at high temperature with solutions and gases from great depths. This theory for the origin of granitic batholiths is known as *granitization*.

Crystallization of Magmas

The great variety in mineral and chemical composition of igneous rocks suggests that the crystallization of magma is not a simple process. Pioneering experiments by N. L. Bowen in the early 1900s (Bowen, 1922) demonstrated that minerals crystallize sequentially as the temperature drops in a silicate magma and that solid crystals can react with the liquid phase of the magma to form new minerals during the crystallization process.

To explain the crystallization process, let us assume that we have a silicate melt of basaltic composition at about 1500°C. As the temperature is slowly lowered, crystals begin to separate from the liquid. There are two crystallization sequences that are observed as the melt cools. The first sequence can be illustrated by the crystallization of plagioclase, which, as indicated in Chapter 2, forms a solid-solution series between calcium-rich and sodium-rich compositions. The first plagioclase crystals to form are higher in calcium content than the calcium content of the liquid phase. As the mixture continues to cool, the crystals that form have progressively less calcium and more sodium than the original plagioclase crystals. In addition, the early-formed crystals react with the liquid by exchanging sodium from the liquid for calcium in the solid crystal. The internal structure of the mineral remains the same during this process. The crystallization of plagioclase follows what is called a *continuous reaction series*, in which the liquid and the crystals continuously change in composition until no liquid remains.

The ferromagnesian minerals follow a second type of crystallization sequence. In this series, olivine is the first ferromagnesian mineral to crystallize. As the temperature decreases, no change in the olivine crystals occurs until a critical temperature is reached. At this point, augite rather than olivine begins to crystallize and the early-formed olivine crystals react with the liquid to form augite. These reactions are different from the continuous reaction of plagioclase because entirely new minerals with different internal structures form at specific temperatures. For this reason the ferromagnesian crystallization sequence is called a *discontinuous reaction series*. The same type of

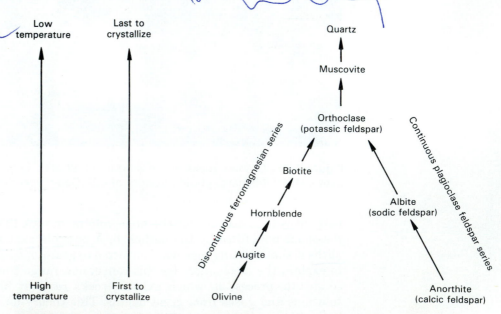

Figure 3-30 Bowen's reaction series describes the sequence of minerals that crystallize from a silicate magma.

reaction occurs between augite and the liquid to form hornblende at a lower temperature.

The entire sequence of mineral crystallization is known as *Bowen's Reaction Series* (Figure 3-30). Although the series demonstrates the trend of mineral changes in magmatic crystallization, most individual magmas would not progress through the entire sequence to quartz and muscovite. A magma of basaltic composition, for example, may be totally solidified before its composition could significantly change.

Bowen's work did suggest a way in which magmas could evolve from mafic to silicic composition. This mechanism, known as *fractional crystallization*, involves a separation of crystals from the remaining liquid. If crystals of olivine and augite were separated from a basaltic magma, by gravitational settling for example, they would be prevented from further reactions with the liquid phase. The remaining liquid would be enriched in silica and would progress further along the reaction series. By fractional crystallization, a small amount of granite could be produced from a basaltic magma if crystals were progressively separated from the magma as they formed.

ENGINEERING AND IGNEOUS ROCKS

Igneous rocks vary greatly in suitability for various types of engineering projects. An engineering site investigation must answer two questions: first, what rock types are present and how are they distributed, and second, how have the rocks been changed or altered since formation? The first question deals directly with the topic of this chapter. The geologists and engineers working on the project must determine the origin of the igneous rock, its contacts with adjoining rock types and their conditions, and the mineralogy of the rocks.

Unaltered intrusive igneous rocks generally are very suitable for most types of engineering projects. The interlocking network of mineral crystals gives the rock great strength. These rocks thus provide adequate support for building or dam foundations, can remain stable at high angles in excavations, and require minimal support in tunnels. Because of the dense interlocking of crystals within the rock, very little water can move through. Therefore, unaltered intrusive rocks are well suited for construction of reservoirs because of the low potential for leakage.

The engineering properties of extrusive rocks are much less uniform. Sequences of extrusive rocks contain pyroclastic materials and lahar deposits, which are much weaker than crystalline rocks. These rocks may be susceptible to slope failures in excavations and also provide more variable and generally weaker foundation support. In general, the water-bearing capacity of extrusive rocks is much greater than intrusive rocks. Sequences of basaltic lava flows on the Columbia Plateau and elsewhere are known for their great supplies of groundwater. This same property can render the rocks unsuitable for reservoir or tunnel construction. The presence of lava tubes in a series of basalt lava flows would prove to be an especially serious problem.

The second question that was posed concerns the geologic history of the igneous rocks since the time of their formation. Several types of changes have significant engineering implications. Any process that tends to fracture the rocks will cause weakening of a large body, or mass, of rock with respect to engineering suitability. Under the forces imposed by the interaction of lithospheric plates, fracturing of crustal rocks is quite common. These processes will be discussed in Chapter 9. It is sufficient to note at this point that a network of fractures within a rock mass can greatly increase the potential for failures of natural or excavated slopes and also increase the construction problems of dams, tunnels, and other structures.

Although tectonic fracturing is important, there are several other mechanisms that cause fracturing. Fracturing of the extrusive rocks, as we have seen, occurs during cooling, in addition to any later processes that may take place. An additional class of fracturing mechanisms becomes possible when the rock mass is at or near the earth's surface. These processes are collectively called *weathering.* A detailed discussion of weathering is found in Chapter 10 to supplement the brief comments included here. Rocks can be fractured in the near-surface environment by freezing and thawing as well as by other means. Thus a mass of igneous rocks generally is more fractured near the surface than at depth.

Weathering produces other changes in the rock as well as fracturing. Chemical reactions between the minerals within the rock and air and water gradually form new minerals. Clay minerals are a common product of these alteration processes. The result is a significant loss of strength as the feldspars and ferromagnesian minerals are converted to clay. In warm, humid climates, igneous rock bodies may be mantled with tens of meters of weathered material. The engineering properties of this material are totally different from the properties of unaltered rock.

CASE STUDY 3-1
VOLCANIC HAZARDS: MOUNT ST. HELENS

The eruption of a volcano is much more than a geologic curiosity. The 1980 eruption of Mount St. Helens provides a particularly well-documented example of the variety of phenomena accompanying such an event.

After several months of preliminary signals, Mount St. Helens experienced a cataclysmic eruption on May 18, 1980. The eruption began in an unpredicted fashion; the north side of the mountain gave way in a massive debris avalanche that buried the upper part of the North Fork of the Toutle River valley beneath 3 billion cubic meters of rock, ash, pumice, snow, and ice (Figure 3-31). The sudden release of pressure caused by the debris avalanches unleashed a lateral blast of hot gases and ash. This unexpected directed blast of pyroclastic material devastated the forests to the north of the mountain for a distance of 25 km. Trees within the blast zone were blown over like toothpicks (Figure 3-32). Subsequent to the lateral blast, ash was erupted vertically into the atmosphere, where it began to drift as a dense cloud eastward across the United States.

Among the most damaging long-term effects of the Mount St. Helens eruption were lahars mobilized from the debris avalanches; within hours after these huge landslides had come to rest, waves of fluidized debris moved down the major rivers draining the volcano. The largest lahar swept down the North Fork of the Toutle River, the Cowlitz River, and eventually into the Columbia River (Figure 3-33). The channels of the Toutle and Cowlitz were filled by the surging mudflows, which occasionally overflowed onto the flood plain, where roads and buildings were buried by the debris (Figure 3-34).

Deposition within the channel itself was also a serious problem. In some places the capacities of the channels were reduced to only one-tenth of their preeruption value. Thus, the threat of flooding exists to a much greater degree than before the eruption. Soon after the eruption, flood-abatement programs were initiated. These projects included dredging and levee improvement in flood-prone areas. The Mount St. Helens experience is an impressive warning to all those who live near potentially active volcanoes.

Figure 3-31 Debris-avalanche deposits on the north flank of Mount St. Helens and extending into the upper portions of the North Fork Toutle River valley. Deposits are about 200 m thick in the area shown. (R. M. Krimmel; photo courtesy of U.S. Geological Survey.)

Figure 3-32 Blown-down timber in the Green River valley from the Mount St. Helens eruption. (Washington Dept. of Natural Resources.)

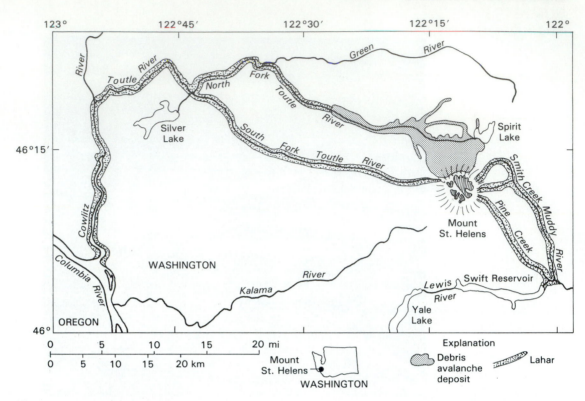

Figure 3-33 Map of the Mount St. Helens region showing the extent of debris avalanches and lahars. (Modified from U.S. Geological Survey Circular 850-B.)

Figure 3-34 Building partially buried by lahar deposits, North Fork Toutle River valley. (R. L. Schuster; photo courtesy of U.S. Geological Survey.)

SUMMARY AND CONCLUSIONS

Igneous rocks can be divided into extrusive rocks that crystallized at the earth's surface and intrusive rocks that crystallized below the surface. Texture is the major criterion used to assign rocks to one of these categories. Texture and composition form the basis for the classification of igneous rocks. Composition refers to the amount of silica in the rock, ranging from the low-silica mafic rocks to the silicic rocks. Each major compositional group (silicic, intermediate, and mafic) has both intrusive and extrusive rock types. Color is a good indication of composition because the silicic minerals like quartz and feldspar are mostly light in color.

Volcanoes erupt several types of materials and produce a variety of surface landforms. Most active volcanoes are associated with tectonic-plate margins, although intraplate volcanism is known to occur; mantle plumes are thought to be the cause of some of this activity.

Volcanic products include gas, lava, and pyroclastic material. Lava flows range from high-temperature, low-viscosity flows of basaltic composition to highly viscous, silicic flows. When gas is explosively released during an eruption, pyroclastic materials are abundant. Large clasts fall in the immediate vicinity of the vent, whereas volcanic ash can be transported thousands of kilometers by winds. Occasionally, destructive pyroclastic flows are generated during eruptions. These nuée ardentes flow down the mountainside at high velocities and with great force. Lava and pyroclastic materials sometimes mix with snow, ice, or water to form lahars, flows that are similar to mudflows in terms of velocity and mechanics of movement.

Volcanic products can be extruded from fissures or central vents. Fissure eruptions produce highly fluid lavas of basaltic composition that flow great distances. By successive eruptions, vast plateaus composed of thick vertical sequences of flood basalts are built up. Eruptions from centralized vents, on the other hand, tend to produce conical mountains composed of lava and/or pyroclastic materials. Shield volcanoes are composed of lava flows only, whereas composite volcanoes contain alternating layers of lava and pyroclastic materials. Cinder cones form when the eruption is limited to pyroclastic material.

Intrusions are formed by migration of magma from a magma chamber to a subsurface location higher in the crust. The resulting plutons are classified by size, dimensions, and relation to the surrounding country rock. Tabular plutons include sills and dikes. Nontabular plutons usually contain larger volumes of intrusive rock. These range from moderate-size laccoliths and stocks to giant batholiths that form the cores of major mountain ranges. Some batholiths probably formed by melting or alteration of existing rocks rather than by intrusion of magma from a separate magma chamber.

Igneous rocks crystallize in a complex fashion. The order in which minerals crystallize from a silicate melt is known as Bowen's Reaction Series. This sequence includes both a discontinuous and a continuous series, the latter represented by plagioclase, in which the early-formed crystals and the liquid change composition gradually while retaining the internal structure of plagioclase. The ferromagnesian minerals follow the discontinuous series, in which reactions between crystals and liquid produce entirely new minerals at specific temperatures.

Engineering conditions in igneous-rock terrains depend upon the composition and type of rock as well as the amount and type of alteration that has occurred since intrusion or extrusion. Weathering and fracturing are two types of alteration that can weaken the rock and render it less suitable for engineering projects.

REFERENCES AND SUGGESTIONS
FOR FURTHER READING

BOWEN, N. L. 1922. The Reaction Principle in Petrogenesis. *Journal of Geology*, 30:177–198.

HAY, E. A., and A. L. MCALESTER. 1984. *Physical Geology: Principles and Perspectives*, 2d ed. Englewood Cliffs, N.J.: Prentice-Hall, Inc.

JUDSON, S., M. E. KAUFFMAN, and L. D. LEET. 1987. *Physical Geology*, 7th ed. Englewood Cliffs, N.J.: Prentice-Hall, Inc.

U.S. GEOLOGICAL SURVEY. 1980. *Hydrologic Effects of the Eruptions of Mount St. Helens, Washington*, 1980. U.S. Geological Survey Circular 850.

PROBLEMS

1. If an igneous rock has phaneritic texture, what origin can be presumed for the rock?
2. List all the criteria you might use to identify a volcanic rock.
3. What criteria are used to classify the igneous rocks?
4. How can the distribution of volcanoes on the earth be explained by the theory of plate tectonics?
5. If someone told you that you were walking across rocks that cooled from pahoehoe lavas, what would you know about the characteristics of the lava at the time of eruption?
6. Nuée ardentes and lahars both involve flow of material from a volcano. What are the differences between the two processes?
7. How can you distinguish a shield volcano from a composite volcano?
8. What factors determine whether a dike or sill will form?
9. Summarize Bowen's Reaction Series.
10. What engineering problems should you expect in volcanic rock terrains?

CHAPTER
4
SEDIMENTARY ROCKS AND PROCESSES

Early in the development of geology as a science, James Hutton and others recognized that rocks exposed on the continents, particularly in mountainous areas, are continually broken down into small particles under the relentless attack of the atmosphere. These finer constituents of the rock are gradually transported to lower elevations. This phase of the geologic cycle leads to the formation of sedimentary rocks, the second major rock group. The paths leading from existing rocks to sediments and finally to sedimentary rocks may be long and complicated. Interpretation of the origin and history of sedimentary rocks is aided, however, by using the Principle of Uniformitarianism. Careful observation of modern sedimentary environments will tell us a great deal about ancient sedimentary rocks.

Although sedimentary rocks constitute less than 10% of the rocks in the crust, their importance is emphasized by the fact that they cover about 70% of the surface of the continents. As a result, sedimentary rocks are the most common type encountered in construction projects. In addition, sedimentary rocks contain the deposits of petroleum, coal, and other energy resources as well as a host of other metallic and nonmetallic mineral deposits that are so necessary to our society.

The benefits of utilizing geology in engineering practice have never been better illustrated than by the work of the English civil engineer William Smith (1769–1839). Smith compiled the first real geologic map, which showed the surficial sedimentary rock units in southern England, as a result of his observations of rocks and fossils in connection with the construction of canals and roads. Since the time of Smith, countless studies have demonstrated the relationship and importance of sedimentary rocks to agriculture, water and mineral resources, and construction.

ORIGIN OF SEDIMENTARY ROCKS

The processes involved in the formation of sediment from existing rocks, in addition to those processes relating to the transportation and deposition of sediment, will be described in considerable detail in Part III of this text. For this reason, we briefly introduce these concepts here and concentrate on the characteristics and properties of the sedimentary rocks that result from these processes.

The disintegration of existing igneous, sedimentary, and metamorphic rocks is accomplished by a number of physical and chemical *weathering* processes. The result of this activity is the production of detached rock fragments, or clasts, which can be transported by various components of the hydrologic cycle from elevated areas of the continents to points of lower elevation (Figure 4-1). The agents that effect the *erosion* and *transportation* of sediments include gravity, wind, running water (Figure 4-2), and ice. The ultimate repository of sediments is the ocean basin, but by the time sediments reach the oceans, they have been reduced to extremely small particles. There are many intermediate points of *deposition* along the path from mountain peak to ocean basin. These sites, or *depositional environments*, can include almost any part of the continent or ocean floor. On the continents, depositional environments include alluvial (stream), lacustrine (lake), paludal (swamp), desert (Figure 4-3), and glacial types. Along the continental margins a variety of shallow marine depositional environments can be recognized (Figure 4-4). Finally, sediment can be deposited in deep marine settings out of the influence of continents.

The nature of the depositional environment determines the characteristics of the resulting sedimentary rock. Some of the parameters that influence

Figure 4-1 The weathering of rocks in mountainous regions commonly begins by the formation of large rock fragments, or talus, that accumulate on slopes. (J. T. Hack; photo courtesy of U.S. Geological Survey.)

rock type are listed in Table 4-1. These factors leave their geologic signature upon the sediments and resulting rocks that form in a particular environment. One of the most frequent objectives of studies of sedimentary rocks is to reconstruct the conditions of an ancient depositional environment from the rocks that remain.

A final step is required to convert an aggregate of particles into a sedi-

Figure 4-2 The river plain in the foreground is composed of sediments transported by the stream from uplands in the background.

Figure 4-3 Deserts are depositional environments for sand transported by strong winds. (E. D. McKee; photo courtesy of U.S. Geological Survey.)

Figure 4-4 Coral reefs constitute important shallow marine depositional environments in tropical areas. (J. I. Tracey, Jr.; photo courtesy of U.S. Geological Survey.)

mentary rock. This step is called *lithification*, and it can be accomplished in several ways. The most basic lithification process is *compaction*. Sediment deposited in a loose condition is gradually compacted to a denser state as additional material is added from above. Glaciers and tectonic forces can impose additional stress upon the sediment. If compaction is sufficient, the sediment becomes a rock. A process that accompanies compaction is the expulsion of water from the void spaces between particles as they are forced closer together. Clay-rich sediments undergo a much greater degree of compaction than sands.

TABLE 4-1
Important Factors of Depositional
Environments

Type of transporting agent (water, wind, ice).
Flow characteristics of depositing fluid (velocity, variation in velocity).
Size, shape, depth of body of water, and circulation of water (in lacustrine and marine basins).
Geochemical parameters (temperature, pressure, oxygen content, pH).
Types and abundances of organisms present.
Type and composition of sediments entering environment.

Interparticle attractive forces between clay particles also aid in the lithification process.

Even after sediment is compacted to a state of maximum density, pore space remains between particles. Over long periods of time these pore spaces may be gradually filled by precipitation of solid material from groundwater circulating through the sedimentary sequence. The filling of void spaces by chemical precipitation is appropriately called *cementation*. Cementation is one of the most effective lithification processes because the chemical cement bonds the particles in the rock together. Most commonly, the cementing agents are either silica (SiO_2) or calcium carbonate ($CaCO_3$). Rocks cemented by silica are among the hardest and strongest sedimentary rocks known. Other cementing agents have been recognized in some rocks.

In addition to cementation, *crystallization* can also contribute to the lithification of a rock. Crystallization usually refers to crystal growth within the void spaces in a rock. Under high pressures, however, some of the original minerals in the rock may chemically react to form crystals of new minerals that are more stable under the prevailing conditions.

The actual point at which a sediment becomes a rock is a very difficult judgment to make. This problem arises because lithification is very gradual and the boundary between rock and sediment is completely gradational. Accordingly, some poorly lithified sedimentary rocks may be classified as soils for engineering purposes because their mechanical behavior may be closer to particle-aggregate systems than to well-cemented rock.

CHARACTERISTICS OF SEDIMENTARY ROCKS

Sedimentary rocks exhibit a variety of distinctive features. By carefully observing these characteristics, it is possible to assign the proper rock name to a specimen as well as to interpret the origin and depositional environment of the sediment that was lithified to form the sedimentary rock.

Texture

One of the first observations to make of a sedimentary rock is to determine its texture, which, as is true for the igneous rocks, refers to the size and arrangement of the particles or grains that make up the rock. There are only two main types of texture that we need to be concerned with—*clastic* and *nonclastic*. Clastic rocks are composed of aggregates of individual mineral or rock fragments. When these fragments have been eroded, transported, and then deposited, the origin of the rock can be described as *detrital* (Table 4-2). If the texture of a rock is nonclastic, it means that the grains form an interlocking network similar to igneous rocks with crystalline texture. Nonclastic rocks are formed by chemical precipitation from aqueous solutions. This type of texture can be observed in rocks formed by both inorganic and organic processes (Table 4-2). Organic precipitation is common to organisms that secrete shells composed of calcium carbonate or silica. However, if a rock is composed of an accumulation of shells or fragments of shells, its texture is considered to be clastic.

Once the texture of a rock is identified, it is necessary to observe the size of the rock particles. Geologists often use the Wentworth scale for classification of particle sizes (Table 4-3). The names of several particle sizes are then used to form the rock name for the corresponding rocks. For example, sandstone is mainly composed of particles between $\frac{1}{16}$ and 2 mm in diameter. Because of its resistance to erosion, sandstone often forms cliffs or upland areas (Figure 4-5). Detrital rocks with most particles larger than sand size are called *conglomerate* if the particles are rounded, or *breccia* if the particles are angular (Figure 4-6). Fine-grained clastic rocks are called *shale* or *mudstone* (Figure 4-7). The term shale usually refers to a rock that tends to split into thin slabs parallel to the depositional layering of the sediment.

The particle size of detrital rocks gives an important clue about the depositional environment. In the case of water-laid materials, the ability to transport particles is dependant upon the current velocity of the transporting fluid. Therefore, conglomerates represent high-velocity streams, like those that rush down steep mountain slopes or form at the edge of a melting glacier. In contrast, silt and clay accumulate in low-energy environments like lakes or sea floors.

The chemical and biochemical sedimentary rocks include a wide variety

TABLE 4-2

Classification of Sedimentary Rocks

Origin	Texture	Particle size or composition	Rock name
Detrital	Clastic	Granule or larger	Conglomerate (round grains) Breccia (angular grains)
		Sand	Sandstone
		Silt	Siltstone
		Clay	Mudstone and shale
Chemical Inorganic	Clastic or nonclastic	Calcite, $CaCO_3$	Limestone
		Dolomite, $CaMg(CO_3)_2$	Dolomite
		Halite, $NaCl$	Salt
		Gypsum, $CaSO_4 \cdot 2H_2O$	Gypsum
Biochemical	Clastic or nonclastic	$CaCO_3$ (shells)	Limestone, chalk, Coquina
		SiO_2 (diatoms)	Diatomite
		Plant remains	Coal

SOURCE: From S. Judson, M. E. Kauffman, and L. D. Leet, *Physical Geology*, 7th ed. copyright © 1987 by Prentice-Hall, Inc., Englewood Cliffs, N.J.

TABLE 4-3
Wentworth Scale of Particle Sizes for Clastic
Sediments

Size (mm)	Size name
	Boulder
256	
	Cobble
64	
	Pebble
4	
	Granule
2	
	Sand
1/16 (0.0625)	
	Silt
1/256 (0.0039)	
	Clay

SOURCE: From C. K. Wentworth, A Scale of Grade and Class Terms for Clastic Sediments. *Journal of Geology* 30:381, copyright © 1922, University of Chicago Press.

Figure 4-5 Resistant sandstone exposed in Arches National Park, Utah. Natural arches are formed primarily by wind erosion. (Photo courtesy of R. A. Kehew.)

of materials. Chemical precipitation from water is responsible for deposits of limestone (Figure 4-8), dolomite, salt, and gypsum. This process is common in seawater but also occurs in lakes, streams, caves (groundwater), and springs.

Deposits of marine origin are the most common type of limestone. Most of Florida is underlain by fossiliferous limestone deposited under shallow marine conditions on a carbonate shelf. Areas of carbonate rocks frequently are plagued by rapid subsidence of the land surface to form *sinkholes,* which are caused by the dissolution of the soluble limestone by fresh water falling as precipitation to the ground surface (Chapter 11). Although dissolution of the rock may take thousands of years, collapse of the surficial soil into the underground cavity may be instantaneous. A solution cavity filled with younger

(a)

(b)

Figure 4-6   (a) Conglomerate is a detrital sedimentary rock composed of large, rounded particles. (b) Breccia is similar in grain size to conglomerate but differs because the grains are angular.

soil materials is shown in Figure 4-8. The actual and potential damage caused by subsidence is a major threat to development in carbonate terrains.

Cave and hot-spring deposits often include a form of limestone known as travertine. The spectacular deposits at Mammoth Hot Springs in Yellowstone National Park (Figure 4-9) are a good example of travertine precipitation.

Minerals precipitate from water because the concentration in the liquid phase is greater than the solubility of the solid phase. The concentration at the point of equilibrium between dissolution and precipitation is a function of temperature, pH, and other factors. Some minerals are so soluble that they rarely precipitate. For example, halite requires a very high concentration of

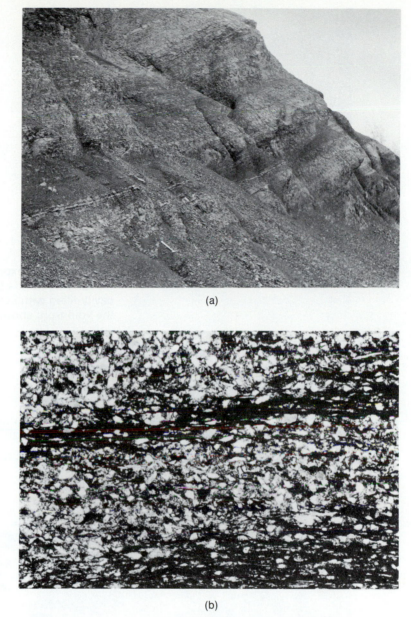

(a)

(b)

Figure 4-7   (a) An outcrop of shale showing the characteristic layering. (B. Bryant; photo courtesy of U.S. Geological Survey.) (b) Under the microscope, the mineral composition of shale can be observed. Minerals in the light-colored bands are mostly quartz, whereas the dark-colored layers are made up of platy crystals of muscovite. (W. M. Cady; photo courtesy of U.S. Geological Survey.)

sodium and chloride to initiate precipitation. The necessary concentrations are sometimes achieved by evaporation of water from a lake or marine basin. The Great Salt Lake in Utah, for example (Figure 4-10), is a remnant of a much larger lake, Lake Bonneville, that existed during a cooler and wetter climatic period during the Pleistocene Epoch. Climatic warming led to a gradual shrinking of the lake by evaporation. As the lake water became concentrated, pre-

Figure 4-8 An exposure of limestone in a quarry in Florida showing a solution cavity filled with soil that collapsed into the void from above.

Figure 4-9 These terraces at Mammoth Hot Springs at Yellowstone National Park were formed by precipitation of travertine from the emerging spring water.

cipitation of halite and other salts formed the Bonneville salt flats, the floor of ancient Lake Bonneville. Salt deposits of this type are called *evaporites*. Evaporites can also form in marine basins that are deeper than the surrounding parts of the sea. The isolation and depth of these basins limits normal oceanic circulation, and the necessary evaporative concentration can occur.

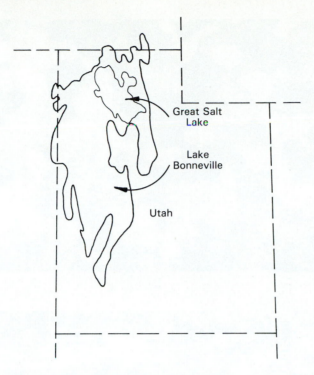

Figure 4-10   The Great Salt Lake is a small remnant of ancient Lake Bonneville. Evaporites were deposited as evaporation caused the lake to shrink in size.

Sorting

The range of particle sizes in a sedimentary rock is known as *sorting*. Rocks that contain a narrow range of sizes are described as well sorted; those that consist of a wide size range are termed poorly sorted (Figure 4-11). The value of sorting in interpretation of sedimentary rocks lies in its use as an environmental indicator. Poor sorting in a water-laid sedimentary rock suggests rapid fluctuations in current velocity and rapid deposition. Wind-deposited sediments, on the other hand, tend to be well sorted because wind is capable of transporting only small particles. Thus the small particles are separated from the large particles and accumulate in one location when the windstorm ceases. Sediment transported directly by a glacier can often be recognized by its very poor sorting. Although glaciers move very slowly, they are equally capable of picking up huge boulders and clay particles as they advance over an area. Very little segregation of the particles occurs during the movement of the glacier.

Sedimentary Structures

Sedimentary structures set sedimentary rocks apart from other rock types. These structures are characteristic features imparted to the rock during the processes of sediment transport and deposition.

The most fundamental sedimentary structure is *bedding*. Individual layers, or beds, of sediment stacked upon each other are the most recognizable indication of sedimentary rock (Figure 4-12). Bedding is the result of variation in the sedimentation process. A slight change in current velocity is all that is necessary to change the size of material being transported and deposited. A

(a)

(b)

Figure 4-11 (a) Photomicrograph of well-sorted quartz sandstone. (W. R. Hansen; photo courtesy of U.S. Geological Survey.) (b) A poorly sorted sandstone under the microscope. Particles range in size from silt to fine gravel. (J. Gilluly; photo courtesy of U.S. Geological Survey.)

wave rushing up the face of a beach will deposit a thin, but distinctive, layer of sediment. The sediment deposited by the next wave will be of a slightly different character. Large exposures of sedimentary rock like the Grand Canyon record much more drastic changes in depositional environments—alternating terrestrial and marine conditions, for example. The layers of bedding are the evidence of these changes.

Differences in the process of sedimentation can lead to different types of bedding. Three main types include *parallel bedding, cross bedding,* and *graded bedding* (Figure 4-13). Parallel bedding can be produced by transport of sediment by currents or by deposition in standing bodies of water where sediment particles fall through a column of water to their resting points on the bottom.

Cross bedding indicates deposition of sediment that moves in waves near the sediment-fluid interface. These conditions are often met by rivers and wind. As the fluid moves above its bed, the coarsest sediment is transported in un-

Figure 4-12 Bedding stands out at great distances in the well-exposed sedimentary rocks of the Grand Canyon. (J. R. Balsley; photo courtesy of U.S. Geological Survey.)

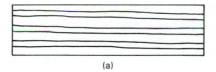

(a)

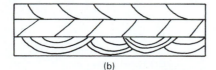

(b)

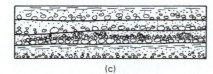

(c)

Figure 4-13 Types of bedding: (a) parallel bedding; (b) cross bedding; and (c) graded bedding.

dulating wavelike forms in the direction of flow. Small waves are called *ripples* (Figure 4-14) and larger sediment waves are called *dunes*. The waveforms move downstream as particles are eroded from the upstream side of a ripple or dune and then cascade down the downstream face. Cross beds are the preserved deposits of the downstream faces of the waves. The repeated migration of sediment waves across the bed results in successive sets of cross beds. Each set

Figure 4-14 Sand ripples on a modern stream bed. The internal structure of ripples and dunes is often preserved as cross bedding.

has upper and lower boundaries parallel to the bed separated by beds inclined to the bottoms at angles ranging from 5 to 40° (Figure 4-15). Large-scale cross bedding is typical of subaerial sand dunes in desert environments. The Navajo sandstone on the Colorado Plateau (Figure 4-16) is interpreted to be a rock formed from eolian deposits.

Graded bedding is similar to parallel bedding except that the grain size within the bed decreases systematically from the bottom toward the top. This gradual change in grain size is produced by a special type of current in which the sediment is carried partially in *suspension* rather than being transported along the bed. Such currents, called *turbidity currents,* are dense, rapidly moving currents that flow down submarine slopes near the edges of continents.

Figure 4-15 Cross bedding in sand. The inclined layers are remnant downstream faces of migrating dunes.

Figure 4-16 Checkerboard Mesa, at Zion National Park, Utah, displays large-scale cross bedding typical of wind-deposited sandstone. (Photo courtesy of R. A. Kehew.)

Graded bedding is produced by the order in which particles settle out of suspension. As the current flows out on to an area of more gentle slopes, the flow velocity gradually decreases. As this occurs, the larger particles begin to fall to the bed, so that large particles are concentrated at the bottom. The process continues, leading to a gradual decrease in grain size from bottom to top. Sequences of turbidities indicate that, in some marine environments, turbidity currents occur at regular intervals. Graded beds therefore alternate with nongraded beds, representing the intervals between turbidity currents (Figure 4-17).

Figure 4-17 Alternating beds of sandstone with graded bedding and shale in a turbidite sequence. (D. G. Howell; photo courtesy of U.S. Geological Survey.)

Other types of sedimentary structures are also recognized in sedimentary rocks. The entire form of small sediment waves, or ripples, is sometimes preserved at the top of a bed of sediment. The resulting sedimentary structures are known as *ripple marks* (Figure 4-18). Sand ripples can often be observed on tidal flats or on stream bottoms. *Mud cracks* (Figure 4-19) are polygons formed at the surface of a fine-grained material as it gradually dries and shrinks. The presence of these structures in an ancient sedimentary rock would be an indication of periodic exposure of the sediment surface, an important clue in deducing the depositional environment of the rock.

Color

The color of sedimentary rocks is often diagnostic of the geochemical environment at the time of formation. Shades of red or brown indicate formation of the sediment in an environment with abundant free oxygen. Under these conditions, iron exists in the ferric, or oxidized state. A small amount of ferric iron is sufficient to impart reddish or yellowish color to the deposit. Sediments that accumulated in environments lacking oxygen are usually darker in color. The somber gray and green shades of these rocks are attributed to the presence of iron in the ferrous, or reduced state. Marine and lacustrine environments frequently have sufficiently low dissolved-oxygen levels to ensure that iron will exist in the reduced form.

If sediment rich in organic matter is deposited in a reducing environment, the resulting sediments will be black in color. Petroleum and coal are good

Figure 4-18 Current ripple marks on bed of sandstone tilted upward by tectonic forces. (R. H. Campbell; photo courtesy of U.S. Geological Survey.)

Figure 4-19 Mud cracks indicate that this bed dried in the open
air soon after deposition. (R. G. Wolff; photo courtesy of U.S.
Geological Survey.)

examples. The lack of oxygen is critical to the preservation of these materials
because bacteria will decompose the organic matter if an oxygen supply exists.

Fossils

The presence of organic remains in sedimentary rocks is a major concern to
the science of geology. It was primarily by the study of fossils that the geologic
time scale was devised.

Living organisms consist of soft and hard parts. The soft parts, including
cell tissue and internal organs, decay rapidly in the presence of oxygen when
the organism dies. Hard parts—bones and shell, for example—can be preserved for long periods of time under favorable conditions in sedimentary deposits (Figure 4-20). Shells are usually composed of calcium carbonate or silica,
whereas bones consist of a phosphatic material.

In anaerobic environments soft organic remains can be preserved, although they may be greatly altered from their original state. These environments account for deposits of petroleum, which are derived from the remains
of microscopic marine organisms, and coal, the product of terrestrial plants.

Fossils also include material that is not necessarily organic in origin.
Petrified wood, for example, is composed of silica that has been precipitated
by groundwater circulating through sedimentary materials. In this process,
silica precipitation gradually replaces the organic material, while retaining
the original cellular structure (Figure 4-21).

Stratigraphy

As William Smith demonstrated, isolated exposures of sedimentary rocks can
be matched, or correlated, to other outcrops of rocks of the same age by careful
study of the physical characteristics of the rock as well as its fossil content.

Figure 4-20   Coquina, a rock composed almost entirely of fossils of marine organisms. (C. Teichert; photo courtesy of U.S. Geological Survey.)

Figure 4-21   Petrified stumps preserved in their original growth position by the replacement of organic matter by dissolved silica carried by groundwater. The stumps have been exposed by erosion of the surrounding materials.

With systematic observations, formations of sedimentary rock can be traced throughout large areas. The branch of geology that deals with the distribution and origin of sedimentary rocks is called *stratigraphy*. The root of this word is *strata*, which refers to the layers of sedimentary rocks. Stratigraphers attempt to reconstruct ancient landforms and depositional environments by studying the sedimentary rocks and fossils that remain from those times. Geologic maps and cross sections constructed from stratigraphic studies (Figure 4-22) are very useful in highway route planning and other engineering projects because they enable a prediction of the sequence of rocks that may underlie a particular location.

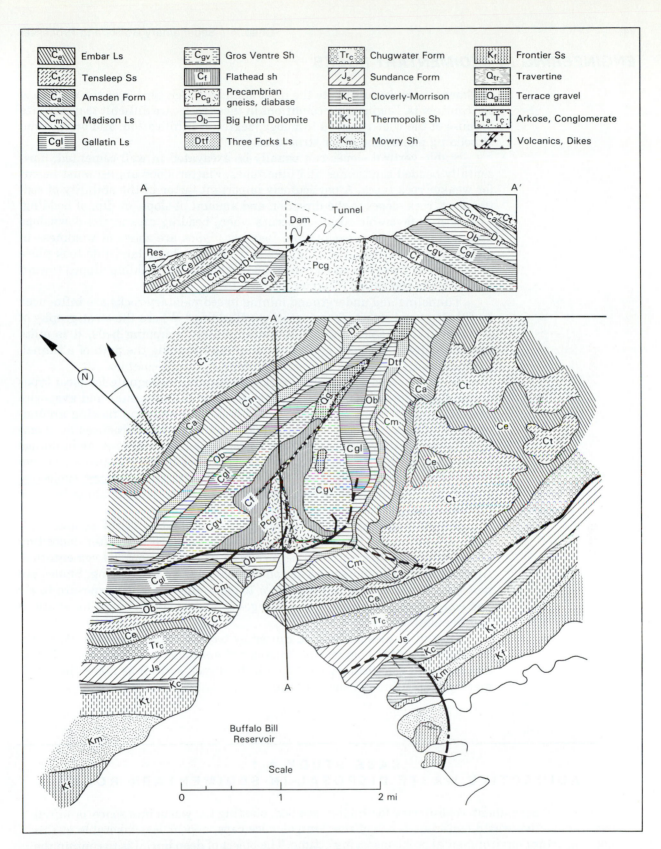

Figure 4-22 Geologic map and cross section of the Cody Highway Tunnels area, Wyoming. (From W. F. Sherman, 1964, "Engineering Geology of the Cody Highway Tunnels, Park County Wyoming." In *Engineering Geology Case Histories* No. 4, edited by P. D. Trask and G. A. Kiersch.)

ENGINEERING IN SEDIMENTARY ROCKS

Sedimentary rocks constitute the most abundant rock type over the surface of the continents. Engineering conditions in these rocks are difficult to generalize because of the wide range in lithology, degree of lithification, and orientation of bedding planes and other structures.

Stable vertical slopes can usually be excavated in well-cemented, horizontally bedded sandstones and limestones. Flatter slope angles must be cut for weaker rock types. A particularly important factor in the stability of sedimentary rock slopes is the direction and amount of slope, or dip, of bedding. The most unfavorable situation occurs where bedding dips in the downslope direction of a slope or excavation. Bedding planes are zones of weakness in sedimentary rock masses and failure may occur. A huge landslide took place at the Vaiont Reservoir in Italy in 1963 partly because bedding dipped toward the center of the valley (Chapter 13).

Tunneling and underground mining in sedimentary rocks are influenced by lithology and structure (orientation of bedding). Where the stratigraphy of sedimentary rocks consists of horizontal or gently dipping beds, it is quite simple to predict the rock types to be encountered along the path of a tunnel. Difficulties arise in areas where the structure is more complex.

Well-cemented sedimentary rocks are generally adequate for most types of building foundations. Special problems occur in limestones and evaporite deposits because these rocks are soluble under the action of flowing groundwater. The soils and rocks overlying underground cavities produced by chemical dissolution may collapse into the voids, damaging or destroying buildings constructed at the surface (Chapter 11). This phenomenon is also a problem above shallow underground mines. Dams and reservoir sites are subject to similar limitations. Undesirable leakage of water may occur along bedding planes or through solution cavities in the rock.

Like the igneous rocks, the engineering properties of the sedimentary rocks are influenced by geologic events that take place long after deposition of the sediments. Strength can be increased by compaction and cementation. Alternatively, sedimentary rocks can be weakened by weathering. Shales are particularly susceptible to breakdown to clay minerals upon exposure to air and water. These relatively rapid changes can lead to problems in excavations and foundations.

Discontinuities in the rock caused by tectonic events are also important in determining the engineering behavior of a sedimentary rock mass. Faults and fractures significantly weaken a body of rock as well as allow the movement of water through the material.

CASE STUDY 4-1
RADIOACTIVE WASTE DISPOSAL IN SEDIMENTARY ROCKS

Selection of permanent repositories for high-level radioactive waste materials is one of the most important environmental problems facing our society. The current approach to disposal of these wastes is to locate a suitable site for deep underground burial. Other disposal alternatives, such as emplacement on or beneath the sea bed, blasting the waste into space, or burial in polar ice caps, appear less favorable at this time. The object of deep burial is to contain the wastes in a geologic environment that will remain stable for at least 10,000 years, allowing natural radioactive decay to reduce the hazard of the wastes. Isolation of the waste for this

Figure 4-23 Potentially acceptable sites for the first high-level radioactive waste repository in the United States. (From U.S. Dept. of Energy, Draft Environmental Assessment—Deaf Smith County Site, Texas.)

period of time must preclude any contact with groundwater that could leach soluble components and transport wastes back to the ground surface or to shallow subsurface rock units.

Public Law 97-425, the Nuclear Waste Policy Act of 1982, established the process and schedule for repository site selection. Under this act, the U.S. Department of Energy (DOE) selected nine potential repository sites in 1983 for further investigation (Figure 4-23). Seven of the nine sites are located in basins containing thick sequences of sedimentary rocks. The remaining two sites, Hanford and Yucca Mountain, are located in basalt lava flows and volcanic tuff, respectively.

In each of the seven sedimentary sites, evaporite deposits constitute the target horizon for the repository, which will be constructed at a depth of between 300 and 1200 m below land surface. Two types of salt deposits are under consideration. Bedded salt (Figure 4-24), in which the evaporite strata occur in beds conformable with the overlying and underlying

formations, is characteristic of the Deaf Smith, Swisher, Davis Canyon, and Lavender Canyon sites. The Vacherie, Richton, and Cypress Creek sites are associated with evaporite bodies known as *salt domes* (Figure 4-25). Salt domes develop from bedded salt units because of the low density and plastic flow of the material. Under the pressure of overlying rock and higher temperatures at depth, salt begins to gradually deform or flow toward a point in the bed that slowly thickens. Because of the low density of salt, thickened zones bulge upward, forming a dome that may extend to the land surface. Sinking occurs adjacent to the dome because of the loss of salt due to flow toward the growing dome. More sediment is deposited in the low-lying areas during dome growth, so that rock formations thicken outward from the center of the dome (Figure 4-25). With continued dome expansion, salt may break through overlying rock units and rise until a stable condition of equilibrium is reached and no further growth occurs. Dissolution of salt at the mar-

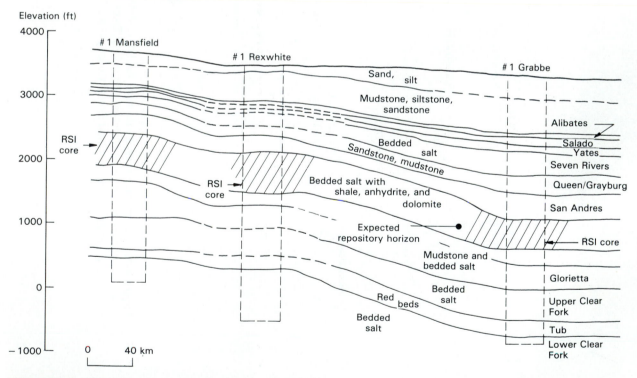

Figure 4-24 Geologic cross section of the bedded salt and associated sedimentary rock units of the Deaf Smith site. (From U.S. Dept. of Energy, Draft Environmental Assessment—Deaf Smith County Site, Texas).

gins of the dome by groundwater flowing in the adjacent rock formations leaves behind less soluble material as a caprock (Figure 4-25). The formation of a caprock isolates the soluble salt to some extent from groundwater in adjoining rocks.

Salt is considered to be a possible material for disposal of high-level radioactive wastes because it is generally dry and impermeable. In addition, the plastic flow of salt causes the natural sealing of any opening that could form. Thus salt will gradually flow in around a waste container placed in an excavation, forming a tight seal. This property also provides protection against migration of waste along rock fractures that could form during earthquakes. Finally, salt allows radioactive heat dissipation and shielding of radioactivity to a greater degree than other rock types.

After selecting the nine potential repository sites, DOE grouped the sites into five settings with similar geologic and hydrologic conditions. For each group that contained more than one site, a preferred site was chosen for that setting. This choice was made by conducting an environmental assessment for each of the nine original sites. The assessment included all known information about the geology of the site and a prediction of the impacts that construction of the site would have upon the environment.

The next step of the selection process was composed of a comparison of the five preferred sites in order to compile a final list of three sites for recommendation to the President for extensive field and laboratory testing, including the construction of test shafts down to the level of the proposed repository. The three final sites include Hanford, Yucca Mountain, and Deaf Smith. After site characterization is complete at these sites, a final choice will be made for the repository. This process is expected to be completed by the end of the century.

The completed radioactive waste repository may look something like the drawing in Figure 4-26. In this plan for the Deaf Smith site, waste would be stored in rooms excavated at depths of approximately 800 m below land surface. The surface facilities for the repository would occupy an area of about 1.6 km^2, whereas the area of the underground openings would be about 7.5 km^2. The design lifetime of the repository is 25 to 30 years. After that time, the site will be closed and sealed.

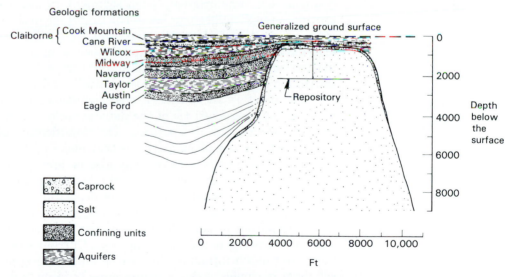

Figure 4-25 Geologic cross section of the Vacherie Dome site. Notice caprock above and on sides of dome. Sedimentary rock beds are thin above dome and thicken away from dome. Aquifers are rock units that contain economically significant groundwater supplies, and confining units are beds that do not contain economically significant groundwater resources. (From U.S. Dept. of Energy, Draft Environmental Assessment—Vacherie Dome Site, Louisiana.)

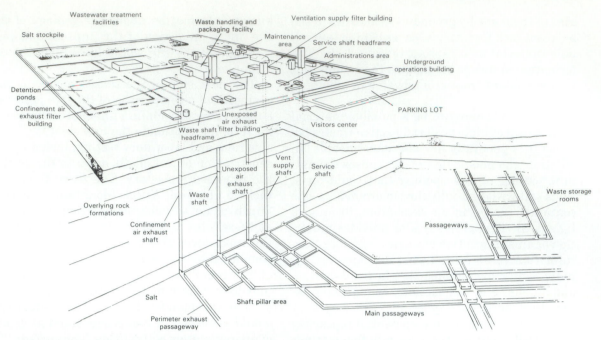

Figure 4-26 Preliminary design of a radioactive waste repository at the Deaf Smith site. (From U.S. Dept. of Energy, Draft Environmental Assessment—Deaf Smith County Site, Texas.)

SUMMARY AND CONCLUSIONS

Sedimentary rocks originate from sedimentary processes that have been operating throughout geologic time. The first stage in the evolution of these rocks is the weathering and erosion of existing rocks. The products produced by this activity are then transported and deposited in a variety of terrestrial and marine environments. After deposition, compaction, cementation, and crystallization cause lithification of the initially soft sediment.

Sedimentary rocks exhibit either clastic or nonclastic textures. Clastic textures are characteristic of the detrital rocks, those that have been transported to their point of deposition. The classification of the detrital rocks depends upon the size of the particles that make up the rock.

Although clastic textures may also be present in the chemical and biochemical sedimentary rocks, many of these rocks are composed of grains that chemically precipitated at the site of deposition. Nonclastic textures, characterized by interlocking grain networks, result from this process.

The nature of each depositional environment imparts numerous characteristic features to the resulting rocks. Current strength, mode of transport, and water depth influence the sorting and bedding of the rock. Transport of sediment in ripples or dunes causes cross bedding to develop. Parallel bedding forms both by current action and by deposition in standing water. Turbidity currents transport high concentrations of suspended sediment that is deposited in graded beds.

The color and fossil content of sedimentary rocks are partially determined by geochemical factors. The presence or absence of oxygen is an extremely important variable in this regard.

Study of the stratigraphy of sedimentary rocks, using all the characteristics that have been described, has produced important information about the geologic past as well as practical data of great value to engineering projects.

REFERENCES AND SUGGESTIONS FOR FURTHER READING

BLATT, H., G. MIDDLETON, and R. MURRAY. 1980. *Origin of Sedimentary Rocks,* 2d ed. Englewood Cliffs, N.J.: Prentice-Hall, Inc.

GEOLOGICAL SOCIETY OF AMERICA. 1964. *Engineering Geology Case Histories Numbers 1– 5,* edited by P. D. Trask and G. A. Kiersch. Boulder, Colo.

JUDSON, S., M. E. KAUFFMAN, and L. D. LEET. 1987. *Physical Geology,* 7th ed. Englewood Cliffs, N.J.: Prentice-Hall, Inc.

U.S. DEPARTMENT OF ENERGY. 1984. Draft Environmental Assessment—Deaf Smith County Site Texas (DOE/RW-0014).

U.S. DEPARTMENT OF ENERGY. 1984. Draft Environmental Assessment—Vacherie Dome Site, Louisiana (DOE/RW-0016).

PROBLEMS

1. How do sedimentary rocks fit into the geologic cycle?
2. What characteristics of the depositional environment influence the properties of sedimentary rocks?
3. How do postdepositional changes convert sediments to sedimentary rocks?
4. What chemical conditions influence the precipitation of nonclastic sedimentary rocks?
5. How do cross beds form?
6. What does color indicate about the geochemical conditions of the depositional environment?
7. What methods are available to gather data necessary to construct a stratigraphic cross section in an area? Why are cross sections useful in highway construction and other engineering projects?
8. What properties influence the strength or possible failure of slopes underlain by sedimentary rocks?
9. What types of sedimentary rocks are considered to be potentially suitable for high-level nuclear-waste disposal? Why?

METAMORPHIC ROCKS AND PROCESSES

The igneous and sedimentary rocks that form at or beneath the earth's surface are not the final result of the dynamic activity of the earth's crust. When exposed to the atmosphere, the minerals in these rocks may chemically alter to new minerals that are more stable under the temperatures and pressures at the earth's surface. Similarly, when igneous and sedimentary rocks become buried deep within the crust because of the movements of lithospheric plates, they are subjected to elevated temperatures and pressures. Just as the rocks are unstable at low temperatures and pressures, they are also unstable at the temperatures and pressures existing at depths well below the surface. Under these conditions, minerals in the rock recrystallize to new minerals, and new rock is formed. The processes that cause these changes are grouped under the term *metamorphism*. Study of these rocks yields valuable information about the effects of these conditions on rock and about the geologic history of a region.

METAMORPHIC PROCESSES

The normal increases in temperature and pressure with depth in the earth are sufficient to initiate metamorphic activity. In addition, tectonic stresses caused by the collision of two lithospheric plates, for example, can generate intense pressures oriented in a particular direction (Figure 5-1). The heat and pressure in deep crustal environments are two of the main components of metamorphic processes. The third component is the effect of liquid or gaseous solutions that move through the rocks and promote the chemical reactions that form new minerals.

Heat is perhaps the most important metamorphic agent. In Figure 5-2 the rate of increase of temperature with depth, or the *geothermal gradient*, suggests that the temperature in the crust at a depth of 15 km is approximately 300°C. This temperature is sufficient for recrystallization of some minerals to begin. Because the geothermal gradient is highly variable from place to place, the depth associated with a particular temperature is not constant.

The effect of pressure varies at different depths in the crust. At shallow depths, rocks are relatively cold and brittle, so they can be fractured and crushed when subjected to high pressures. At greater depths, rocks are much softer because of the high temperatures. Under the action of pressure, they tend to deform by plastic flow, like modeling clay squeezed between the fingers. In the region of plastic deformation, pressure influences the types of new minerals formed. Typically, the atoms within the mineral structure are more closely packed together when the mineral crystallizes under high pressure.

The recrystallization of minerals during a metamorphic event is largely a solid-phase process. The rock does not actually melt and then recrystallize.

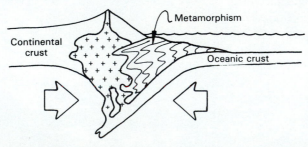

Figure 5-1 Metamorphism at a convergent plate margin. Rocks metamorphosed may include sediments or other continental or oceanic crustal material.

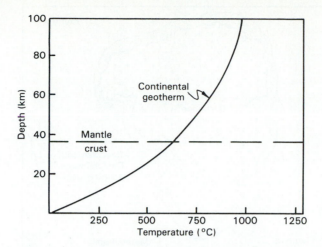

Figure 5-2 Representative continental geotherm. (From G. C. Brown and A. E. Mussett, *The Inaccessible Earth*, copyright © 1981 by George Allen & Unwin, Ltd., London.)

The solid-phase reactions between minerals are greatly facilitated by the movement of small amounts of liquid or gaseous solutions through the rock. These solutions, which travel through the pores and cracks of the rock, add and remove various ions and molecules as the reactions occur. In this way new chemical constituents can be brought in contact with mineral grains so that they may diffuse through the mineral structures during recrystallization.

Types of Metamorphism

Metamorphism is associated with several types of geologic events. The intrusion of a pluton, for example, brings magma into contact with existing crustal rocks (Figure 5-3). *Contact metamorphism* is the name given to the alteration of the surrounding *country* rock by the intruding magma. Heat is the most significant influence in contact metamorphism. The effect of pressure is much less significant. The extent of contact metamorphism, the zone of altered rock called the *metamorphic aureole,* rarely extends more than several hundred meters outward from the magmatic body. Thus contact metamorphism is limited in areal extent to a thin shell around the pluton. More extensive effects may occur if solutions and vapors given off by the magma penetrate into the country rock along fractures. These *hydrothermal solutions* carry volatile components that separate from the cooling magma. Important vein-type ore deposits can be formed by hydrothermal solutions that migrate during contact metamorphic events.

A second type of metamorphism is characterized by much more extensive zones of rock alteration than contact metamorphism zones. Its name, *regional metamorphism*, suggests a process that may transform rocks through huge portions of the crust. The conditions that produce regional metamorphism involve the effects of both temperature and pressure. Thus regional metamorphism must occur deep within the crust, at least at depths of 10 km or more.

Geologic studies of metamorphic rocks indicate that the mineral content of rocks in regionally metamorphosed areas varies systematically. The specific group of minerals present in rock can be used to infer a certain *metamorphic grade* at that particular point. High grade, for example, means that the rock was subjected to very high temperatures and pressures. Different metamorphic minerals can be produced from the same original rock under various meta-

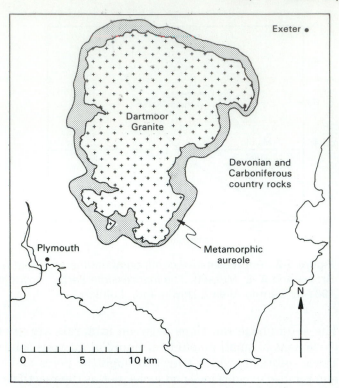

Figure 5-3 The contact metamorphic aureole surrounding the
Dartmoor Granite, Devon, England. (From R. Mason, *Petrology of
the Metamorphic Rocks*, copyright © 1978 by George Allen &
Unwin, Ltd., London.)

morphic grades. Figure 5-4 shows the conditions of temperature and pressure
under which the minerals kyanite, sillimanite, and andalusite form. All three
metamorphic minerals have the same chemical composition (Al_2SiO_5) but
different internal structures. Regional metamorphism produces areas in which
metamorphic grade is very high. Grade decreases in all directions from the
most intensely altered zones (Figure 5-5).

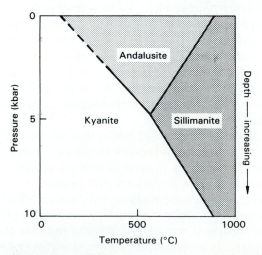

Figure 5-4 Metamorphic minerals formed from Al_2SiO_5 under
varying conditions of temperature and pressure. (From S. Judson,
M. E. Kauffman, and L. D. Leet, *Physical Geology*, 7th ed.,
copyright © 1987 by Prentice-Hall, Inc., Englewood Cliffs, N.J.)

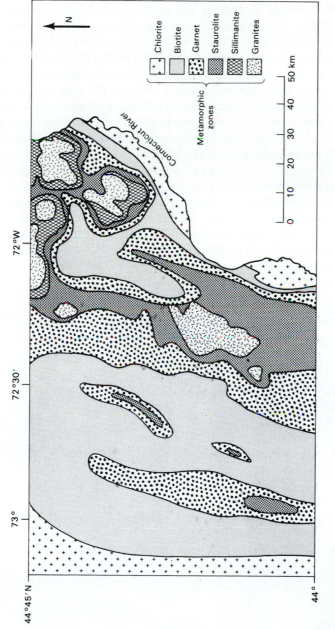

Figure 5-5 Regional metamorphism in northern Vermont. Metamorphic grade decreases outward from granitic plutons. (From F. J. Turner, *Metamorphic Petrology*, 2d ed., copyright © 1981 by Hemisphere Publishing Corp., New York.)

Regional metamorphism is often associated with the central cores of mountain ranges, whose rocks are exposed after the removal by erosion of thousands of meters of overlying rocks. This relationship suggests that regional metamorphism may be a part of the mountain-building process. A possible scenario for regional metamorphism involves the intense lateral pressure generated at convergent plate margins. Thick wedges of sediment deposited adjacent to continents or island arcs are downwarped as the plates collide. Granite batholiths are often found at the centers of these metamorphosed zones surrounded by metamorphic rocks that decrease in grade radially outward (Figure 5-5). Rocks called *migmatites* (Figure 5-6) exhibit characteristics of both igneous and metamorphic rocks. Migmatites can often be observed to merge laterally with granite of igneous appearance in one direction and with unmistakable metamorphic rocks in the opposite direction. All these relationships suggest that metamorphism of downwarped crustal rocks may be so intense that remelting occurs in the zone of highest temperature and pressure. Thus granitic batholiths surrounded by metamorphic rocks may represent an episode of plate collision in the geologic past.

The dominant effects of pressure lead to the recognition of a third type of metamorphism. In the upper part of the crust, the intense pressure associated with folding and faulting is sufficient to crush and pulverize the minerals along a fault plane. The alteration of rocks under these conditions is called *dynamic metamorphism*. The zones of crushing are limited to the close proximity of planes of shearing between adjacent masses of brittle rock. Temperature is a minor factor in dynamic metamorphism.

Figure 5-6   Migmatite, a metamorphic rock containing thin layers of light-colored granitic igneous rock. (C. C. Hawley; photo courtesy of U.S. Geological Survey.)

CHARACTERISTICS OF METAMORPHIC ROCKS

A major division of metamorphic rocks can be made according to texture. The aspect of texture that is most important is called *foliation*, the parallel orientation of mineral grains within a metamorphic rock. Foliation gives metamorphic rocks a banded or layered appearance because foliation resembles the bedding of sedimentary rocks (Figure 5-7). In reality, however, foliation is produced by an entirely different mechanism from bedding. Tectonic stresses acting on rocks during metamorphism are usually applied in one principal direction. The stress, or pressure, acting in other directions is considerably lower. When minerals recrystallize under these pressure conditions, elongated mineral grains that form in the direction of lowest pressure are favored (Figure 5-8). The long axes of the mineral grains are therefore perpendicular to the direction of greatest pressure, and the parallel orientation of these grains, the micas for example, gives the rock foliation.

Metamorphic rocks can be classified as foliated or nonfoliated. Foliated rocks are associated with regional metamorphism where tectonic stress is a major factor. Nonfoliated rocks are produced during contact metamorphism or

Figure 5-7 Foliation, the orientation of platy minerals that gives metamorphic rocks a layered appearance.

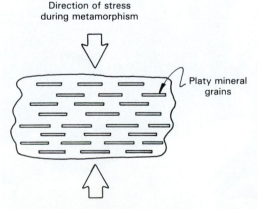

Figure 5-8 Orientation of platy mineral grains with respect to direction of highest pressure during metamorphism.

where platy minerals cannot crystallize because of a lack of necessary chemical components.

The presence of foliation also imparts to a rock a tendency to split or break along the foliation planes rather than across the planes or at some other orientation. This inclination to break along planes of weakness is called cleavage, although it must be kept in mind that the planes of weakness are not of the same origin as cleavage in a mineral.

Several common foliated metamorphic rocks are named because of the type of cleavage that they possess. If the cleavage planes are very thin and the rock is fine grained, the cleavage is known as *slaty cleavage*. The resulting rock is *slate* (Figure 5-9), a rock that has been used as roofing material for

Figure 5-9   Well-developed slaty cleavage in slate outcrop near Kodiak, Alaska. (G. K. Gilbert; photo courtesy of U.S. Geological Survey.)

centuries because of its smooth and regular cleavage surfaces. Slate is usually produced by low-grade metamorphism of shale, primarily consisting of directed pressure and fairly low temperatures. With increasing metamorphic grade, slate can be converted to *phyllite*, so named because of its phyllitic cleavage. Phyllite differs from slate in that it is somewhat coarser grained. In addition, the flaky mineral grains of phyllite possess a distinctive silky luster.

Schist (Figure 5-10) is a very common foliated rock in which the grains are coarser than in slate or phyllite. In addition, the surfaces of cleavage planes are relatively rough because of coarse grain size. Schist represents a still-higher metamorphic grade. Many types of igneous and sedimentary rocks can be metamorphosed to form schist. Because of the variety of mineralogy in schists, the rock name is formed by using the name of the most abundant mineral or minerals. Thus, a biotite schist would be mainly composed of biotite with lesser amounts of other minerals.

The foliated rock that develops under high-grade metamorphic conditions is known as *gneiss*. Gneiss is a coarse-grained, coarsely banded rock (Figure 5-11). The foliation consists of alternate bands of light- and dark-colored minerals. The light-colored layers are mainly composed of quartz and feldspar whereas the dark layers contain biotite, hornblende, augite, and other minerals. Gneisses are formed from silicic igneous rocks as well as various types of sedimentary rocks.

Of the nonfoliated metamorphic rocks, *quartzite* and *marble* are the most common. Quartzite is the name given to metamorphosed quartz sandstone (Figure 5-12). Recrystallization of quartz in the original sandstone fills voids between the original grains and increases the strength and density of the rock. Quartzites are among the strongest and hardest rock types. Because quartzite consists mainly of nonplaty quartz grains, foliation is lacking, although some quartzite may appear foliated because of relict bedding structures in the rock.

Figure 5-10 An outcrop of schist. The well-developed foliation is evident. (W. B. Hamilton; photo courtesy of U.S. Geological Survey.)

Figure 5-11 Gneiss, a coarse-grained metamorphic rock, with broad light- and dark-colored bands of foliation. (R. L. Parker; photo courtesy of U.S. Geological Survey.)

Figure 5-12 Quartzite, a very hard metamorphic rock in which recrystallization forms a dense network of quartz grains. (J. B. Hadley; photo courtesy of U.S. Geological Survey.)

Figure 5-13 Metamorphism of limestone or dolomite produces marble, a rock commonly used as building stone. The coarse, sugary appearance on the broken face of the specimen is characteristic.

Figure 5-14 Fault breccia composed of dolomite fragments. (C. Deiss; photo courtesy of U.S. Geological Survey.)

Marble (Figure 5-13) is recrystallized limestone or dolomite. The recrystallization process produces large interlocking grains of calcite or dolomite. Impurities in the rock give marble a number of possible colors. Many of these varieties are highly sought after for use as decorative building stone.

Rocks formed by dynamic metamorphism may be foliated or nonfoliated. If rocks adjacent to a fault zone are crushed by pressure and shear displacement, a structureless *fault breccia* (Figure 5-14) may be produced. If the zone of shear is subjected to an intense degree of crushing, a fine-grained rock called *mylonite* may be formed. Under these conditions, recrystallization and foliation are produced.

ENGINEERING IN METAMORPHIC ROCK TERRAINS

The engineering characteristics of metamorphic rocks can be generalized into two basic types. Nonfoliated rocks possess similar engineering properties to intrusive igneous rocks. In an unaltered and unfractured condition, they can be considered to be strong materials, with few limitations for foundations, tunnels, and dams. Vertical excavation slopes will remain stable. Foliated metamorphic rocks, however, are more similar to sedimentary rocks because of their tendency to fail along specific planes. Foliation planes in this instance are similar to bedding planes. The orientation of foliation planes with respect to a natural slope or excavation, therefore becomes critical to the stability of the material.

In a way similar to the igneous and sedimentary rocks, the ultimate behavior of a metamorphic rock mass depends upon the degree and orientation of fractures and the weathering characteristics. These properties must be ascertained prior to construction of each individual engineering project.

CASE STUDY 5-1
FAILURE OF THE ST. FRANCIS DAM

Construction in metamorphic rock terrains requires careful mapping of rock types and foliation directions. Foliation imparts a directional weakness to the rock that can be critical to the design and operation of engineering projects. On March 12, 1928, only months after construction was completed, the St. Francis Dam in California (Figure 5-15) collapsed, releasing a wall of water 30 to 50 m high that caused more than 400 deaths and great destruction in the valley below.

The geologic setting of the dam site is shown in Figure 5-16. On the east side of the valley, schists with foliation planes inclined toward the center of the valley formed the foundation material for the dam. Beneath the center of the

dam, the schists were in fault contact with sedimentary rocks. The fault zone itself contained brecciated rocks produced during fault movement.

Although the precise cause of the dam collapse has never been proven, it is highly likely that water from the impounded reservoir seeped into the rocks beneath the dam, thus weakening the rock materials. On the east side, penetration of water into the schists may have weakened the resistance to failure along the inclined foliation planes in this abutment (Outland, 1977). At a critical point, failure of the rock abutment may have initiated the collapse of the dam.

(a)

(b)

Figure 5-15 St. Francis Dam before (a) and after (b) failure. (Photos courtesy of Los Angeles Department of Water and Power.)

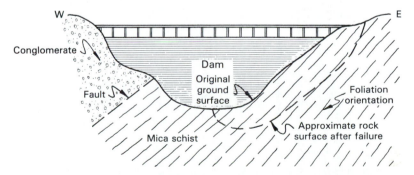

Figure 5-16 Geologic cross section of the St. Francis Dam. Failure probably occurred in the schists beneath the east side of the dam. (Modified from C. F. Outland, *Man-made Disaster*: *The Story of St. Francis Dam*, copyright © 1977 by the Arthur H. Clark Co., Glendale, Calif.)

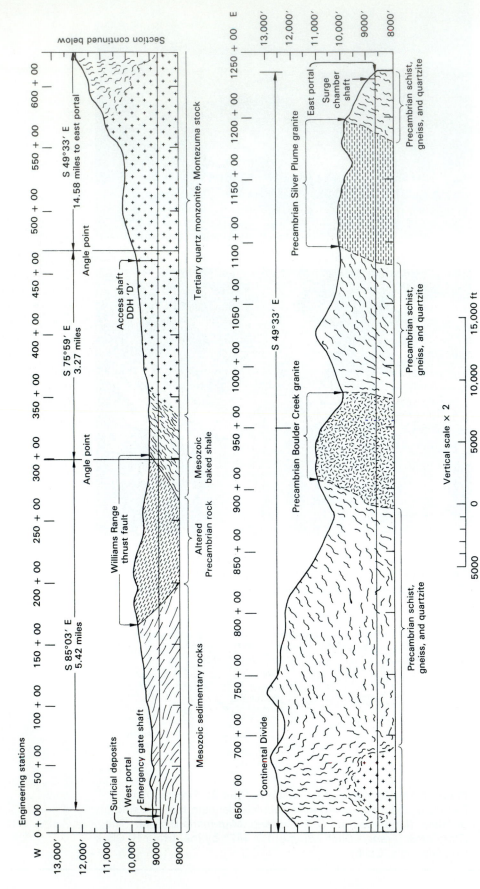

Figure 5-17 Generalized geologic section along the route of the Harold D. Roberts Tunnel, Colorado. (From L. A. Warner and C. S. Robinson, 1981, *Geology of the Eastern Part of the Harold D. Roberts Tunnel, Colorado* (*Stations 690 + 00 to 1238 + 58*), U.S. Geological Survey Professional Paper 831-D. Diagram reprinted with permission of the Board of Water Commissioners, City and County of Denver.)

114

CASE STUDY 5-2
TUNNELING PROBLEMS IN METAMORPHIC ROCKS

Tunneling often presents some of the most complex and dangerous problems in civil engineering. These problems usually result from unknown or unexpected adverse geologic conditions encountered along the path of the tunnel.

The Harold D. Roberts Tunnel was constructed to divert water from the west slope of the Continental Divide, through the Front Range of the Rocky Mountains, for municipal use in the city of Denver, Colorado. Problems experienced during the construction of this 37.5-km, concrete-lined tunnel were directly related to distribution of rock types along the center line of the tunnel (Warner and Robinson, 1981). The geologic setting of this part of the Front Range consists of a highly complex series of igneous, metamorphic, and sedimentary rocks (Figure 5-17). The Precambrian metamorphic rocks, including schist, gneiss, quartzite, and migmatite, provided some of the most troublesome tunneling conditions along the route. Foliated and nonfoliated metamorphic rocks are present in various sections of the route, with biotite responsible for most of the foliation.

One of the major problems was the inflow of groundwater at high pressure from fractured rocks at certain locations. Flows as high as 32 L/s (500 gal/min) were measured. In order to test for zones of water inflow, which could not be predicted in advance, feeler holes were drilled ahead of the advancing tunnel face. When a flow was detected, grout (cement), was injected into the rock at high pressure to seal the fractures. Relatively strong, nonfoliated gneiss frequently yielded high inflows of water. Apparently, this rock behaved as a brittle material that could maintain open fractures, unlike the weaker foliated schist, in which fractures were more likely to seal themselves under the immense weight of the mountain range.

The cost of a tunnel is directly related to the number of steel supports that are needed to prevent collapse of the rock walls into the opening. These supports were installed as needed during construction at 0.6- to 0.18 m centers along the tunnel. The percentage of a particular segment of tunnel that required steel supports ranged from 7.3 to 100, depending upon the rock type.

When steel supports are needed at a section, the contractor must blast and excavate a larger

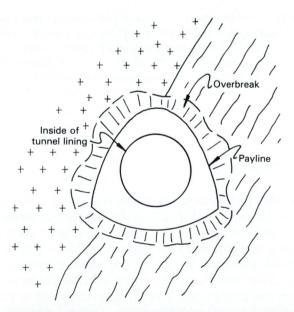

Figure 5-18 Cross section of a tunnel showing the overbreak, or area excavated beyond the payline to accommodate steel supports. The payline is the outer boundary of the area estimated for excavation prior to construction.

cross-sectional area of rock than the estimated cross-sectional area, which is the area inside the *payline* (Figure 5-18). The extra area excavated is called *overbreak*. Overbreak increases the cost of a tunnel to the contractor, because more cement is needed to construct the finished tunnel lining. In Figure 5-19, mean overbreak, roughness factor, and eccentricity are plotted against percent support for a section of the Harold D. Roberts Tunnel. Roughness factor and eccentricity are parameters related, respectively, to irregularities and to differences in direction of the amount of overbreak. The graph clearly shows that, of the rock types compared, the foliated gneiss and schist had the highest values of the overbreak parameters and required the highest percentage of support. In this case, foliated metamorphic rocks were more costly to tunnel construction than other rock types.

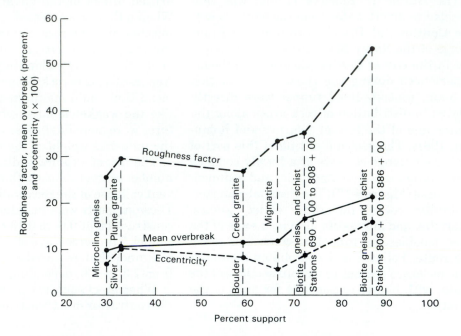

Figure 5-19   Plot of various overbreak parameters versus percent support, showing that overbreak and therefore cost of tunneling increases in areas of foliated metamorphic rocks. (From L. A. Warner and C. S. Robinson, 1981, *Geology of the Eastern Part of the Harold D. Roberts Tunnel, Colorado* (*Stations 690 + 00 to 1238 + 58*), U.S. Geological Survey Professional Paper 831-D.)

SUMMARY AND CONCLUSIONS

Metamorphic processes produce new rocks under the influences of heat, pressure, and hydrothermal solutions. Contact metamorphism is produced by the emplacement of magma bodies within the crust. Under the action of heat from the cooling magma, rocks are altered within a thin zone around the pluton. Nonfoliated metamorphic rocks are most commonly produced by contact metamorphism because of the minor effect of pressure.

Pressure plays a much more important role in the formation of foliated rocks during regional metamorphism. Tectonic stress, the result of lithospheric plate interactions, combined with heat from deep burial produces metamorphism on a regional scale. Metamorphic grade often decreases outward from a core of intensely altered rocks, where remelting may occur. This process may explain the huge granitic batholiths of the crust.

Dynamic metamorphism produces alteration of rocks by pressure and shearing in zones parallel to faults. These effects are limited to the brittle, upper zones of the crust.

Foliated metamorphic rocks develop by recrystallization of long, platy minerals orientated normal to the direction of greatest pressure. Foliation is associated with various types of rock cleavage. With increasing metamorphic grade, slate, phyllite, schist, and gneiss can be formed.

Nonfoliated rocks, such as quartzite and marble, involve recrystallization without the development of mineral-grain alignment.

The orientation of foliation is an important factor in construction in areas of metamorphosed rocks. The strength of a rock mass is much lower in the direction of the foliation than in other directions.

REFERENCES AND SUGGESTIONS FOR FURTHER READING

MASON, ROGER. 1978. *Petrology of the Metamorphic Rocks.* London: George Allen & Unwin, Ltd.

OUTLAND, C. F. 1977. *Man-made Disaster: the Story of St. Francis Dam.* Glendale, Calif.: The Arthur H. Clark Co.

TURNER, F. J. 1981. *Metamorphic Petrology,* 2d ed. New York: McGraw-Hill.

WARNER, L. A., and C. S. ROBINSON. 1981. *Geology of the Eastern Part of the Harold D. Roberts Tunnel, Colorado (Stations 690 + 00 to 1238 + 58).* U.S. Geological Survey Professional Paper 831-D.

PROBLEMS

1. Where and why does metamorphism occur in the earth?
2. What are the differences between regional and contact metamorphism?
3. Explain how metamorphism can alter a rock chemically as opposed to mineralogically.
4. How is foliation produced in a metamorphic rock?
5. How do the characteristics of the original rock influence the properties of the metamorphic rock?
6. What are some of the engineering problems associated with metamorphic rocks?

CHAPTER
6
MECHANICS OF EARTH MATERIALS

It is not enough for engineers to have a basic understanding of the physical, chemical, and mineralogical characteristics of rocks, for in dealing with rock as an engineering material, engineers are involved with the *mechanics* of rock and other earth materials. Mechanics refers to the response of materials to applied loads.

GENERAL TYPES OF EARTH MATERIALS

Rocks, Soils, and Fluids

From an engineering perspective, earth materials can be subdivided into three categories: rocks, soils, and fluids. Rocks are solid, dense aggregates of mineral grains. Igneous and metamorphic rocks are composed of interlocking mineral grains that formed by crystallization from a magma or recrystallization of an existing rock. The degree of interlocking is one of several factors that determine how strong a rock will be. Chemically precipitated sedimentary rocks can also be interlocking aggregates of mineral grains. Clastic sedimentary rocks, on the other hand, are composed of particles that were derived from a preexisting rock, transported by wind, water, or ice, deposited at a particular location, and bound together by various types and amounts of cementing agents. *Soils*, from the engineering standpoint, are similar to transported sedimentary rocks in that they consist of rock particles and minerals derived from preexisting rocks. Soils, however, lack strong cementing material between grains. For this reason, they usually can easily be excavated without heavy machinery or blasting.

The third major category of earth materials is composed of fluids. The earth's fluids include such liquids as water, magma, and petroleum, but also gases as, for example, natural gas beneath the land surface and the gases of the atmosphere above the surface.

Phase Relationships

All rocks and soils contain some void space between particles. This void space may be occupied by liquids as well as by gas. For example, a sandstone rock unit several thousand meters below the surface may contain petroleum and natural gas. Near the surface, however, void spaces are usually filled or partially filled with water. Gases like nitrogen, oxygen, and carbon dioxide fill the remainder. Rocks and soils, therefore, must be considered as three-phase systems in many instances.

The volume of void space in an earth material can range from a very small percentage of the solid volume to several times the volume of the solid material. The relative amount of void space, along with the type and amount of fluid occupying the space, have an important influence on the mechanical behavior of the material.

The relative amount of void space is quantified by the use of the parameters *porosity* and *void ratio* (Figure 6-1). The porosity, n, is defined as

$$n = \frac{V_v}{V_T}$$ Eq. 6-1

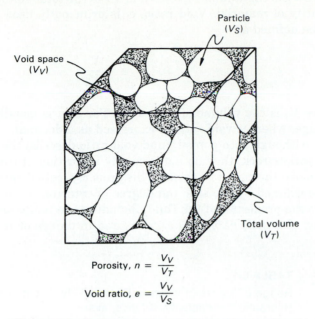

Figure 6-1 Definition of porosity and void ratio of an earth material. Void space (V_v) includes liquid and gas phases.

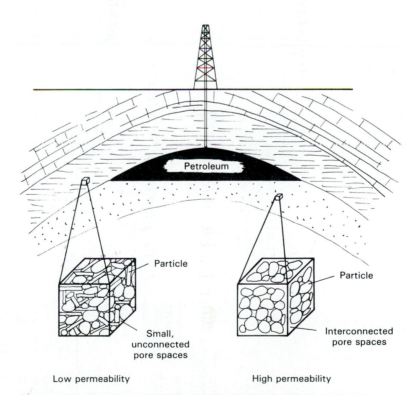

Figure 6-2 Rocks with large, interconnected pore spaces allow the rapid migration of fluids like petroleum or water, but rocks with low permeability transmit fluids very slowly.

where V_v is the volume of voids and V_T is the total volume of a representative quantity of material. Void ratio, e, is principally used by soils engineers. It can be defined as

$$e = \frac{V_v}{V_s} \qquad \text{Eq. 6-2}$$

where V_s is the volume of soil solids. Porosity is usually expressed as a percentage, whereas void ratio is expressed as a decimal.

Although both porosity and void ratio describe the amount of void space in a soil or rock, they do not give any indication of the rate at which fluids will move through a material. This property, called *permeability* (Figure 6-2), is determined by the size and degree of interconnection of the voids, as well as by the properties of the fluid, including the temperature, density, and viscosity. *Intrinsic permeability* measured in darcys or m² (Table 6-1) refers to

TABLE 6-1

Range of Values of Intrinsic Permeability, k (cm², darcys) and Hydraulic Conductivity, K (cm/s, m/s)

SOURCE: From R. A. Freeze and J. A. Cherry, *Groundwater*, copyright © 1979 by Prentice-Hall, Inc., Englewood Cliffs, N.J.

the material properties that influence fluid movement. The characteristics that determine intrinsic permeability include the size, shape, and packing of grains, as well as the degree of cementation and fracturing. Intrinsic permeability is of great importance to petroleum geologists and engineers. A rock may be saturated with oil, but if the permeability of the rock is low, oil would not move rapidly enough to a well for the well to be considered economically feasible.

A composite parameter, *hydraulic conductivity* (m/s), is used by groundwater hydrologists and soils engineers to measure the ability of a rock or soil to transmit water (Table 6-1). Hydraulic conductivity, K, is defined as

$$K = k \frac{\rho g}{\mu} \qquad\qquad \text{Eq. 6-3}$$

where k is the intrinsic permeability, ρ and μ are the density and viscosity of water, and g is the acceleration of gravity. The properties of the material and the fluid are combined in hydraulic conductivity because water at relatively constant values of density and viscosity is usually the only liquid encountered when dealing with the upper part of the earth's crust.

STRESS, STRAIN, AND DEFORMATIONAL CHARACTERISTICS

Stress

When a load is applied to a solid, it is transmitted throughout the material (Figure 6-3). The load subjects the material to pressure, which equals the amount of load divided by the surface area of the external face of the object over which it is applied. Pressure, therefore, is force per unit area. Within the material, the pressure transmitted from the external face to an internal location is called *stress*. Stress, at any point within the material, can also be defined as force per unit area.

Stresses can be classified according to their orientation within a body (Figure 6-4). Stresses of equal magnitude that act toward a point from opposite directions are called *compressive stresses*. When the stresses are directed away from each other, *tensile stress* is present. The third type of stress, *shear stress*,

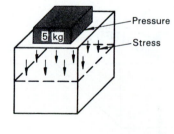

Figure 6-3 A weight resting on a block causes pressure on the external surface of the block and stress on internal planes in the body.

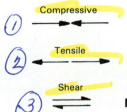

Figure 6-4 Classification of stress.

includes stresses that are offset from each other and act in opposite directions, as in a couple.

On any plane passed through a solid body, there are stresses acting normal to the plane, either compressional or tensional, as well as shear stresses acting parallel to the plane (Figure 6-5). In mechanics, it is possible to resolve the stresses acting at any point within an object into three mutually perpendicular *principal stresses* that are called the maximum, minimum, and intermediate principal stresses (Figure 6-5). The planes perpendicular to the directions of the three principal stresses have no shear stresses acting upon them.

At shallow depths beneath the earth's surface, the vertical and lateral (horizontal) stresses present are due to the weight of the overlying rocks, soil, and air. Thus in Figure 6-6, the vertical stress, σ_v, acting on a horizontal plane at a depth h can be calculated as

$$\sigma_v = \gamma h + P_a$$ Eq. 6-4

where γ is the unit weight of rock material and P_a is atmospheric pressure acting on the surface above the rock. The vertical stress is calculated like hydrostatic stress in a liquid, although the lateral stress in the earth may or may not be equal to the vertical stress. In most subsurface applications, the vertical stress is considered to be the maximum principal stress. The minimum and intermediate principal stresses are assumed to be horizontal.

Deformation—Response to Stress

The application of stress to a body of rock or soil causes the material to yield or deform. The amount of deformation is called *strain*. The type and amount of strain that a particular material experiences depends on the type of stresses applied, as well as the depth and temperature. An understanding of deformation is important to geologists concerned with the behavior of earth ma-

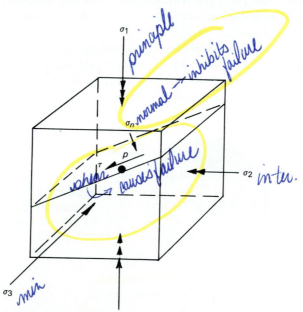

Figure 6-5 Maximum (σ_1), intermediate (σ_2), and minimum (σ_3) principal stresses acting at a point (P) within a body produce shear (τ) and normal (σ_n) stresses acting on a plane containing point P passed through the body.

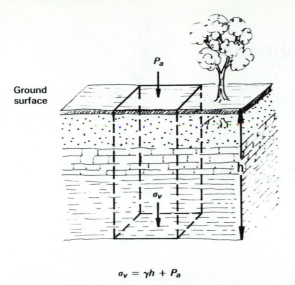

$$\sigma_v = \gamma h + P_a$$

Figure 6-6 The vertical stress acting on shallow horizontal planes in the earth is the sum of the unit weight (γ) of the material times the depth (h) and the atmospheric pressure (P_a). Where layers of different material are present, vertical stress would be the sum of the unit weight of each layer times its thickness plus the atmospheric pressure.

terials during geologic processes, and engineers, who are more interested in deformation of material under the loads applied by engineering structures. Before considering natural materials, we will first review the deformation of so-called ideal materials.

Ideal Materials. Fundamental types of deformation can be demonstrated by the use of simple mechanical systems. As shown in Figure 6-7, the three basic types of material behavior are *elastic, viscous,* and *plastic.* Perfect elastic behavior, the type demonstrated by a spring, results in a linear plot of stress versus strain. The slope of the line relating stress and strain is an important material property called the *modulus of elasticity.* It specifies how much strain will occur under a given stress. In this case, strain (ϵ) is measured as

$$\epsilon = \frac{\Delta L}{L} \qquad \qquad \text{Eq. 6-5}$$

where ΔL is the change in length of the spring and L is the original length. Another characteristic of elastic materials is that they tend to return to their original condition when the stress is removed. Thus elastic strain is recoverable. Some rocks approach ideal elastic behavior during deformations of small magnitudes. If the modulus of elasticity is known, it is therefore possible to predict the amount of deformation that will occur under an applied load.

Certain fluids display the type of behavior known as *viscous* (Figure 6-7). Stress is directly proportional to *strain rate* for those fluids. The constant of proportionality is the viscosity of the liquid. Even solid earth materials may exhibit viscous behavior under certain circumstances. The model for viscous behavior is a dashpot, in which a piston is moved through a cylinder containing a viscous liquid. The liquid flows through a space between the piston and the wall of the cylinder as the piston is pulled.

Plastic behavior, as demonstrated in Figure 6-7 by a block pulled along

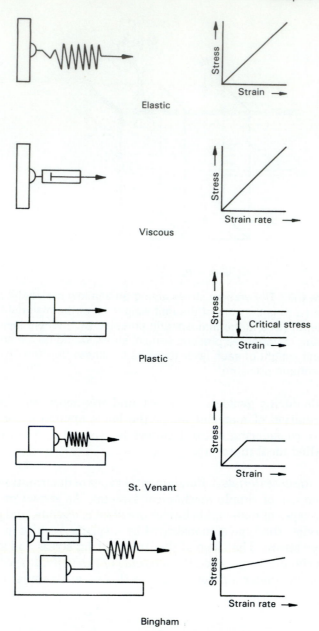

Figure 6-7 Types of idealized material deformation as illustrated
by simple mechanical systems.

a surface, involves continuous deformation after some critical value of stress
has been achieved. There is no deformation until the critical stress has been
reached. Many rocks display plastic deformation under stress, but they differ
from ideal plastic behavior in that other types of deformation occur before
plastic behavior begins.

Models that more closely approximate the behavior of real earth materials are provided by combining basic models of ideal behavior. The St. Venant
model, for example, which combines elastic and plastic behavior, approximates
the deformation of many rocks under stress. The Bingham model combines all
three ideal types of ideal behavior. This response has been used as a model
for mud flows and similar phenomena. These processes will be described in
Chapter 13.

Stress-Strain Behavior of Rocks.   Although rocks can be compared to the idealized models just discussed, their actual behavior is more complex. One common method of testing rock behavior is the *unconfined compression test*, in which a cylindrical sample of rock is subjected to an axial load applied to the ends of the sample (Figure 6-8). As the axial stress is increased during the test, the changes in length of the sample can be measured. From this value, the strain at any instant can be determined using Equation 6-5.

In Figure 6-9, a generalized stress-strain curve for rocks is shown. Unlike the ideal models, the curve is nonlinear and has three distinct segments. The first segment relates to the closing of microscopic pores and void spaces in the rock as stress is applied at low levels. In the middle section of the curve, the nearly linear response approximates elastic behavior. The slope of this segment can be used to calculate a modulus of elasticity for the rock, even though the rock is not perfectly elastic (Figure 6-7). The final segment of the curve is most similar to plastic behavior. At the point of *failure*, where the sample crumbles and loses all resistance to stress, the curve terminates.

Different types of rocks vary considerably in their stress-strain behavior. In Figure 6-10, two types of response are shown. Rocks that display mostly elastic behavior until rupture are called *brittle*. Those that exhibit a significant amount of plastic response before failure are termed *ductile*. Ductile rocks usually deform elastically to a point called the *elastic limit*, where strain becomes more plastic with increasing stress.

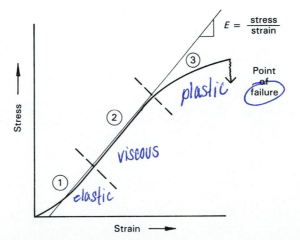

Axial load

Sample

Deformation gauge

Figure 6-8   Schematic diagram of the unconfined compression test. An axial load is applied to a rock or soil sample and the resulting strain is measured.

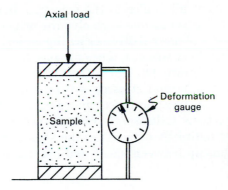

$E = \dfrac{\text{stress}}{\text{strain}}$

Point of failure

Stress

Strain

Figure 6-9   A generalized stress-strain curve for rocks. The modulus of elasticity is measured by the slope of the tangent to segment 2 of the curve.

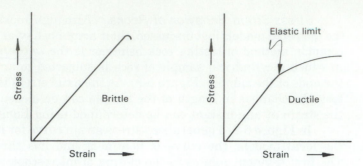

Figure 6-10 Stress-strain curves for brittle and ductile rocks.

Strength.   We have described the failure of a brittle rock as the point when the rock loses all resistance to stress and crumbles. In a plastic material, a specific point of failure is harder to identify because deformation continues indefinitely at a constant level of stress. Failure, in this case, has to be defined as a certain amount, or percentage, of strain. For whatever criteria established for failure, *strength* is defined as the level of stress at failure.

When the stress at a certain point in the ground or in a laboratory test sample is resolved into its principal stresses, a *stress differential* is apparent between the three principal stresses. In an unconfined compression test, for example, the axial load becomes the greatest principal stress and the other two principal stresses are zero. When a stress differential exists, shear stresses develop on planes at all angles to the principal stress directions within the body (Figure 6-11). It is possible to resolve the principal stresses acting on the sample into shear and normal stresses acting on any plane within the sample (Figure 6-12). The shear stress, τ, tends to cause failure, and the normal stress, σ_n, tends to resist failure. On two critical planes within the rock the combination of shear and normal stresses will produce the greatest tendency for failure. If the shear stress on these planes exceeds the *shear strength* of the rock in those directions, failure of the rock will occur. Therefore, the unconfined compression test indirectly determines the shear strength of the rock.

The relationship between shear and normal stresses during a strength

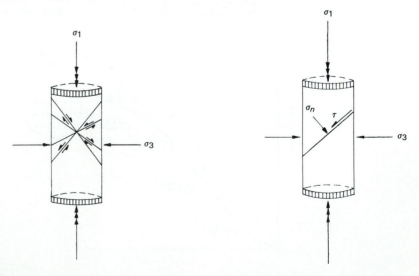

Figure 6-11 Shear stresses on planes inclined to the principal planes produced by a stress differential.

Figure 6-12 Shear and normal stresses acting on a plane within a test sample.

test, and at failure, is critical to understanding the deformation behavior of the material. A test that can be used to determine this relationship directly for rocks or soils is called, appropriately, the *direct shear test* (Figure 6-13). In the direct shear test, the failure plane is specified by the construction of the test cell so that variable shear and normal stresses can be applied to test for shear strength along the plane where the cell separates. The test is usually conducted by applying a constant value of normal stress and then increasing the shear stress until failure occurs. Successive tests are then repeated at a higher normal stress. If the cell is filled with dry sand for a strength test, the results will be as shown in Figure 6-14. The line on the graph shows the relationship between shear stress and normal stress at failure. The constant slope indicates that shear strength is directly proportional to the normal stress applied to the failure plane. This result is obtained because the strength of the sand is controlled by frictional contact between sand grains, and the shear stress necessary to cause failure is directly dependent upon the normal stress applied to the failure plane. The angle that the plotted line makes with the horizontal axis (ϕ) is a basic property of the specific sand tested called the *angle of internal friction.* It depends upon the grain-size distribution, shape, and packing of the sand grains in the sample tested. The equation of the line relating shear strength, S, and normal stress, σ_n, is

$$S = \sigma_n \tan \phi$$ Eq. 6-6

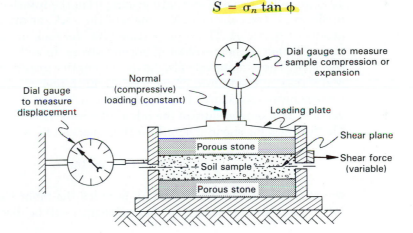

Figure 6-13 The direct shear test.

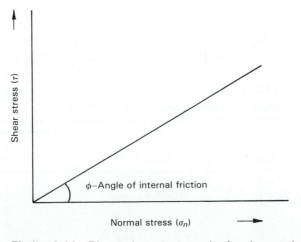

Figure 6-14 Direct shear test results for dry sand.

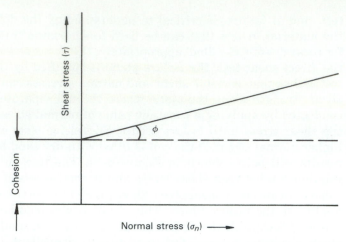

Figure 6-15 Direct shear test results for rocks and some cohesive soils.

The shear stress versus normal stress curve for dry sand passes through the origin because frictional contact between grains is the only component of strength in this material. For rocks, as well as for some soils, there are additional components of strength arising from the interlocking nature of grains in the rock, the cement in the pores of the rock, or attractive forces between grains or particles. These properties give the rock or soil a certain inherent strength that is independent of normal stress. In soils, this strength is called *cohesion*. These materials also have a strength component proportional to normal stress on the failure plane. The strength-test results, then, for rocks and some cohesive soils, are of the type shown in Figure 6-15. The intercept on the vertical axis is a measure of the inherent strength, or cohesion, of the material. The formula for shear strength in this case can be written as

$$S = C + \sigma_n \tan \phi \qquad \qquad \text{Eq. 6-7}$$

where C is cohesion. This equation is called the Mohr-Coulomb equation. The influence of fluid pressure on shear strength will be discussed in Chapter 13.

ENGINEERING CLASSIFICATION OF INTACT ROCK

Stress-strain behavior is utilized in the engineering classification of rock developed by D. U. Deere and R. F. Miller (Deere, 1968). One component of the classification is the unconfined compressive strength (Table 6-2). Five cate-

TABLE 6-2

Strength Classification of Intact Rock

Class	Description	Unconfined compressive strength (kg/cm^2)
A	Very high strength	>2250
B	High strength	1125–2250
C	Medium strength	562–1125
D	Low strength	281–562
E	Very low strength	<281

gories are established for an intact sample of rock. The modulus of elasticity forms the other aspect of the classification. This classification applies only to rock that is internally continuous, or intact. Most masses of rock contain discontinuities that strongly influence engineering behavior. Rock-mass properties will be discussed in a subsequent section.

In Figure 6-16, the definition of the particular modulus value used is illustrated. E_{t50} is the modulus obtained by taking the slope of a line tangent to the stress-strain curve at 50% of the unconfined compressive strength. The classes of modulus of elasticity are listed in Table 6-3.

Together, the modulus of elasticity (E_{t50}) and the unconfined compressive strength (σ_a) can be used to calculate a modulus ratio, which is defined as

$$M_R = \frac{E_{t50}}{\sigma_a} \qquad \text{Eq. 6-8}$$

Modulus-ratio values are plotted on a diagram (Figure 6-17) divided into fields of high modulus ratio (>500:1), medium modulus ratio (500:1–200:1), and low modulus ratio (<200:1). The position of plotted points on the modulus ratio diagram gives a visual comparison of rocks and indicates the strength and modulus values, the rock properties that control engineering behavior.

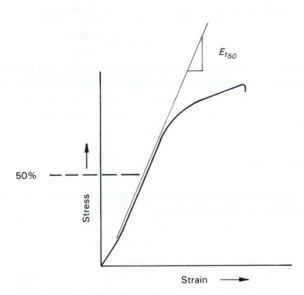

Figure 6-16   Definition of the tangent modulus (E_{t50}) used in the Deere and Miller classification.

TABLE 6-3

Modulus of Elasticity Classification of Rocks

Description	E_{t50} (kg/cm² × 10⁵)
Very stiff	8–16
Stiff	4–8
Medium stiffness	2–4
Low stiffness	1–2
Yielding	0.5–1
Highly yielding	0.25–0.5

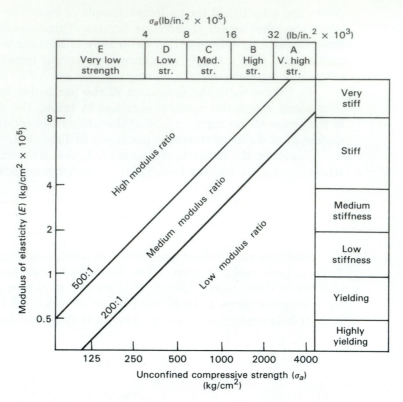

Figure 6-17   Classification diagram for the Deere and Miller classification. This classification applies to intact rock. (Modified from D. U. Deere in *Rock Mechanics in Engineering Practice*, edited by K. G. Stagg and O. C. Zienkiewicz, copyright © 1968 by John Wiley and Sons, Ltd., London.)

Igneous Rocks

The strength of igneous rocks is high when the rock is composed of a dense network of interlocking crystals. This condition is usually present in the intrusive rocks, which had sufficient time of crystallization to develop a three-dimensional crystal network. As shown in Figure 6-18, intrusive rocks generally have a high modulus of elasticity and a medium modulus ratio.

The extrusive igneous rocks have a much wider range in strength and modulus. Lava flows may develop strength and modulus properties that are similar to the intrusive rocks, but vesicular extrusive rocks are usually weaker. The pyroclastic rocks extend the field of the extrusive rocks into the area of low strength and modulus because of their low density and high porosity caused by their formation by airfall or flow processes. Most extrusive rocks, however, fall into the medium-modulus-ratio category.

Sedimentary Rocks

Sedimentary rocks have extremely variable strength and modulus properties (Figure 6-19). For the clastic rocks, these values depend on the rock characteristics acquired in the depositional environment, as well as all changes that have affected the rock in the lithification process. Important depositional factors include grain-size distribution, sorting, rounding, and mineral composition. Lithification processes, such as compaction and cementation, usually in-

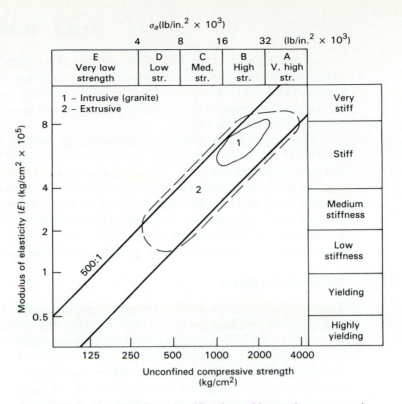

Figure 6-18 Engineering classification of intact igneous rock. (Modified from D. U. Deere in *Rock Mechanics in Engineering Practice*, edited by K. G. Stagg and O. C. Zienkiewicz, copyright © 1968 by John Wiley and Sons, Ltd., London.)

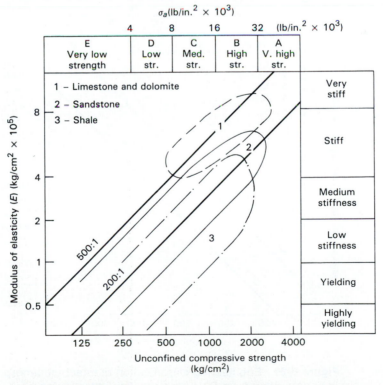

Figure 6-19 Engineering classification of intact sedimentary rock. (Modified from D. U. Deere in *Rock Mechanics in Engineering Practice*, edited by K. G. Stagg and O. C. Zienkiewicz, copyright © 1968 by John Wiley and Sons, Ltd., London.)

crease the strength of the rock. Clastic rocks of medium and coarse grain size usually fall into the moderate-modulus-ratio class over a wide range of strength and modulus values. Shale exhibits a strong tendency for plastic deformation. As a result, the modulus of elasticity is low and the modulus ratio overlaps into the low zone.

Nonclastic rocks differ in engineering properties according to the composition of the rocks. Limestone and dolomite generally have medium to high strength and modulus ratio. The evaporite rocks exhibit weak, plastic behavior. The tendency for plastic deformation at low strength in the evaporite rocks explains why these rocks are considered to be a possible option for high-level nuclear-waste repositories. Because of their stress-strain behavior, any cavities or cracks in the rocks are sealed by plastic flow. This property is important in preventing migration of waste products from the disposal site.

Metamorphic Rocks

Metamorphism increases the strength of some sedimentary rocks by compaction and recrystallization. Thus (Figure 6-20), the modulus ratio of quartzite is limited to the high-strength side of the diagram relative to the corresponding sedimentary rock sandstone. Marble is an exception to the trend of increased strength following metamorphism. When limestone or dolomite recrystallizes, the crystals in the resulting marble are larger. The strength of a rock generally is proportional to the surface area of contact between grains, and a fine-grained rock has a higher contact area than a coarse-grained rock. Therefore, limestone and dolomite lose strength when they are metamorphosed. Because only the

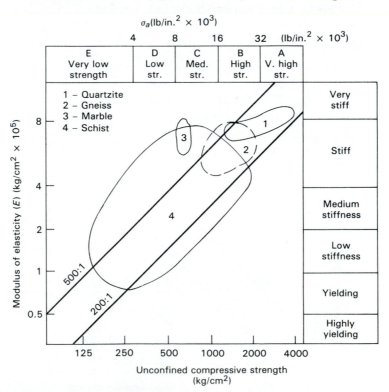

Figure 6-20 Engineering classification of intact metamorphic rock. (Modified from D. U. Deere in *Rock Mechanics in Engineering Practice*, edited by K. G. Stagg and O. C. Zienkiewicz, copyright © 1968 by John Wiley and Sons, Ltd., London.)

grain size changes during metamorphism, and the mineral composition remains the same, the modulus-ratio field of marble falls within the high-modulus-ratio section of the chart.

Schists have a wide variation in strength and modulus because of their strongly oriented foliation, which produces planes of weakness parallel to the foliation. The strength of a schist depends upon the direction from which stresses are applied. If the stress configuration tends to produce failure in the direction of the foliation, the strength is minimized. A much greater strength develops if the stresses tend to cause failure perpendicular to the foliation planes. Gneisses are limited to the high strength and modulus regions of the chart because the coarse bands of alternating light- and dark-colored minerals in gneiss do not result in highly developed planes of weakness in the rock.

ROCK-MASS PROPERTIES

The strength and deformational properties of rocks that are described in this chapter usually are determined by laboratory tests on rock samples. These samples are necessarily small, intact specimens taken from large bodies of rock at a field location. Although the test results obtained from intact samples are useful for comparison of properties between various rock types, the strength values cannot be directly applied to the overall rock mass in the field situation. The reason for this apparent discrepancy is that the behavior of a rock mass under load in the field is partially controlled by the strength developed along discontinuities in the rock and by weathering characteristics, rather than by the strength of the intact portions of the rock itself. Discontinuities are present in almost every type of rock and they act to lower the strength of the rock mass. Under stress, the rock will fail along existing planes of weakness rather than develop new fracture planes within intact, solid rock. Therefore, it is very important to determine the properties of the rock mass as well as the properties of intact rock within the rock mass.

Some of the types of rock discontinuities are shown in Table 6-4. Depositional discontinuities are acquired by sedimentary rocks as their constituent

TABLE 6-4

Rock Discontinuities

Sedimentary
  Bedding planes
  Sedimentary structures (mud cracks,
    ripple marks, cross beds, and so on)
  Unconformities
Structural
  Faults
  Joints
  Fissures
Metamorphic
  Foliation
  Cleavage
Igneous
  Cooling joints
  Flow contacts
  Intrusive contacts
  Dikes
  Sills
  Veins

---

particles are deposited as sediment. These discontinuities include bedding planes between units of sediment of different lithology. The units are called beds when they are greater than 10 mm in thickness and laminations when they are less than 10 mm thick. Sedimentary structures, including mud cracks and ripple marks, also represent discontinuities in the rock. In addition to knowing the types of discontinuities, engineers must also determine their spacing, orientation, and roughness. The orientation is particularly important in considering the ability of a rock slope to support itself under the load of the rocks and soils that make up the slope. Failure in this situation occurs when part of the rock mass breaks away along a discontinuity and moves downslope. Slope movements, which will be discussed in more detail later, comprise a very significant category of geologic hazards.

Quantification of rock-mass properties is very difficult because of the number of variables involved. One index that is used is called the *rock quality designation* (RQD). During site investigation for an engineering project, test holes are drilled to determine subsurface rock formations. One sampling technique is to obtain a *core*, which is a small cylindrical sample of the entire vertical interval drilled. The core can then be used for laboratory tests of the rocks penetrated. The length of an individual piece of core is dependent upon the degree and orientation of fractures in the rock. Cores of highly fractured rocks will be recovered as small, broken chips, whereas cores of relatively intact rock will consist of long, unbroken segments. RQD is defined as the percentage of core recovered in pieces 10 cm or longer in length. The index is expressed as a percentage of the total depth drilled (Table 6-5). Thus in 10 m of drilling, recovery of 9.2 m in pieces 10 cm or longer in length would have an RQD of 92% and be described as excellent quality.

TABLE 6-5

Rock Quality Designation (RQD)

RQD (%)	Description of rock quality
0–25	Very poor
25–50	Poor
50–75	Fair
75–90	Good
90–100	Excellent

Several rock-mass classifications have been devised for specific applications. Few of these, if any, are suitable for all purposes. The variables usually considered in a classification include the strength of intact rock within the rock mass, RQD, state of weathering of the rock mass, fracture spacing, continuity and in-filling, orientation of fractures, and groundwater conditions. Numerical values can be assigned for each variable and summed to obtain an overall rock-mass classification. Problems arise in assigning numerical values to largely descriptive variables such as weathering characteristics.

ENGINEERING PROPERTIES OF SOIL

Index Properties and Classification

The engineering approach to the study of soil focuses on the characteristics of soils as construction materials and the suitability of soils to withstand the load applied by structures of various types. For these purposes, the characteristics that specify the size and type of particles in the soil and the condition of the soil in its natural setting are most important to utilize in classification. The properties that are used in engineering soil classification are called *index properties.* These properties, which are measured by *classification tests,* are important because they relate to shear strength, compressibility, and other aspects of soil behavior.

Table 6-6 lists index properties and their respective classification tests. An important division of soils for engineering purposes is the separation of coarse-grained, or *cohesionless* soils, from fine-grained, or *cohesive* soils. Fine-grained soils containing clay behave much differently than coarse-grained soils. For coarse-grained soils, the index properties refer to the size and distribution of particles in the soil. These characteristics are evaluated by *mechanical analysis,* a laboratory procedure that consists of passing the soil through a set of sieves with successively smaller openings (Figure 6-21). The

TABLE 6-6

Index Properties of Soils

Soil type	Index property
Coarse-grained (cohesionless)	Particle-size distribution Shape of particles Clay content In-place density Relative density
Fine-grained (cohesive)	Consistency Water content Atterberg limits Type and amount of clay Sensitivity

Figure 6-21   Sieves used for determining grain-size distribution of soil. (Photo courtesy of Soiltest Inc.)

size of sieve opening determines the size of the particles that may pass through. After the test, the particles retained on each sieve are converted to a weight percentage of the total and then plotted against particle diameter as determined by the known sieve opening size. The result is a *grain-size distribution curve* (Figure 6-22), plotted as the cumulative weight percent versus particle size. The cumulative weight percent is expressed as the percentage finer than a corresponding particle size. For example, if 20% of the soil by weight were retained on all the sieves with openings of 1 mm or larger in diameter, a point at 80% finer by weight would be plotted at the 1-mm size.

The shape of the grain-size distribution curve is a very important soil characteristic. A curve that covers several log cycles of the graph, as shown on the left in Figure 6-23, contains a variety of sizes. This type of soil would be called *well graded*. The opposite type of soil, composed of a very narrow range of particle sizes, would be classified as *poorly graded* (Figure 6-23). These terms can be easily confused with sorting, the geologic designation for grain-size distribution. Geologists are concerned with various depositional processes, such as river flow, that tend to separate particle sizes during transportation and deposit particles in beds composed of particles of similar sizes. For this reason, geologists would refer to a poorly graded soil with a narrow range of

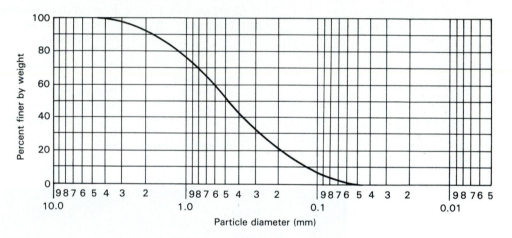

Figure 6-22 A cumulative grain-size distribution curve drawn from data obtained in a mechanical analysis of a soil sample. (From D. F. McCarthy, *Essentials of Soil Mechanics and Foundations*, copyright © 1977 by Reston Publishing Co., Inc., Reston, Va.)

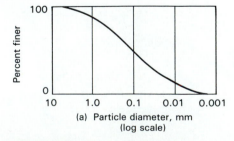
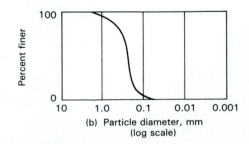

Figure 6-23 Grain-size distribution curves of a well-graded soil (left) and a poorly graded soil (right). (From D. F. McCarthy, *Essentials of Soil Mechanics and Foundations*, copyright © 1977 by Reston Publishing Co., Inc., Reston, Va.)

sizes as well sorted. The opposite, a well-graded soil, would be considered poorly sorted by geologists.

The other index properties describing coarse-grained soils include *particle shape, in-place density*, and *relative density*. These properties are related because particle shape influences how closely particles can be packed together. In-place density refers to the actual density of the soil at its particular depth in the field, whereas relative density is the ratio of the actual density to the maximum possible density of that soil. Relative density is a good indication of possible increases in density, or compaction, that may occur if load is applied to the soil. The compaction of soil under the load of structures, called *settlement*, can be very damaging when it occurs in excessive or variable amounts beneath a building.

The index properties of fine-grained, or cohesive, soils are somewhat more complicated than the index properties of cohesionless soils because of the influence of clay minerals. The type and amount of clay minerals is, therefore, very significant. With respect to engineering behavior, *consistency* is the most important characteristic of cohesive soils. It refers to the strength and resistance to penetration of the soil in its in-place condition. Consistency is determined by the arrangement of soil particles, particularly clays, in the soil, which is called its *fabric*. Although many types of fabric are possible, soils in which edge-to-face contact of clay particles exists (Figure 6-24), or *flocculated* fabrics, are much stronger than the parallel arrangement of particles in *dispersed* fabrics. Flocculated fabrics change to dispersed fabrics during *remolding*, which involves disturbance and alteration of the soil by natural processes or during various lab tests. The consistency of a soil can be determined by field tests in which the soil is evaluated in place, or by lab tests on samples that have been carefully handled to avoid remolding. The unconfined compression test is often used as an indication of consistency. In practice, the relative terms soft, medium, stiff, very stiff, and hard are applied to describe consistency. The relationship between these terms, unconfined compressive strength values, and the results of simple tests made with the hand and other objects are shown in Table 6-7.

The ratio of unconfined compressive strength in the undisturbed state to strength in the remolded state defines the index property called *sensitivity*. Thus sensitivity, S_t, can be expressed as

$$S_t = \frac{\text{strength in undisturbed condition}}{\text{strength in remolded condition}} \qquad \text{Eq. 6-9}$$

Soils with high sensitivity (Table 6-8) are highly unstable and can be converted to a remolded or dispersed state very rapidly, with an accompanying drastic loss of strength. The disastrous slope movements that follow these transformations are discussed in Chapter 13.

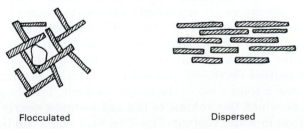

Flocculated   Dispersed

Figure 6-24 Clay particles displaying flocculated and dispersed fabrics.

TABLE 6-7

Consistency and Strength for Cohesive Soils

Consistency	Shear strength (kg/cm^2)	Unconfined compressive strength (kg/cm^2)	Feel or touch
Soft	<0.25	<0.5	Blunt end of a pencil-size item makes deep penetration easily
Medium (medium stiff or medium firm)	0.25–0.50	0.50–1.0	Blunt end of a pencil-size object makes 1.25-cm penetration with moderate effort
Stiff (firm)	0.50–1.0	1.0–2.0	Blunt end of pencil-size object can make moderate penetration (about 0.6 cm)
Very stiff (very firm)	1.0–2.0	2.0–4.0	Blunt end of pencil-size object makes slight indentation; fingernail easily penetrates
Hard	>2.0	>4.0	Blunt end of pencil-size object makes no indentation; fingernail barely penetrates

SOURCE: From D. F. McCarthy, *Essentials of Soil Mechanics and Foundations*, copyright © 1977 by Reston Publishing Co., Inc., Reston, Va.

TABLE 6-8

Sensitivity of Clays

Type	Sensitivity value
Nonsensitive	2–4
Sensitive	4–8
Highly sensitive	8–16
Quick	>16

The *water content*, an important influence upon bulk properties and behavior of a soil, is defined as

$$W = \frac{W_w}{W_s} \qquad \text{Eq. 6-10}$$

where W is the water content, W_w is the weight of water in a volume of soil, and W_s is the total weight of the soil particles in the same volume. In the remolded state, the consistency of the soil is defined by the water content. Four consistency states are separated by the water content at which the soil passes from one state to another (Figure 6-25). These water content values are known as the *Atterberg limits*. For example, the *liquid limit* is the water content at which the soil-water mixture changes from a liquid to a plastic state. As the water content decreases, the soil passes into a semisolid state at the *plastic limit*, and a solid state at the *shrinkage* limit. The shrinkage limit defines the point at which the volume of the soil becomes nearly constant with further decreases in water content. The Atterberg limits can be determined with simple laboratory tests. The use of the Atterberg limits in predicting the behavior

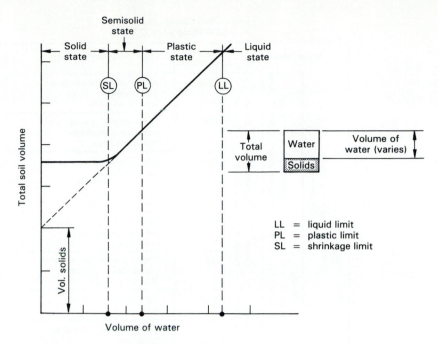

Figure 6-25 Atterberg limits. As the volume of water (water content) in a volume of soil decreases, the soil passes through four states, separated by the Atterberg limits. (From D. F. McCarthy, *Essentials of Soil Mechanics and Foundations*, copyright © 1977 by Reston Publishing Co., Inc., Reston, Va.)

of natural, in-place soil is limited by the fact that they are conducted on re-molded soils. The relationship between moisture content and consistency defined by the Atterberg limits may not be the same in the undisturbed state. Therefore, the Atterberg limits are mainly used for classification rather than for the prediction of soil behavior under field conditions.

The most useful engineering classification of soils is the *Unified Soil Classification System* (Figure 6-26). This classification gives each soil type a two-letter designation. For coarse-grained soils, the first letter, either *G* for gravel or *S* for sand, refers to the dominant particle size in the soil. The second letter is either *W*, for well graded, *P*, for poorly graded, or *M* or *C*, for coarse-grained soils that contain more than 12% of silt or clay, respectively. The first letter of the designation for fine-grained soils is *M* or *C*, for silt or clay. The second letter, either *H* (high) or *L* (low) refers to the plasticity of the soil as defined in Figure 6-27. The chart is a plot of *plasticity index* (PI) against liquid limit (LL). The plasticity index is defined as

$$PI = LL - PL \qquad\qquad \text{Eq. 6-11}$$

The difference between the liquid and plastic limits (PI) is a measure of the range in water contents over which the soil remains in a plastic state. The plotted position of a soil with respect to the A-line on the plasticity chart determines whether the soil receives the letter *H* for high plasticity or the letter *L* for low plasticity. Highly organic soils, which form a final category in the classification, are also subdivided into high- and low-plasticity types. Once a soil has been classified by the unified system, predictions can be made of the soil's permeability, strength, compressibility, and other properties.

Unified Soil Classification System
(ASTM Designation D-2487)

Major division			Group symbols	Typical names
Coarse-grained soils More than 50% retained on No. 200 sieve	Gravels 50% or more of coarse fraction retained on No. 4 sieve	Clean gravels	GW	Well-graded gravels and gravel-sand mixtures, little or no fines
			GP	Poorly graded gravels and gravel-sand mixtures, little or no fines
		Gravels with fines	GM	Silty gravels, gravel-sand-silt mixture
			GC	Clayey gravels, gravel-sand-clay mixtures
	Sands More than 50% of coarse fraction passes No. 4 sieve	Clean sands	SW	Well-graded sands and gravelly sands, little or no fines
			SP	Poorly graded gravels and gravel-sand mixtures, little or no fines
		Sands with fines	SM	Silty sands, sand-silt mixtures
			SC	Clayey sands, sand-clay mixtures
Fine-grained soils 50% or more passes No. 200 sieve	Silts and clays Liquid limit 50% or less		ML	Inorganic silts, very fine sands, rock flour, silty or clayey fine sands
			CL	Inorganic clays of low to medium plasticity, gravelly clays, sandy clays, silty clays, lean clays
			OL	Organic silts and organic silty clays of low plasticity
	Silts and clays Liquid limit greater than 50%		MH	Inorganic silts, micaceous, or diatomaceous fine sand or silts, elastic silts
			CH	Inorganic clays of high plasticity, fat clays
			OH	Organic clays of medium to high plasticity
Highly organic soils			Pt	Peat, muck, and other highly organic soils

Figure 6-26 Engineering classification of soils by the Unified Soil Classification System. (Modified from D. F. McCarthy, *Essentials of Soil Mechanics and Foundations*, copyright © 1977 by Reston Publishing Co., Inc., Reston, Va.)

Shear Strength

The shear strength of a soil determines its ability to support the load of a structure or remain stable upon a hillslope. Soils engineers must therefore incorporate soil strength into the design of embankments, road cuts, buildings, and other projects. The division of soils into cohesionless and cohesive types for classification is also useful for discussing strength. In the Mohr-Coulomb theory of failure, shear strength has two components—one for inherent strength due to bonds or attractive forces between particles, and the other produced by frictional resistance to shearing movement (Eq. 6-7). The shear strength of cohesionless soils is limited to the frictional component. Cohesive

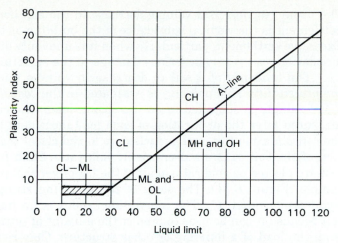

Figure 6-27 Chart for determining classification of fine-grained soils by the Unified Soil Classification System. (From D. F. McCarthy, *Essentials of Soil Mechanics and Foundations*, copyright © 1977 by Reston Publishing Co., Inc., Reston, Va.)

soils, on the other hand, develop strength through the attractive forces between clay particles—the cohesion—in addition to the frictional resistance to failure. The difference between the strength-test results for cohesionless and cohesive soils is shown in Figure 6-28. The shear strength of a cohesive soil is constant with increasing normal stress in the type of test where soil water is prevented from draining out of the test cell (undrained). In comparison, the shear strength of a cohesionless soil increases with increasing normal stress.

When a landslide occurs, it is an indication that stress within the soil mass exceeded the shear strength of the soil at the time of failure. The load imposed by buildings and other structures on most soils is rarely great enough to overcome the ultimate shear strength, or *bearing capacity*, of the soil. Case Study 6-1 describes an example of a building that actually did cause failure of the soil upon which it was constructed.

Settlement and Consolidation

Despite the lack of large-scale bearing-capacity failures caused by the load upon the soil exerted by structures, the soil does deform under the load applied.

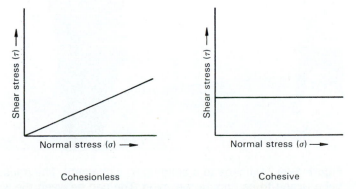

Figure 6-28 Shear strength of cohesionless and undrained cohesive soils. The plotted lines indicate the value of shear stress at failure (shear strength).

The usual response is a volume decrease in the soil beneath the foundation. Settlement is the vertical subsidence of the building as the soil is compressed. Excessive settlement, particularly when it is unevenly distributed beneath the foundation, can result in serious damage to the structure (Figure 6-29).

The tendency of a soil to decrease in volume under load is called the *compressibility.* This property is evaluated in the *consolidation test,* in which a soil sample is subjected to an increasing load. The change in thickness is measured after the application of each load increment (Figure 6-30). During the consolidation test, soil particles are forced closer together. Thus compression of the soil causes a decrease in void ratio. Data from the consolidation test are plotted as void ratio against the log of vertical pressure applied to the sample (Figure 6-31). The slope of the resulting straight line is called the *compression index*, C_c. The compression index is an important property of a soil because it can be used to predict the amount of settlement that will occur from the load of a building or other structure. Clay-rich soils and soils high in organic content have the highest compressibility. The compressibility of cohesionless soils is generally less than that of cohesive soils.

The compression of a saturated clay soil is a slightly different process

Figure 6-29 Damage to a building by settlement. The house has been raised to its original elevation. The gap between the wall and the foundation shows the amount of settlement that occurred. (Photo courtesy of USDA Soil Conservation Service.)

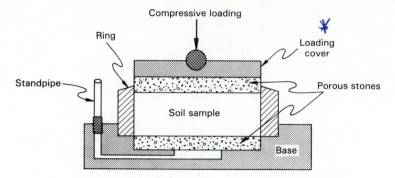

Figure 6-30 Schematic diagram of a consolidation test to determine soil compressibility. (From D. F. McCarthy, *Essentials of Soil Mechanics and Foundations*, copyright © 1977 by Reston Publishing Co., Inc., Reston, Va.)

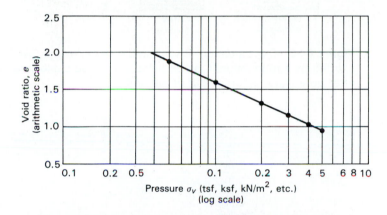

Figure 6-31 Consolidation test data plotted as decrease in void ratio vs. log of vertical pressure. (From D. F. McCarthy, *Essentials of Soil Mechanics and Foundations*, copyright © 1977 by Reston Publishing Co., Inc., Reston, Va.)

from the compression of an unsaturated material. When the pores of a soil are filled with water, compression is impossible because the void volume is totally occupied. The increase in load upon the saturated clay during a consolidation test or actual field situation is initially balanced by an increase in fluid pressure in the pores of the soil (Figure 6-32). Higher fluid pressure in the clay causes the pore water to flow out of the clay toward any direction where fluid pressure is lower. Only after pore fluid is removed, can the soil compress to a lower void ratio. This process is called *consolidation*. It is an important phenomenon to consider when constructing a building on a saturated clay because consolidation is very slow. The movement of pore water out of the soil is a function of the permeability, and clay has the lowest permeability of any soil. Therefore, it may take years for the soil to reach equilibrium under the load imposed. The settlement occurring during this period can be dangerous to the building. A famous example of consolidation is the Leaning Tower of Pisa (Figure 6-33). Tilting of the tower is the result of nonuniform consolidation in a clay layer beneath the structure. This ongoing process may eventually lead to failure of the tower.

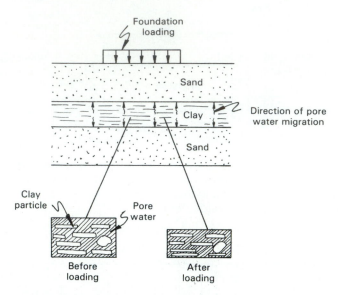

Figure 6-32 Consolidation of saturated clay under the load exerted by a foundation. Clay particles are forced closer together after pore water is forced out of the unit.

Figure 6-33 The Leaning Tower of Pisa (tower at right), whose tilt is the result of nonuniform consolidation beneath the structure. (Photo courtesy of the Italian Government Travel Office, E.N.I.T.)

Clay Minerals

Clay minerals are so important to soils engineering that we must take a closer look at their properties. Clays are good examples of *colloids, particles so small that their surface energy controls their behavior*. In particular, the electrical charge on the particle surface influences the interaction between adjacent clay particles. In Figure 6-34 the effect of the clay-surface charge upon the pore-water solution surrounding it is shown. Surface charges on clay particles are negative under most conditions. In order to balance these charges, cations from the pore-water solution and water molecules are attracted to the particle surface. The attraction of water is possible because of the polar nature of the water molecule. The positive side of the water molecule is, therefore, attracted to the negative clay surface. The negative charges on the surface of the particle in combination with the attracted cations and water molecules are called the *diffuse double layer*. Water molecules in diffuse double layers behave somewhat differently than water that is beyond the double layer in pore spaces. In addition, the positively charged sides of the water molecules cause repulsive forces between the double layers of adjacent clay particles.

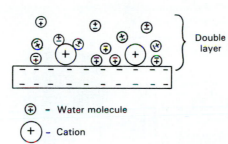

Double layer

⊕ – Water molecule

+ – Cation

Figure 6-34 Attraction of cations and water molecules to the charged surface of a clay particle.

When clays are initially deposited by settling to the bottom in a lake or the ocean, for example, both attractive and repulsive forces between clay particles are present. The relative strength of these forces determines which of the fabrics shown in Figure 6-24 will develop during sedimentation. Repulsive forces are stronger when the electrolyte content (ionic concentration) of the solution is weak. Therefore, *dispersed fabrics* are more common in clays deposited in fresh water and *flocculated fabrics* characterize seawater deposition.

The electrostatic properties of clays explain many aspects of their engineering behavior. Because of the tendency of clays to absorb water, the water content of clay soils is frequently near or above the plastic limit. In the plastic state, the strength of clay soils is relatively low. When load is applied to clay soils, clay particles can be forced closer together or forced to reorient themselves, causing a new fabric to develop. The large decreases in void ratio that accompany these adjustments to load are responsible for the high compressibility of clay. As we have seen, when pore water is expelled during consolidation, the decrease in void ratio may be a very slow process. Because of these properties, clay soils can create troublesome engineering problems. Unstable slopes and settlement of foundations are examples. Site investigations for heavy structures must not neglect the presence of clay soils at depth; even though surficial soils may have low compressibility, a clay bed in the subsurface may lead to settlement of the structure (Figure 6-32).

CASE STUDY 6-1
BEARING CAPACITY FAILURE IN WEAK SOIL

A rare example of a true bearing-capacity failure of a soil beneath a foundation is provided by the Transcona, Manitoba, grain elevator (White, 1953). This structure (Figure 6-35) was built in 1913 by the Canadian Pacific Railway. During the initial filling of the elevator, the underlying soils failed along a nearly circular surface, causing a rotation of the elevator to an angle of 27° from the vertical (Figure 6-36). Despite the sinking and rotational movement, the elevator was largely undamaged. By excavating beneath the side that was rotated upward and by underpinning the structure with piles, engineers were able to restore the elevator to

an upright position, where it now rests at a depth of 7 m lower than the depth of the original foundation. Years later, when testing and analysis of the soils were done, it was determined that the pressure of the fully loaded elevator exceeded the shear strength of the soil. The soils in the area are clay-rich sediments deposited in a Pleistocene glacial lake. These materials have a very low strength and high compressibility. Failures similar to the Transcona grain elevator are less likely today because of the great progress in the field of soil mechanics with respect to testing and analysis of soils.

Figure 6-35   The Transcona grain elevator as it appears today.

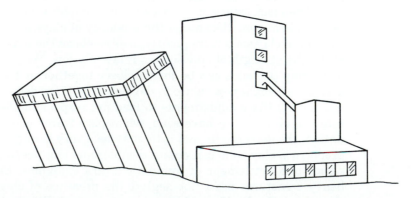

Figure 6-36   Rotation of the grain elevator by bearing-capacity failure of the weak foundation soil. (Sketched from a photo by L. S. White.)

CASE STUDY 6-2
THE GROS VENTRE SLIDE: ROLE OF DISCONTINUITIES IN ROCK-MASS STABILITY

A good example of the effect of discontinuity orientation upon slope movements is provided by the Gros Ventre slide, the largest historic rock slide in the United States (Figure 6-37). Sedimentary rock formations are oriented as shown in the Gros Ventre Valley near Jackson, Wyoming (Alden, 1928). On the south slope of the valley, bedding planes are tilted, or dip, in the same direction as the slope. The strength of the rock mass, therefore, is controlled by the strength developed along the sloping bedding planes rather than the strength of the rock itself. After a period of heavy rains in 1925, the slope failed because the stress along one or more of the bedding planes was greater than the shear strength. As a result, about 40 million m^3 of rock slid rapidly into the valley along bedding-plane discontinuities.

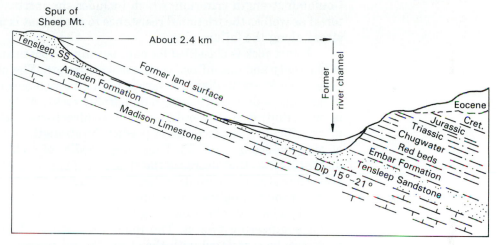

Figure 6-37 The Gros Ventre rockslide. (From Alden, 1928; used by permission of the American Institute of Mining, Metallurgical, and Petroleum Engineers.)

SUMMARY AND CONCLUSIONS

Earth materials are three-phase materials with possible solid, liquid, and gas components. The volume of the void spaces in the rock or soil, along with the size and interconnectedness of pores, determines the intrinsic permeability of the material. A related parameter, hydraulic conductivity, includes the properties of the fluid in its definition.

Stress is present beneath the surface of the earth as a result of the weight of overlying rock and soil as well as any load applied at land surface. The stress at any point can be resolved into maximum, minimum, and intermediate components, acting in mutually perpendicular directions at the point.

Strain results from the application of stress to a material. Earth materials exhibit complex deformational responses to stress. Models can be used to approximate the stress-strain relationships of different types of materials. Simple models, including elastic, plastic, and viscous behavior, can be combined to develop models that are more relevant to earth materials. Natural materials, however, are more complex in behavior and can only be approximated by models.

The stresses applied to a rock sample may be high enough to cause failure. Failure is controlled by the shear strength of the material because the rock fails under the action of shear stresses that develop on planes inclined at various angles to the principal stresses. Shear and normal stress at failure are plotted on diagrams to show the variations of shear strength under various stress conditions.

The relationship between shear and normal stresses at failure varies for different types of material. The strength of a sand is controlled by friction between grains along the failure plane, so the shear strength is directly proportional to normal stress on the failure plane. Rocks and some soils have additional sources of strength, resulting from cementation between grains, cohesion between particles, and the interlocking grain structure of minerals that crystallized from a magma or solution. These materials follow the Mohr-Coulomb strength criterion, which includes the intrinsic strength of the material as well as the frictional resistance to shear that is proportional to normal stress along the failure plane.

Intact rock is classified for engineering purposes on the basis of modulus of elasticity and unconfined compressive strength. The plot of these quantities on a modulus-ratio diagram serves to distinguish several main subdivisions of igneous, sedimentary, and metamorphic rocks. Although this classification is important, it applies only to small, nonfractured rock samples. It is much more difficult to characterize or predict the strength and deformation behavior of rock masses in the field. These large bodies of rock almost always contain discontinuities that decrease the strength of the overall rock mass. The types and orientations of discontinuities must be identified in order to conduct meaningful stability analysis or design.

The objective of engineering soils classification is to predict soil behavior in an engineering project. The index properties, which are used to classify soils, can be correlated with the strength and compressibility of the soil. For coarse-grained soils, the relevant index properties are grain-size distribution, grading, in-place density, and relative density. Cohesive soils are more difficult to characterize because of the interaction between clay minerals and water. The index properties used include clay content, consistency, water content, sensitivity and the Atterberg limits. The index properties are used to classify the soil in the Unified Soil Classification System.

The performance of a soil as a foundation support material is dependent upon its shear strength and compressibility. The bearing capacity is rarely exceeded as it was beneath the Transcona grain elevator. Even though a soil may not fail, damage to structures will be experienced if settlement takes place. Settlement is the subsidence of a foundation because of compression of the underlying soil. Consolidation is the time-dependent compression of clay soil as pore water is slowly expelled from the loaded soil.

Many of the engineering problems associated with clay can be explained by its colloidal properties. The surface charge of clay particles attracts water and cations from the pore solution. The attraction or repulsion of adjacent particles during deposition determines the initial fabric. During consolidation, reorientation of particles often occurs. The low strength of clay is indicative of high Atterberg limits and high water contents.

REFERENCES AND SUGGESTIONS
FOR FURTHER READING

ALDEN, W. C. 1928. Landslide and flood at Gros Ventre, Wyoming. *Transactions, AIME* 76:347–362.

COSTA, J. C., and V. R. BAKER. 1981. *Surficial Geology: Building with the Earth.* New York: John Wiley & Sons, Inc.

DEERE, D. U. 1968. "Geological considerations." In *Rock Mechanics and Engineering Practice*, edited by K. G. Stagg and O. C. Zienkiewicz. London: John Wiley & Sons, Ltd.

FREEZE, R. A., and J. A. CHERRY. 1979. *Groundwater.* Englewood Cliffs, N.J.: Prentice-Hall, Inc.

HOLTZ, R. D., and W. D. KOVACS. 1981. *An Introduction to Geotechnical Engineering.* Englewood Cliffs, N.J.: Prentice-Hall, Inc.

McCARTHY, D. F. 1977. *Essentials of Soil Mechanics and Foundations.* Reston, Va.: Reston Publishing Co., Inc.

PECK, R. B., and F. G. BRYANT. 1953. The bearing capacity failure of the Transcona elevator. *Geotechnique* 3:201–208.

WHITE, L. S. 1953. Transcona elevator failure: eyewitness account. *Geotechnique* 3:209–214.

PROBLEMS

1. What are the differences among porosity, void ratio, and permeability?
2. Give as many examples as you can of materials that exhibit elastic, plastic, and viscous deformation.
3. How does the response of rocks to stress differ from models of ideal behavior?
4. How are stress and strength related?
5. How is shear strength measured?
6. What are the two major components of shear strength?
7. Define modulus ratio and explain what it is used for.
8. Discuss the limitations and problems of extrapolating mechanical properties measured on small rock samples to field situations.
9. How does the engineering classification of soils differ from the engineering classification of rocks?
10. Give some examples of the significance of the properties of clay minerals in soils engineering.
11. What are the Atterberg limits and what are they used for?
12. What natural geologic processes are controlled by the shear strength of soils?
13. What soil characteristics influence compressibility?
14. Why is consolidation a time-dependent process?

CHAPTER
7
EARTHQUAKES

The internal activity of the earth is inextricably linked to its external environment. Although these relationships are not always readily apparent, their importance should not be overlooked. The most obvious reminders of the earth's dynamic interior are volcanoes and earthquakes. Protection of the human population against the hazards of these processes requires an understanding of the earth's interior. But how do we study such an inaccessible region? In the next several chapters we will examine the evidence relating to the earth's internal processes and consider their impact upon engineering. Even though a rock mass may have deformed millions of years ago at a great depth, the results of that deformation are highly relevant to the utilization of the surface region in which we live.

At 12 min after 5 in the morning on April 18, 1906, a great earthquake struck San Francisco. During the next 60 s, violent ground shaking devastated portions of the city and caused panic and terror. The ground actually was observed to be moving in waves in some places; train tracks were bent like wire, and whole buildings or chimneys collapsed to the ground (Figure 7-1). The aftermath of the earthquake was even more destructive because fires broke out throughout the city and firefighters were left helpless to fight them because the city water main was ruptured by the shaking. In all, at least 700 people were killed and 250,000 were left homeless.

The 1906 earthquake was the result of a sudden rupture along a 400-km-long segment of the San Andreas fault, the boundary between two lithospheric plates. Since that time, the United States has experienced only one other great earthquake, the Alaska earthquake of 1964. Fortunately, it occurred in a sparsely populated area. In recent years building codes have been continually modified in order to minimize damage from earthquakes. In nearly every metropolitan area, however, many buildings predate modern building practices. If a large earthquake strikes a large American city in the near future, the consequences could be truly catastrophic.

Figure 7-1 Damage to the San Francisco City Hall by the 1906 earthquake. (W. C. Mendenhall; U.S. Geological Survey.)

OCCURRENCE

Earthquakes are caused primarily by the rupture of rocks in the earth's crust along discontinuities called *faults*. Faults may coincide with major boundaries between lithospheric plates; in fact, most earthquakes take place along faults in the vicinity of plate boundaries (Figure 7-2). Earthquakes also occur far from plate boundaries. Often the faults associated with these intraplate earthquakes are deeply buried and not visible at land surface.

Elastic Rebound Theory

During earthquakes, bodies of rocks on opposite sides of faults move relative to each other. Several types of displacement, or *slip*, are recognized (Figure 7-3). *Strike slip* is mainly a horizontal type of displacement, whereas *dip slip* involves predominantly vertical movement.

A longstanding theory in geology, formulated by the analysis of surface rupture after the 1906 earthquake, holds that rocks deform in an elastic manner until brittle failure takes place (Figure 7-4). Elastic strain that had been accumulating on either side of the fault is suddenly released causing the rocks

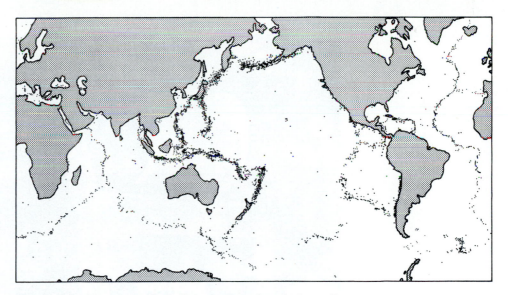

Figure 7-2 Locations of shallow earthquakes during a 7-year period. The epicenters are mainly concentrated on tectonic plate margins. (Data from ESSA, U.S. Coast and Geodetic Survey.)

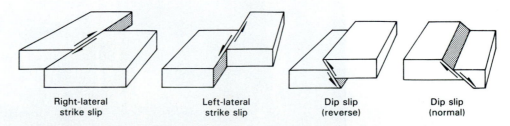

| Right-lateral strike slip | Left-lateral strike slip | Dip slip (reverse) | Dip slip (normal) |

Figure 7-3 Types of displacement during fault movement. (From R. D. Borcherdt, ed., 1975, *Studies for Seismic Zonation of the San Francisco Bay Region*, U.S. Geological Survey Professional Paper 941-A.)

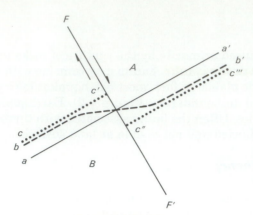

Figure 7-4 Elastic rebound theory. Line a–a' is a real or imaginary linear feature that crosses a fault (F–F') such as a road or fence. Elastic strain produces distortion of the line to the form shown as line b–b'. At the time of rupture, block A slides rapidly past block B and the elastic strain energy is released. The original line is now represented by segments c–c' and c"–c"'. (From M. P. Billings, *Structural Geology*, 3d ed., copyright © 1972 by Prentice-Hall, Inc., Englewood Cliffs, N.J.)

on either side of the fault to abruptly slide past each other. Elastic strain energy is released in the form of *seismic waves*, which radiate outward through the rocks in all directions from the *focus,* or point of rupture. The energy that can be obtained from elastic strain can be painfully experienced by holding a rubber band against someone's arm, stretching it outward, and then releasing it suddenly from the opposite end. The propagation of seismic waves during fault

Figure 7-5 A seismograph is the instrument used to measure seismic waves. (Photo courtesy of U.S. Geological Survey.)

movement is a consequence of elastic rebound, and the hypothesis is known as the *elastic rebound theory*.

Although the elastic rebound theory satisfactorily explains earthquakes produced by rupture of the brittle rocks near the earth's surface, the theory becomes more difficult to apply to those earthquakes generated at depths greater than a few tens of kilometers beneath the surface. At these depths the pressure exerted by the overlying rocks makes elastic behavior less likely. A more comprehensive theory for the origin of earthquakes in regions of plastic deformation is thus needed. Intensive research underway at present may lead to an improved explanation for deep earthquakes.

Detection and Recording

During rupture at the focus of an earthquake, the elastic strain energy stored in the rocks is suddenly released in the form of *seismic waves*. These waves are propagated outward from the focus through the rocks in all directions. This seismic energy causes rapid vibrations in the rocks in the subsurface as well as the land surface itself. The instrument designed to measure and record the characteristics of seismic waves is called a *seismograph* (Figure 7-5). It consists of a weight suspended on a string or wire that acts as a pendulum. The inertia of the mass resists its motion relative to the rigid frame to which it is attached. When the pendulum is connected either mechanically or electrically to a pen in contact with a rotating drum and chart, a record of the wave vibrations, called a *seismogram*, is produced.

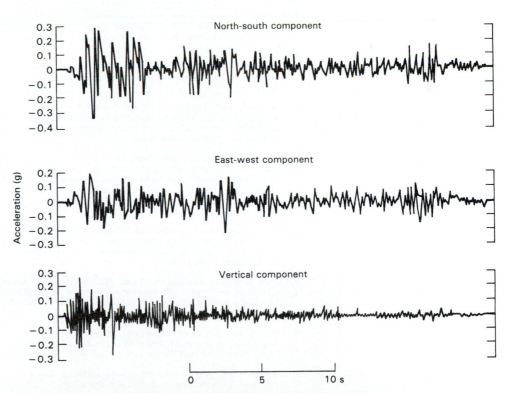

Figure 7-6 Accelerogram of the 1940 Imperial Valley, California, earthquake recorded at El Centro, California. (From W. W. Hays, 1980, *Procedures for Estimating Earthquake Ground Motions*, U.S. Geological Survey Professional Paper 1114.)

The pendulum of a seismograph is constrained to move in only one direction. Therefore, a complete seismological station must contain three seismographs. One measures the vertical component of motion and the other two measure horizontal motion in directions at right angles to each other.

Ground motion from an earthquake usually is most severe near the *epicenter*, the point on the earth's surface directly above the focus. Instruments called *accelerographs* are used to measure the strong motion of an earthquake in the vicinity of the epicenter. Accelerographs differ from seismographs in that they begin recording only during strong earth movements. Seismographs record seismic waves continuously.

MODIFIED MERCALLI INTENSITY SCALE OF 1931

(Abridged)

I. Not felt except by a very few under especially favorable circumstances.

II. Felt only by a few persons at rest, especially on upper floors of buildings. Delicately suspended objects may swing.

III. Felt quite noticeably indoors, especially on upper floors of buildings, but many people do not recognize it as an earthquake. Standing motor cars may rock slightly. Vibration like passing of truck. Duration estimated.

IV. During the day felt indoors by many, outdoors by few. At night some awakened. Dishes, windows, doors disturbed; walls made cracking sound. Sensation like heavy truck striking building. Standing motor cars rocked noticeably.

V. Felt by nearly everyone; many awakened. Some dishes, windows, etc., broken; a few instances of cracked plaster; unstable objects overturned. Disturbance of trees, poles and other tall objects sometimes noticed. Pendulum clocks may stop.

VI. Felt by all; many frightened and run outdoors. Some heavy furniture moved; a few instances of fallen plaster or damaged chimneys. Damage slight.

VII. Everybody runs outdoors. Damage negligible in buildings of good design and construction; slight to moderate in well-built ordinary structures; considerable in poorly built or badly designed structures; some chimneys broken. Noticed by persons driving motor cars.

VIII. Damage slight in specially designed structures; considerable in ordinary substantial buildings with partial collapse; great in poorly built structures. Panel walls thrown out of frame structures. Fall of chimneys, factory stacks, columns, monuments, walls. Heavy furniture overturned. Sand and mud ejected in small amounts. Changes in well water. Disturbed persons driving motor cars.

IX. Damage considerable in specially designed structures; well designed frame structures thrown out of plumb; great in substantial buildings, with partial collapse. Buildings shifted off foundations. Ground cracked conspicuously. Underground pipes broken.

X. Some well-built wooden structures destroyed; most masonry and frame structures destroyed with foundations; ground badly cracked. Rails bent. Landslides considerable from river banks and steep slopes. Shifted sand and mud. Water splashed (slopped) over banks.

XI. Few, if any (masonry), structures remain standing. Bridges destroyed. Broad fissures in ground. Underground pipe lines completely out of service. Earth slumps and land slips in soft ground. Rails bent greatly.

XII. Damage total. Waves seen on ground surfaces. Lines of sight and level distorted. Objects thrown upward into the air.

Figure 7-7 The Modified Mercalli scale (abridged) for determining earthquake intensity.

Accelerographs measure ground acceleration relative to g, the acceleration of gravity, which has a value of 9.8 m/s². The highest acceleration recorded was 1.25g during the 1971 San Fernando, California, earthquake. Accelerations are measured in three directions at right angles to each other (Figure 7-6). Commonly, accelerographs are installed at different levels in a building or in different geologic settings in order to evaluate the response to seismic waves of different types of structures, soils or rocks.

Seismograph stations are now integrated into regional and worldwide networks. These networks use similar equipment and process data uniformly in order to standardize data gathering and analysis.

Magnitude and Intensity

The strengths and effects of earthquakes are determined in two ways. *Magnitude* is a quantitative measure of the energy released by an earthquake, as obtained from seismograph records. A method for determining magnitude developed by seismologist Charles F. Richter defines magnitude as the logarithm of the largest seismic wave amplitude produced by the trace of a standard seismograph located at a distance of 100 km from the epicenter. Magnitudes obtained by this method are commonly referred to as the magnitude on the *Richter scale*. The highest magnitude values measured have been approximately 8.9. The logarithmic aspect of magnitude means that an earthquake of magnitude 7, for example, produces 10 times the wave amplitude of an earthquake of magnitude 6. Great earthquakes are classified as those with a magnitude of 8 or more.

Unlike earthquake magnitude, *intensity* is a subjective measure that is determined by observing the effects of the earthquake on structures and by interviewing people who experienced the event. The most commonly used intensity scale is the *Modified Mercalli* scale, which has 12 divisions (Figure 7-7). Each division describes a particular level of human response and building performance. Intensity generally decreases with distance from the epicenter (Figure 7-8); however, local geologic conditions also influence the amount of

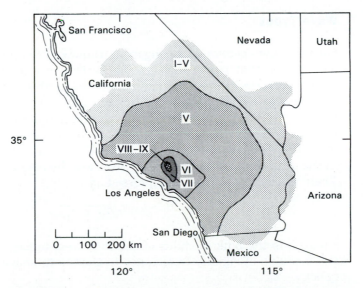

Figure 7-8 Seismic intensity map of the 1971 San Fernando, California, earthquake. (From W. W. Hays, 1980, *Procedures for Estimating Earthquake Ground Motions*, U.S. Geological Survey Professional Paper 1114.)

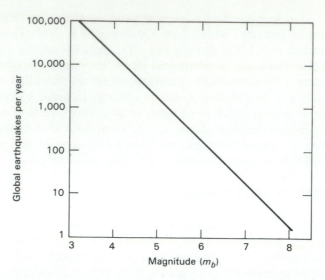

Figure 7-9 Average relationship between magnitude and frequency on a global basis. (From B. A. Bolt, *Nuclear Explosions and Earthquakes: The Parted Veil*, copyright © 1976 by W. H. Freeman and Co., San Francisco.)

ground shaking and structural damage. We will discuss these factors in more detail later.

Earthquakes cannot be predicted with accuracy. On a global basis, however, there is a consistent relationship between magnitude and *frequency* (Figure 7-9). The inverse relationship illustrated in Figure 7-9 shows that whereas on an average there are about 10,000 earthquakes a year of magnitude 4, there are fewer than 10 earthquakes every year above magnitude 7.

SEISMIC WAVES

Wave Types

Seismic waves generated during an earthquake fall into two basic categories—*body waves* that travel through the earth's interior from the earthquake's focus, and *surface waves* that move along the earth's surface from the epicenter. Body waves can be subdivided further into *P* (*primary*) and *S* (*secondary*) *waves*. *P* waves are also known as compressional waves because they travel through rocks as alternate compressions and expansions of the material (Figure 7-10). The movement of individual rock particles is a back-and-forth motion in the direction of propagation. *P* waves are the fastest moving seismic waves. Their velocity (α), in m/s is expressed as

$$\alpha = \sqrt{(\kappa + \tfrac{4}{3}\mu)\rho}$$ Eq. 7-1

where κ is the bulk modulus of the rock (resistance to volume change), μ is the modulus of rigidity (resistance to change in shape), and ρ is the density (g/cc).

S waves are also known as shear waves because of their mode of travel (Figure 7-11). The travel of *S* waves is analogous to the shaking of a rope tied to a wall at the other end. Although the wave motion moves from your hand toward the wall, the movement of the rope is up and down as the wave passes.

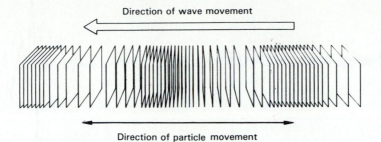

Direction of wave movement

Direction of particle movement

Figure 7-10 The motion of *P* waves. The wave moves through rock as a sequence of compressions and expansions parallel to the direction of wave movement. Particles vibrate in back-and-forth fashion in the direction of wave movement. (After E. A. Hay and A. L. McAlester, *Physical Geology: Principles and Perspectives*, 2d ed., copyright © 1984 by Prentice-Hall, Inc., Englewood Cliffs, N.J.)

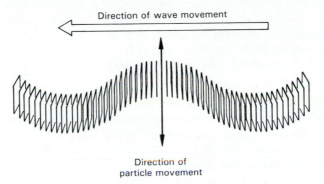

Direction of wave movement

Direction of particle movement

Figure 7-11 Transmission of *S* waves. Particle movement is perpendicular to direction of wave movement through rock. (After E. A. Hay and A. L. McAlester, *Physical Geology: Principles and Perspectives*, 2d ed., copyright © 1984 by Prentice-Hall, Inc., Englewood Cliffs, N.J.)

This shear deformation, which is transverse to the direction of wave propagation, explains why *S* waves cannot be transmitted through a liquid: A liquid has no resistance to shear stresses. The discovery that *S* waves do not travel through the earth's outer core is the main evidence for the hypothesis that this part of the core is liquid.

The velocity of *S* waves (β), which is lower than that of *P* waves, can be expressed as

$$\beta = \sqrt{\frac{\mu}{\rho}}$$

Eq. 7-2

The modulus of rigidity, μ, is equal to zero in liquids; therefore, the S-wave velocity is also zero.

Surface waves, which arrive at seismograph stations after both *P* and *S* waves, include *Love* and *Rayleigh* waves. Surface waves are characterized by long periods (10 to 20 s) and wavelengths of 20 to 80 km. Period measures the amount of time that elapses between the arrival of two successive wave crests and wavelength measures the distance separating identical points on successive waves. Love waves originate from *S* waves that reach the surface at the epicenter. As they move outward, they produce only horizontal ground motion.

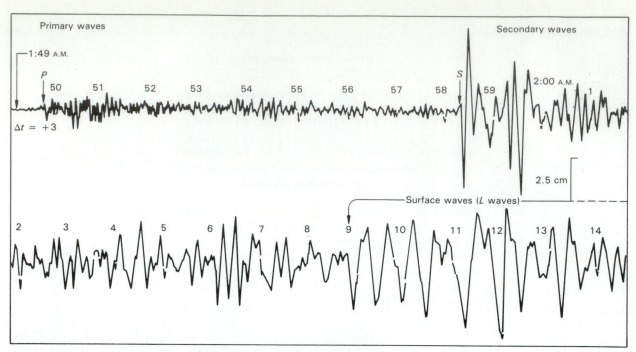

Figure 7-12 Portion of a seismogram showing the arrival of *P*, *S*, and surface waves. (From S. Judson, M. E. Kauffman, and L. D. Leet, *Physical Geology*, 7th ed., copyright © 1987 by Prentice-Hall, Inc., Englewood Cliffs, N.J.)

Rayleigh waves resemble the motion of ocean waves, with predominantly vertical displacement. A seismogram showing the arrival of *P* waves, *S* waves, and surface waves is shown in Figure 7-12.

Time-Distance Relations

The difference in velocity between *P* and *S* waves means that the time interval between the first arrival of *P* waves and *S* waves increases with distance from the earthquake source (Table 7-1). Thus the time interval between *P*- and *S*-

TABLE 7-1

Travel Time and Time Interval for Arrival of Seismic Waves

Distance from source, km	Travel time				Interval between P and S (S − P)	
	P		S			
	min	s	min	s	min	s
2,000	4	06	7	25	3	19
4,000	6	58	12	36	5	38
6,000	9	21	16	56	7	35
8,000	11	23	20	45	9	22
10,000	12	57	23	56	10	59
11,000	13	39	25	18	11	39

SOURCE: From S. Judson, M. E. Kauffman, and L. D. Leet, *Physical Geology*, 7th ed., copyright © 1987 by Prentice-Hall, Inc., Englewood Cliffs, N.J.

wave arrivals measured at a seismograph station is an indication of the distance of the station from the epicenter. Figure 7-13 illustrates how *S-P* time differences vary with distance. The *S-P* time difference can be used for locating the epicenter of a given earthquake if at least three seismograms from different stations are available for that earthquake. The distance from the epicenter corresponding to the *S-P* time difference is used as the radius of a circle to plot a circular arc in the vicinity of the epicenter (Figure 7-14). The point where three circular arcs intersect is the epicenter.

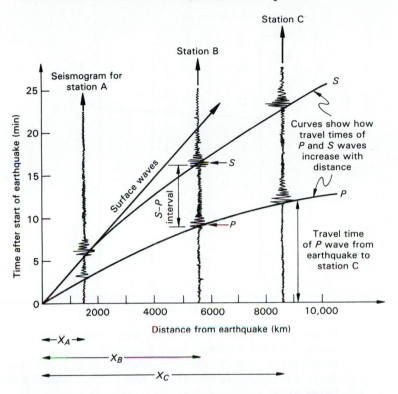

Figure 7-13 Variation of the *S-P* time interval with distance from the epicenter. As the distance from the epicenter increases from seismograph station X_A to X_C, the *S-P* time interval increases. (From F. Press and R. Siever, *Earth*, 3d ed., copyright © 1982 by W. H. Freeman and Co., San Francisco.)

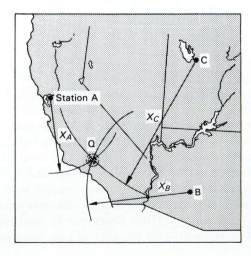

Figure 7-14 The epicenter of an earthquake can be located by plotting three circular arcs with radii determined from the *S-P* time interval as shown in Figure 7-13. The epicenter is located at the intersection of the three arcs. (From F. Press and R. Siever, *Earth*, 3d ed., copyright © 1982 by W. H. Freeman and Co., San Francisco.)

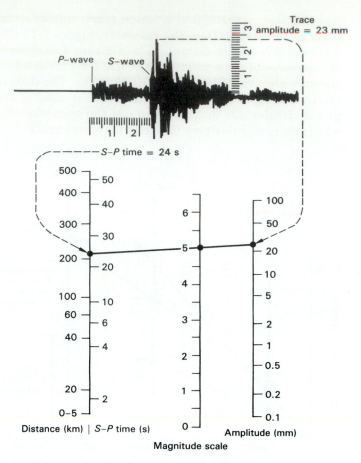

Figure 7-15 Determination of the magnitude from the *S-P* time interval and the maximum trace amplitude measured from a seismogram. (From W. W. Hays, 1980, *Procedures for Estimating Earthquake Ground Motions,* U.S. Geological Survey Professional Paper 1114.)

The *S-P* time interval, along with maximum trace amplitude, is also useful in determining the magnitude (Figure 7-15). If the plotted points representing the trace amplitude and *S-P* time interval are connected with a straight line on a nomogram, the magnitude of the earthquake is indicated where the line crosses the center scale.

EARTHQUAKE HAZARDS

Most earthquakes accompany major movements of crustal blocks along faults. Crustal blocks can also move without generating seismic waves. Such movement, called *aseismic creep,* is desirable from the standpoint that it relieves accumulating strain gradually, thus preventing a buildup of strain leading to a major earthquake. Aseismic creep also has detrimental effects because of the associated displacement that takes place along faults. This displacement can rupture utility lines and wells and cause damage to streets, sidewalks, and other structures. A particularly dramatic example of aseismic creep is the 1963 failure of the Baldwin Hills Reservoir in southern California, an impoundment that stored water for the City of Los Angeles (Figure 7-16). Aseis-

Figure 7-16 Aerial view of the Baldwin Hills reservoir after failure. Aseismic creep along faults beneath the impoundment ruptured the impermeable seal of the bottom. Water then migrated through the dam at an increasing rate until a section of the dam collapsed and the reservoir drained rapidly. (Photo courtesy of Los Angeles Dept. of Water and Power.)

Figure 7-17 Fault scarp formed during the 1959 Hebgen Lake earthquake in Montana. (I. J. Witkind; photo courtesy of U.S. Geological Survey.)

mic displacement along faults located beneath the reservoir caused fracturing of the reservoir's underdrain system and impermeable asphalt membrane. Water leakage through the reservoir floor and dam eventually led to the complete failure of the dam and draining of the impoundment.

Hazards associated with earthquakes include surface faulting, ground shaking, ground failure, and *tsunamis*. Rupture of the ground surface by fault displacement includes both vertical and lateral motion. A *fault scarp* is the steep slope that forms when one crustal block is uplifted relative to the block across the fault (Figure 7-17). Obviously, any structure built directly across a fault that causes surface rupture during an earthquake is likely to be damaged or destroyed (Figure 7-18). By mapping faults known to be active (those in which movement has occurred within a recent period of time), engineers can avoid these hazardous areas for some or all types of structures. Unfortunately, development has occurred over active faults in some areas (Figure 7-19). In

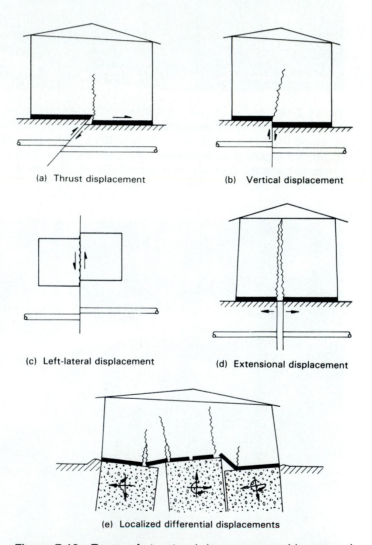

(a) Thrust displacement   (b) Vertical displacement

(c) Left-lateral displacement   (d) Extensional displacement

(e) Localized differential displacements

Figure 7-18 Types of structural damage caused by ground rupture during the San Fernando earthquake. (From U.S. Geological Survey, 1971, *The San Fernando, California, Earthquake of February 9, 1971*, U.S. Geological Survey Professional Paper 733.)

Figure 7-19   Urban development along the San Andreas fault, San Mateo County, California. Arrows show location of fault. (R. W. Wallace; photo courtesy of U.S. Geological Survey.)

addition, surface displacement frequently has occurred along faults that were thought to be inactive.

Ground shaking is usually the greatest threat to buildings and human life associated with earthquakes. The peak acceleration attained by a structure multiplied by its mass defines the dynamic force applied to the structure during an earthquake. Ground acceleration depends on the distance from the epicenter, the focal depth of the earthquake, and the characteristics of the soils and rock beneath the foundation. The response of the structure depends upon its size and design. We will examine these factors more closely in the following sections.

F=ma

Ground failure includes a variety of processes. During intense shaking, soil structure and behavior can be drastically altered. Seismic stresses often cause soils to densify and compact. Certain saturated soils lose strength and liquefy almost instantaneously during seismic shaking. In this state the soil can no longer support buildings, which then may sink and rotate into the soil as complete units (Chapter 11). Mass movements, including slumps, slides, flows, and avalanches can be triggered by earthquakes. Often, these movements involve huge masses of material and are responsible for great damage and loss of life. The Lower Van Norman Dam (Figure 7-20) nearly failed during the 1971 San Fernando, California, earthquake because of a large slump that developed on the upstream side. If the reservoir level had not been maintained at a lower elevation than normal design limits because of concerns for the safety of the dam, it most likely would have totally failed. The imminent failure during the earthquake necessitated the evacuation of 80,000 people down-

Figure 7-20 Lower Van Norman Dam near San Fernando,
California, after damage by the 1971 earthquake. Slumping of
parts of the dam into the reservoir is evident. (Photo courtesy of
U.S. Geological Survey.)

stream. Other examples of mass movements caused by earthquakes are discussed in Chapter 13.

An earthquake hazard that is greatly feared in coastal areas is the tsunami, or *seismic sea wave*. These are high-energy waves generated by earthquakes whose epicenters lie beneath the sea bed. Tsunamis travel rapidly across the ocean basins and cause great damage when they impact coastlines. Tsunamis are described more fully in Chapter 15.

In addition to the hazards just mentioned, other problems are encountered during earthquakes. These include fires, disruption of water supplies, and disease. As the population of seismically active areas continues to increase, the potential for damage and loss of life grows accordingly.

SEISMIC RISKS AND LAND-USE PLANNING

Assessment of seismic hazard for an area is a complex procedure that must take into account the history of seismic events in the area, the location of active faults, the regional and local responses to seismic events, and the seismic characteristics of the surficial geological materials. If the evaluation of all these factors indicates that the area may be subjected to a potentially damaging earthquake, measures can be taken to reduce the risks associated with the event. These measures can include land-use planning and zoning as well as adoption of stringent building codes.

Location of Earthquakes

Locations of earthquake epicenters recorded during a particular period of time provide a good indication of seismically active areas. Such maps (Figure 7-2) show a high concentration of earthquakes along lithospheric plate boundaries. Earthquakes are not limited to plate boundaries, however. Figure 7-21 indicates that damaging earthquakes have struck many areas in the United States during the past 200 years. Table 7-2 provides information on the magnitude and effects of these events.

In addition to determining the location of earthquake epicenters, it is also important to map and identify active faults. A number of specific criteria are used for recognition of an active fault, but in general, there must be evidence of movement at least once within the past 35,000 years. A map of major faults in California is shown in Figure 7-22. If a particular fault segment has not experienced rupture within a certain period of time, this does not necessarily mean that the segment is inactive. Instead, it may indicate that that section of the fault is *locked*, a condition in which large amounts of elastic strain accumulate prior to infrequent, but large-magnitude, earthquakes.

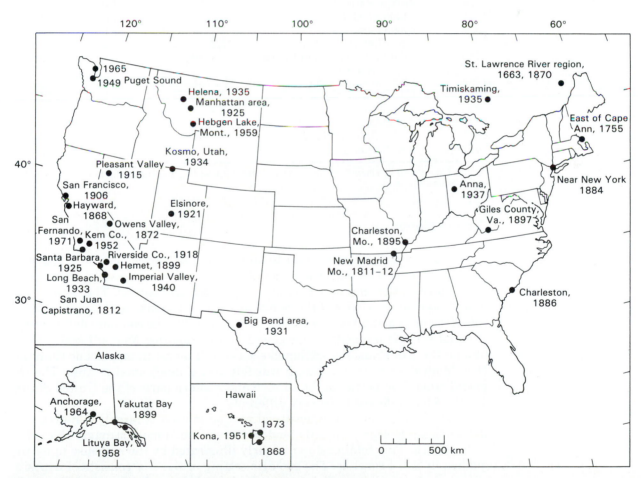

Figure 7-21 Location of past destructive earthquakes in the United States. (From W. W. Hays, 1980, *Procedures for Estimating Earthquake Ground Motions*, U.S. Geological Survey Professional Paper 1114.)

TABLE 7-2

Property Damage and Lives Lost in Notable U.S. Earthquakes

Year	Locality	Magnitude	Damage (million dollars)	Lives lost
1811–12	New Madrid, Mo.	7.5 (est.)	—	—
1865	San Francisco, Calif.	8.3 (est.)	0.4	—
1868	Hayward, Calif.	—	0.4	30
1872	Owens Valley, Calif.	8.3 (est.)	0.3	27
1886	Charleston, S.C.	—	23.0	60
1892	Vacaville, Calif.	—	0.2	—
1898	Mare Island, Calif.	—	1.4	—
1906	San Francisco, Calif.	8.3 (est.)	500.0	700
1915	Imperial Valley, Calif.	—	6.0	6
1925	Santa Barbara, Calif.	—	8.0	13
1933	Long Beach, Calif.	6.3	40.0	115
1935	Helena, Mont.	6.0	4.0	4
1940	Imperial Valley, Calif.	7.0	6.0	9
1946	Hawaii (tsunami)	—	25.0	173
1949	Puget Sound, Wash.	7.1	25.0	8
1952	Kern County, Calif.	7.7	60.0	8
1954	Eureka, Calif.	—	2.1	1
1954	Wilkes-Barre, Pa.	—	1.0	—
1955	Oakland, Calif.	—	1.0	1
1957	Hawaii (tsunami)	—	3.0	—
1957	San Francisco, Calif.	5.3	1.0	—
1958	Khantaak Island and Lituya Bay, Alaska	—	—	5
1959	Hebgen Lake, Mont.	—	11.0	28
1960	Hilo, Hawaii (tsunami)	—	25.0	61
1964	Prince William Sound, Alaska	8.4	500.0	131
1965	Puget Sound, Wash.	—	12.5	7
1971	San Fernando, Calif.	6.6	553.0	65

SOURCE: From Office of Emergency Preparedness, 1972, *Disaster Preparedness*, Report to Congress of the United States, Vol. 3.

Ground Motion

The response of the earth's crust to an earthquake varies both on a regional and local scale. The decrease in amplitude of seismic waves as they move outward from the epicenter is known as *seismic attenuation*. The seismic attenuation characteristics of the San Francisco earthquake and the New Madrid, Missouri, earthquake are shown in Figure 7-23. The contours define zones of equal seismic intensity as measured by the Modified Mercalli scale. Even though the San Francisco earthquake was much larger in magnitude than the New Madrid event (Table 7-2), it was felt over a much smaller area. The regional crustal properties of the eastern and western parts of the United States are therefore inferred to be very different.

On a local scale the intensity of ground motion is largely a function of the surficial geology. Soft soils tend to vibrate much more strongly than stiff soils or rock. This relationship is clearly illustrated by the response to earthquakes in San Francisco. The geologic setting of the city consists of deposits of bay mud and alluvium (river-deposited soils) in low-lying areas (Figure 7-24), and areas of bedrock at shallow depth throughout the rest of the city. Figure 7-25 shows the distribution of intensity during the 1906 earthquake. The areas of high intensity correspond closely to the areas underlain by bay

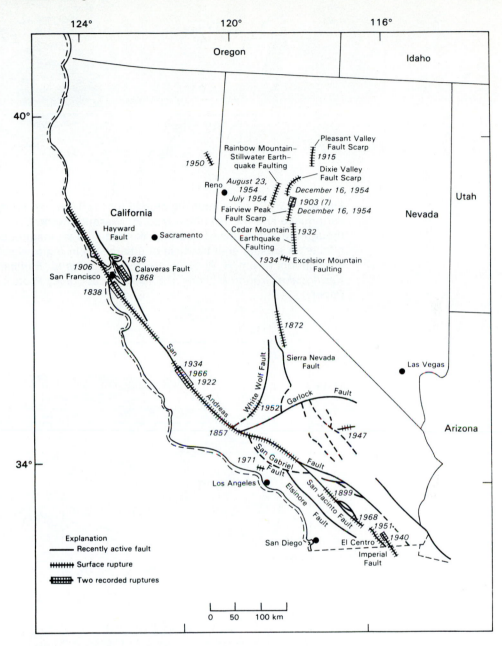

Figure 7-22 Major faults and locations of past ruptures in California and Nevada. (From W. W. Hays, 1980, *Procedures for Estimating Earthquake Ground Motions*, U.S. Geological Survey Professional Paper 1114.)

mud and alluvium. In 1906 many parts of the city were built on poorly compacted fills placed in stream valleys and along the bay. Buildings sited upon these soils were severely shaken and damaged because of the vibrational response of these soils.

The relationship between surficial material and ground motion is further illustrated by the recordings of accelerographs in San Francisco of a distant underground nuclear explosion (Figure 7-26). Horizontal ground motion was much greater for bay mud and alluvium than for bedrock when subjected to the same seismic waves. Soft soils actually amplify incoming seismic waves.

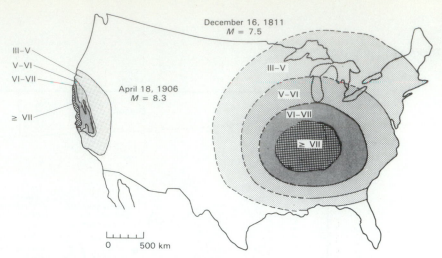

Figure 7-23 Seismic intensity plotted for the 1906 San Francisco earthquake and the 1811 New Madrid, Missouri, event. Even though the San Francisco earthquake had a larger magnitude, it affected a smaller area because of differences in regional seismic attenuation. (From W. W. Hays, 1980, *Procedures for Estimating Earthquake Ground Motions*, U.S. Geological Survey Professional Paper 1114.)

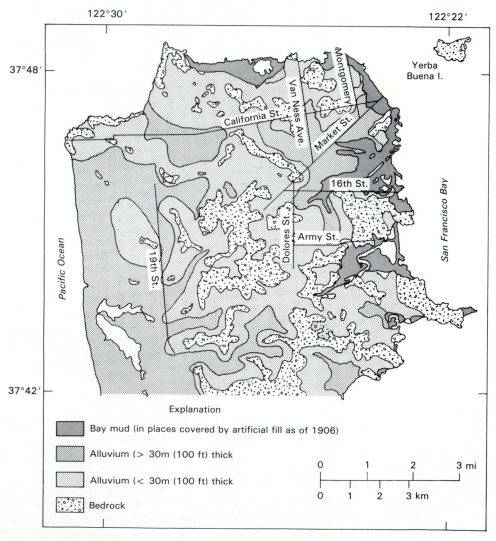

Figure 7-24 Generalized geology of the San Francisco area. (From R. D. Borcherdt, ed., *Studies for Seismic Zonation of the San Francisco Bay Region*, U.S. Geological Survey Professional Paper 941-A.)

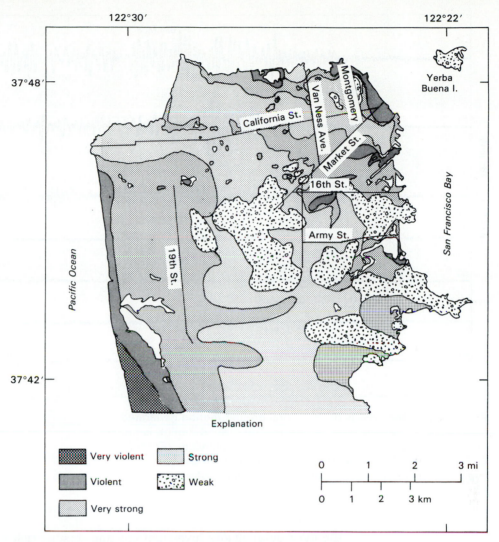

122°30′   122°22′

37°48′

California St.

Van Ness Ave.

Montgomery

Market St.

16th St.

Army St.

19th St.

Yerba Buena I.

Pacific Ocean

San Francisco Bay

37°42′

Explanation

Very violent   Strong

Violent   Weak

Very strong

0   1   2   3 mi

0   1   2   3 km

Figure 7-25 Relative intensity of the 1906 earthquake in San Francisco. The areas of highest intensity correspond to areas of bay mud and alluvium, as shown on Figure 7-24. (From R. D. Borcherdt, ed., *Studies for Seismic Zonation of the San Francisco Bay Region*, U.S. Geological Survey Professional Paper 941-A.)

One of the most effective ways to reduce future losses in areas of high seismic risk is land-use planning. Zoning ordinances of various types can prevent or restrict development in critical areas. An obvious restriction is to prevent development over active faults. Although such a restriction seems to be only common sense, past development has largely ignored the presence of known active faults (Figure 7-19). Further zoning options can establish minimum distances, or *setbacks*, from active faults for buildings such as schools and hospitals, where damage could result in numerous casualties. Similar restrictions could be applied to areas of potential landsliding or severe ground shaking.

The most stringent standards for site location and construction should be established for engineering projects such as dams and nuclear power plants, whose satisfactory performance in an earthquake is necessary to prevent massive losses in human life and property. These types of structures now require

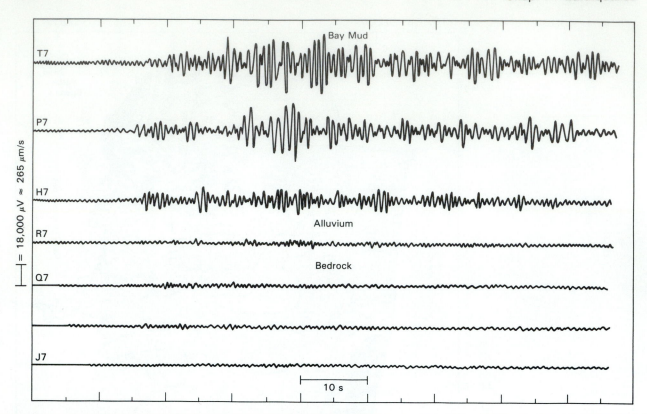

Figure 7-26 Recordings of horizontal ground motion generated
by an underground nuclear explosion. The ground motion is much
stronger for stations located on bay mud and alluvium. (From R.
D. Borcherdt, ed., *Studies for Seismic Zonation of the San
Francisco Bay Region*, U.S. Geological Survey Professional Paper
941-A.)

the most detailed site investigations and seismic risk assessment procedures
possible in most parts of the United States.

EARTHQUAKE ENGINEERING

The effect of an earthquake upon a building is determined by the relationship
and interaction between the characteristics of the seismic waves that reach
the site, the response of the materials beneath the building, and the design
and construction of the building itself.

Horizontal ground motions often present the most severe test of a building's ability to withstand an earthquake without failure. The resulting lateral
accelerations impose inertial forces on a structure. In the absence of strict
building codes, structures rarely are designed to resist lateral forces of the
magnitude imposed by strong ground motion. Figure 7-27 illustrates the nature of these inertial forces. For low, rigid structures (Figure 7-27, left), the
peak lateral force can be calculated as the product of the mass of the building
and the peak lateral acceleration. The lateral force for flexible buildings subjected to short-duration seismic shaking can be less than the product of mass
times acceleration. This force decrease is achieved because part of the energy
is absorbed by the bending action of the frame (Figure 7-27, center). In tall,

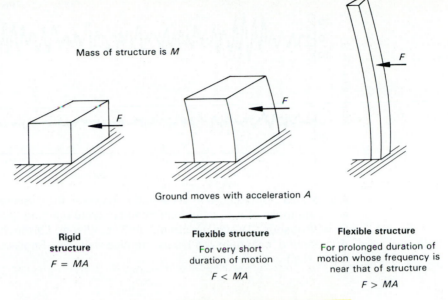

Mass of structure is *M*

Ground moves with acceleration *A*

Rigid structure	**Flexible structure**	**Flexible structure**
F = MA	For very short duration of motion	For prolonged duration of motion whose frequency is near that of structure
	F < MA	*F > MA*

Figure 7-27 Inertia forces developed by different types of buildings when subjected to earthquake shaking. (From H. J. Degenkolb, 1977, *Earthquake Forces on Tall Structures*, Booklet 2717A, Bethlehem Steel Corp., Bethlehem, Pa.)

very flexible buildings, the lateral force can even exceed $m \times a$ because of the increase in inertial force caused by the oscillation of the structure by repeated vibrations (Figure 7-27, right).

One of the most important factors in earthquake damage is the relationship between the vibrational period of a structure and the period of the material upon which it is built. When set into motion, geologic materials, as well as buildings, tend to vibrate at a certain rate. This characteristic rate is defined by the *fundamental period*. The period of a wave or vibration is the time interval between the passage of two corresponding points on the waveform—the time between two successive crests, for example. Fundamental periods of geologic materials range from well below 1 s for bedrock and stiff soils to several seconds or more for deep, soft soils. The fundamental period of a building varies with the height of the structure. Tall buildings have long periods (several seconds), and low buildings have short periods. The effect of building height upon acceleration is illustrated in Figure 7-28. Accelerographs were installed in the basement and penthouse of a 10-story warehouse that was shaken by the 1952 Tehachupi, California, earthquake. Accelerations were much greater in the penthouse than in the basement. In addition, the penthouse accelerations clearly show the effect of the fundamental period of the building, with maximum accelerations occurring every few seconds. No such pattern is evident in the basement accelerogram.

When the period of a soil and a building are similar, the most hazardous situation occurs. Resonance between the building and the soil leads to more violent shaking, and greater damage is usually the result. Tall buildings sustain greater damage when constructed on deep, soft soils because of their similarity in vibrational period. Similarly, small, rigid buildings may perform more poorly when sited upon short-period materials such as bedrock. Figure 7-29 illustrates the effect of building height upon relative seismic damage. The damage intensity for tall structures rises dramatically when the fundamental period of the foundation soil is longer than 1 s.

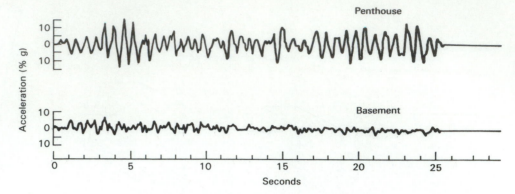

Figure 7-28 Differences in acceleration between the basement and penthouse of a 10-story reinforced-concrete building. (From G. W. Housner, "Design spectrum." In R. L. Wiegel, *Earthquake Engineering*, copyright © 1970 by Prentice-Hall, Inc., Englewood Cliffs, N.J.)

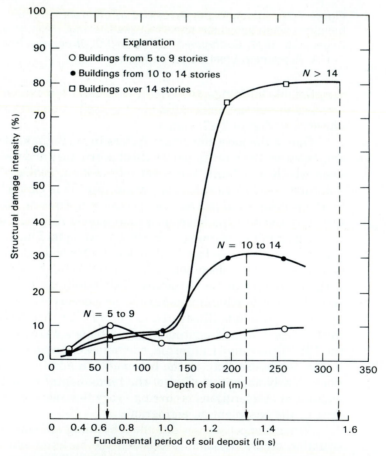

Figure 7-29 Structural damage intensity for different height buildings related to depth of soil and fundamental period of soil. Tall buildings sustain maximum damage when constructed on long-period soil deposits. (From H. B. Seed et al., 1972, Soil Conditions and Building Damage in 1967 Caracas Earthquake, *Journal of the Soil Mechanics and Foundations Division*, American Society of Civil Engineers, SM-8:787–806.

Construction materials and design often play a critical role in the performance of a building during an earthquake (Table 7-3). Small, wood-frame structures are generally the safest type as long as they are anchored securely to their foundations. Improperly anchored houses have been observed to shear apart from their foundations during lateral ground accelerations. Steel-frame or reinforced-concrete construction methods are least hazardous for multistory buildings. Buildings that provide the least degree of protection include those of nonreinforced masonry, brick and mortar, and adobe construction. Structures of this type are literally deathtraps because of their heavy weight and brittle behavior. The Guatemalan earthquake of 1976 provided a good example of the high correlation between adobe construction and nearly total destruction of villages in areas of moderate to high intensity (Figure 7-30).

Although building performance is certainly a major concern, the effects of earthquakes upon *lifeline* systems are also of vital importance. Lifeline systems include electricity and fuel supplies, water and wastewater facilities, transportation lines, and communication networks. Loss of these facilities in a large urban area can lead to greater problems than the earthquake itself. Emergency rescue and relief efforts are severely hampered by damage to energy, transportation, and communication systems. The breakdown of water and sanitary networks can lead to disease and inability to combat fires that may break out. Building codes and regulations must be extended to these types of facilities as well.

TABLE 7-3

Earthquake Ratings for Common Building Types

Simplified description of structural types	Relative damageability (in order of increasing susceptibility to damage)
Small wood-frame structures, i.e., dwellings not over 3000 sq. ft, and not over 3 stories	1
Single or multistory steel-frame buildings with concrete exterior walls, concrete floors, and concrete roof. Moderate wall openings	1.5
Single or multistory reinforced-concrete buildings with concrete exterior walls, concrete floors, and concrete roof. Moderate wall openings	2
Large area wood-frame buildings and other wood-frame buildings	3 to 4
Single or multistory steel-frame buildings with unreinforced masonry exterior wall panels; concrete floors and concrete roof	4
Single or multistory reinforced-concrete frame buildings with unreinforced masonry exterior wall panels, concrete floors and concrete roof	5
Reinforced concrete bearing walls with supported floors and roof of any materials (usually wood)	5
Buildings with unreinforced brick masonry having sandlime mortar; and with supported floors and roof of any materials (usually wood)	7 up
Bearing walls of unreinforced adobe, unreinforced hollow concrete block, or unreinforced hollow clay tile	Collapse hazards in moderate shocks

SOURCE: From D. Armstrong, 1973, The Seismic Safety Study for the General Plan. Sacramento: California Council on Intergovernmental Relations.

NOTE: This table is not complete. Additional considerations would include parapets, building interiors, utilities, building orientation, and frequency response.

Figure 7-30 Devastation of adobe houses in the town of Tecpan, Guatemala, by the 1976 earthquake. (Photo courtesy of U.S. Geological Survey.)

EARTHQUAKE PREDICTION

Prediction of earthquakes is a scientific goal that, if achieved, could prevent widespread loss of life in a major earthquake. Unfortunately, accurate prediction is not currently possible, although intensive research is proceeding in many areas.

Two types of earthquake prediction are theoretically possible. Long-term prediction, in which an earthquake is forecast along a particular fault segment within a period of several years or more, is made by studying seismic gaps and historical records of earthquakes that have occurred along that fault segment. By plotting the numbers of earthquakes within specific time intervals against their magnitude, diagrams similar to Figure 7-9 can be constructed for a local area. From this plot it is possible to determine the *recurrence interval*, or the average time interval between earthquakes of a specific magnitude. Predictions can then be made that an earthquake of that magnitude has a high probability of occurrence within a specified time interval, if the date of the last earthquake of that magnitude is known.

Research leading to short-term predictive ability, involving a time interval small enough for evacuation of an area, for example, has focused on *precursors* that have been observed prior to previous earthquakes. Precursors are physical or chemical phenomena that occur in a typical pattern before an earthquake. These phenomena include the velocity of seismic waves, the electrical resistivity of rocks, preliminary earthquakes (foreshocks), deformation of the land surface, and changes in water level or water chemistry of wells in the area.

Many of these precursors can be explained by a theory called the *dilatancy model*. Under this hypothesis, rocks in the process of strain along a fault show significant dilation or swelling before rupture. This volume increase is caused by the opening of *microcracks,* which are minute failure zones in weaker mineral grains in the rock and along grain boundaries. As the porosity increases during the formation of microcracks, groundwater flows into the highly stressed areas.

These changes in density and water content affect the ability of the rock to transmit seismic waves and conduct electricity. Therefore, seismic-wave velocity and electrical resistivity progressively change as the overall rupture along the fault draws near. Localized changes in land-surface elevation are also related to volume changes at depth. An area of recent uplift along the San Andreas fault near Los Angeles, which has been named the Palmdale Bulge (Figure 7-31), is being monitored in great detail as a possible indicator of a future earthquake.

Volume changes and groundwater movement may be reflected by changes in water levels in wells as well as changes in the chemical composition of groundwater. Radon gas has been observed to increase in wells prior to earthquakes. These increases are perhaps related to the release of radon gas from rocks during the formation of microcracks. The pattern of seismic activity in the vicinity of an area of imminent fault rupture is also significant. This pattern consists of an initial rise in the number of small events, followed by a decline in foreshocks just prior to the major earthquake. The decline may represent a temporary increase in rock strength before the newly formed microcracks are filled with water.

The precursor phenomena can be grouped into stages according to the dilatancy model (Figure 7-32). Stage I consists of a gradual stress buildup along the fault. Stages II and III are correlated with dilatancy and water influx. Stage IV is the major earthquake, and Stage V is the aftermath of the event.

If every earthquake followed the sequence shown in Figure 7-32, with uniform stage duration, earthquake prediction would be a simple matter. Instead, each earthquake is unique in terms of specific precursor behavior patterns and length of precursor stages. Continued research and study of future earthquakes will certainly lead to refinements and calibration of the model.

The strange behavior of animals before an earthquake has been recognized for centuries and now is studied scientifically for use in earthquake prediction. The range of unnatural behavior patterns is extremely diverse. Many animals seem restless and frightened. Snakes have come out of hiber-

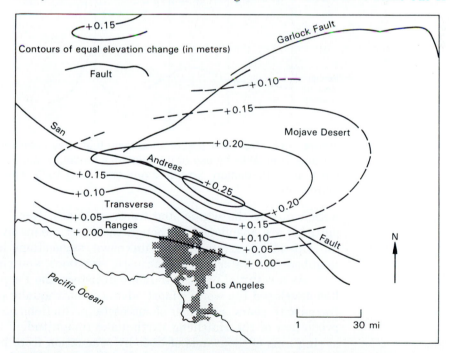

Figure 7-31 Uplift along the San Andreas fault in the vicinity of Palmdale, California. This activity may be a precursor to a future earthquake. (From F. Press and R. Siever, *Earth*, 3d ed., copyright © 1982 by W. H. Freeman and Co., San Francisco.)

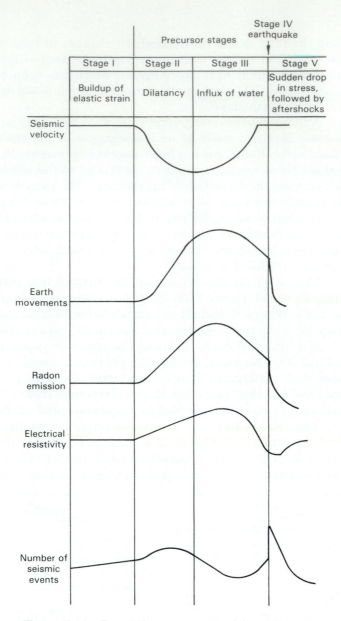

Figure 7-32 Trends in precursor activity that have been related to major earthquakes by use of the dilatancy model. (From F. Press, Earthquake Prediction, *Scientific American*, May 1975, with permission.)

nation during the winter and frozen to death. Normal eating and sleeping patterns are abandoned. Although the exact means by which animals sense an impending earthquake are unknown, suggestions have included changes in subsurface water content, acoustic emissions, and electrostatic effects.

As a result of past earthquake disasters, the People's Republic of China has developed the world's most advanced earthquake prediction program. In the past 15 years, a number of successful predictions have been made. In 1975 prediction of the Haicheng earthquake (magnitude 7.3) was responsible for saving tens of thousands of lives by evacuation from hazardous buildings. A notable failure, however, was the inability to predict the 1976 Tangshan event (magnitude 7.6), in which perhaps as many as three-quarters of a million people were killed.

CASE STUDY 7-1
THE SAN FERNANDO EARTHQUAKE

Among recent earthquakes in the United States, the February 9, 1971, earthquake in San Fernando, California, stands out for several reasons. First, the event struck the outlying vicinity of a major metropolitan area, Los Angeles, California. Thus despite the relatively moderate magnitude of the event (6.6), great damage was caused to the densely populated area. Second, although the earthquake occurred in a region known to be seismically active, displacement took place along several faults that were not considered to be active prior to the earthquake. Third, the network of seismograph stations that recorded the event provided the most detailed and abundant data ever to be gathered for an earthquake. These data provided many interesting facts and interpretations, including the highest ground accelerations ever recorded. These accelerations commonly ranged from $0.5g$ to $0.75g$, with peak recorded values of more than $1g$. Accelerations of this magnitude are particularly significant because they exceed by several times design values used for earthquake-resistant struc-

tures. Finally, the San Fernando earthquake provided a performance test for many modern buildings constructed under recent building codes. Unfortunately, many of the structures did not perform well (Table 7-4).

The San Fernando earthquake was generated by movement along a number of minor faults near a bend in the San Andreas fault. The locations of the main shock and major aftershocks are shown in Figure 7-33. Fortunately, the earthquake took place about 6:00 A.M.; at this hour schools were unoccupied and streets and highways were nearly deserted. Had the earthquake struck just 1 h or 2 h later, the loss of life would have been much greater. Ground shaking from the main shock lasted about 1 min; the duration of strong motion was only about 10 s.

For a moderate event, the destruction inflicted by the San Fernando earthquake was quite striking. Sixty-four people died, and property damage estimates ran into the hundreds of millions of dollars (Table 7-4). These damages would have seemed insignificant had the

TABLE 7-4

Estimate of Damage

Structure	Number damaged	Amount
Schools	180	$ 22,500,000
Hospitals	4	50,000,000
Residential:		
Homes	21,761	
Apartment houses	102	
Mobile homes	1,707	179,500,000
Commercial buildings	542	
Miscellaneous structures	250	
Highways and roads		27,500,000
Dams		36,500,000
Other public structures		145,000,000
Utilities		42,000,000
Personal property		50,000,000
Total		$553,000,000

SOURCE: From R. Kachadoorian, 1971, An estimate of damage in the San Fernando, California, earthquake of Feb. 9, 1971, in *The San Fernando, California, earthquake of February 9, 1971*, U.S. Geological Survey and National Oceanic and Atmospheric Administration, 1971, U.S. Geological Survey Professional Paper 733.

NOTE: Data supplied by the city of Los Angeles, city of San Fernando, county of Los Angeles, Los Angeles Unified School District, Corps of Engineers, Los Angeles Food Control System, State of California, and region 7 of the Office of Emergency Preparedness

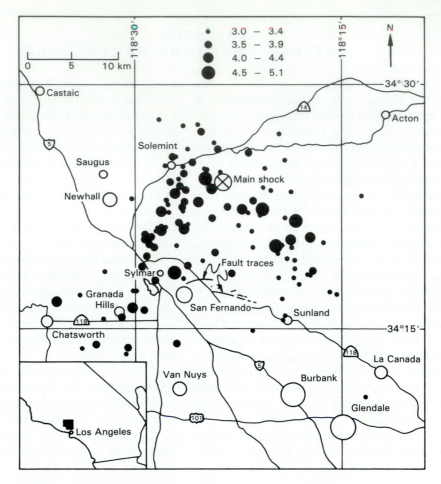

Figure 7-33 Map of main shock and aftershocks greater than magnitude 3.0 for a 3-week period following the San Fernando earthquake. (From U.S. Geological Survey, 1971, *The San Fernando, California, Earthquake of February 9, 1971*, U.S. Geological Survey Professional Paper 733.)

Lower Van Norman Dam failed (Figure 7-20), an event that was narrowly avoided.

Surface ruptures associated with the earthquake were mapped in a zone approximately 15 km in length. The specific types of surface displacements exhibited a complex and variable pattern (Figure 7-18). Damage to buildings, roads, and utilities was intense in these areas (Figure 7-34).

Ground shaking provided the other main cause of damage. High-rise buildings in the area consisted mainly of medical facilities. Of four hospitals damaged, two were total losses. Parts of Veteran's Hospital, a pre-1930 nonreinforced concrete structure, collapsed, killing 45 people (Figure 7-35). The newly constructed $25 million Treatment and Care Facility at the Olive View Hospital also sustained severe damage and had to be demolished. The building was a five-story reinforced concrete structure. First- and second-story columns were particularly hard hit (Figure 7-36). Two other relatively new hospitals were also severely affected.

One hundred eighty school buildings were damaged in the earthquake. Seismic standards were established for schools in California with the enactment of the Field Act in 1933. Schools built before the Field Act performed much less satisfactorily than post–Field Act buildings. Seven pre–Field Act schools in the Los Angeles Unified School District suffered sufficient damage to warrant demolition after the earthquake. Luckily, no schools were occupied at the time of the earthquake.

Other buildings erected prior to 1933 fared

Figure 7-34 Damage to highway by surface rupture from the 1971 San Fernando, California, earthquake. (Photo courtesy of U.S. Geological Survey.)

Figure 7-35 Cleanup following collapse of Veteran's Hospital during the San Fernando earthquake. Forty-five people were killed. (Photo courtesy of U.S. Geological Survey.)

poorly during the shaking. Most of these, including Veteran's Hospital, were constructed of nonreinforced concrete.

Wood-frame buildings were damaged throughout the area of ground shaking, with more severe damage occurring in older houses that lacked lateral bracing and that may have been weakened by rotting or termite damage. Two-story houses sustained more damage than one-story structures (Figure 7-37). Mobile homes were shaken from their foundations in many developments because of inadequate anchoring.

Damage to lifeline systems was extensive in the areas of strongest ground motion. Water-supply systems were disrupted in several cities, including failure of dams, and breaks in water lines, tunnels, and aqueducts. The water distribution system for the city of San Fernando (population 17,000) was completely destroyed. Also affected were gas systems, sewage lines, and highways. Several major highway inter-

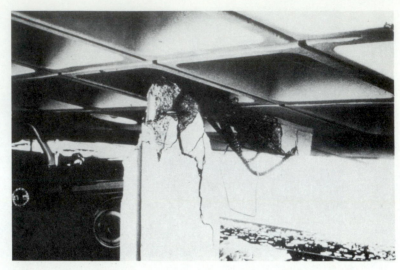

Figure 7-36 Damaged column of reinforced-concrete ambulance port at Olive View Hospital. (National Oceanic and Atmospheric Administration.)

Figure 7-37 Damage to two-story house near San Fernando. (Los Angeles Dept. of Building and Safety.)

changes were blocked by overpass collapse (Figure 7-38).

Overall, the San Fernando earthquake produced widespread damage that could have been inestimably worse if it had struck during midday or if the Lower Van Norman Dam had to-tally failed and released its reservoir upon a large population below. We can only hope that the combination of improved building codes, strict land-use planning, and advances in earthquake prediction will minimize loss of life and damage from future earthquakes.

Figure 7-38 Collapse of an overpass caused by the San Fernando earthquake. (Photo courtesy of U.S. Geological Survey.)

SUMMARY AND CONCLUSIONS

When crustal blocks suddenly slip past one another, elastic strain energy is released in the form of seismic waves. According to the elastic rebound theory, elastic strain energy is gradually accumulated until brittle rupture occurs. Seismic waves are recorded by seismographs or, near the epicenter, as ground accelerations by accelerographs.

Earthquake magnitude is a quantitative measure of the amount of energy released, whereas intensity is a descriptive value assigned by the assessment of earthquake damage at a given location. Magnitude and frequency are inversely proportional; only a few great earthquakes strike each year over the entire earth.

Much information about earthquake processes and the earth's interior is gained by the study of seismic waves. The compressional P waves travel at the highest velocity; shear S waves travel more slowly and terminate abruptly at the earth's liquid outer core. Love and Rayleigh waves travel outward from the epicenter near the earth's surface. The S-P time interval is useful for determining the magnitude of the earthquake as well as the epicentral location.

A variety of hazardous processes are initiated by earthquakes. Surface rupture is common along the fault that generated the earthquake. Ground shaking, which affects a much larger area, is responsible for much of the damage to structures. Shaking also can cause various types of ground failure. When earthquakes take place offshore, tsunamis radiate outward to cause great damage in coastal areas. Finally, earthquakes are responsible for human catastrophies due to disease and fire when urban lifeline systems are severed.

Massive governmental and private efforts have been directed toward reducing earthquake damage. These programs include identification of hazardous areas such as fault zones and prediction of localized ground motion. Soil and rock type govern the response of a particular site to ground shaking. Specifically, buildings constructed on thick, soft soils sustain maximum damage when set into vibration. Risks can be reduced by zoning ordinances that prevent construction in high-hazard areas or specify construction techniques for critical structures.

Earthquake engineering is the analysis of the effect of vibrations upon structures and the design of earthquake-resistant buildings. Each building has a fundamental period of vibration. When this period corresponds closely to the period of the foundation materials, maximum damage is likely. Modern steel, wood-frame, and reinforced-concrete designs perform well during vibration; heavy, rigid buildings including those of nonreinforced concrete, brick, and adobe fare poorly. Increasingly strict building codes eventually will prove worthwhile in savings of life and property.

Earthquake prediction holds promise for future damage reduction. It is hoped that increasing sophistication of seismic detection instruments will allow the recognition of predictable patterns of precursor phenomena. These data, along with the study of animal behavior, someday may yield a method for preventing the great disasters to which some regions are now accustomed.

REFERENCES AND SUGGESTIONS FOR FURTHER READING

BAROSH, P. J. 1969. *Use of Seismic Intensity Data to Predict the Effects of Earthquakes and Underground Nuclear Explosions in Various Geologic Settings*. U.S. Geological Survey Bulletin 1279.

BERLIN, G. L. 1980. *Earthquakes and the Urban Environment*, vols. 1, 2, 3. Boca Raton, Fla.: CRC Press Inc.

BILLINGS, M. P. 1972. *Structural Geology*, 3d ed. Englewood Cliffs, N.J.: Prentice-Hall, Inc.

BLAIR, M. L., and W. E. SPANGLE. 1979. *Seismic Safety and Land-use Planning—Selected Examples from California*. U.S. Geological Survey Professional Paper 941-B.

BOLT, B. A. 1976. *Nuclear Explosions and Earthquakes: The Parted Veil*. San Francisco: W. H. Freeman and Co.

BORCHERDT, R. D., ed. 1975. *Studies for Seismic Zonation of the San Francisco Bay Region*. U.S. Geological Survey Professional Paper 941-A.

COSTA, J. E., and V. R. BAKER. 1981. *Surficial Geology: Building with the Earth*. New York: John Wiley.

DEGENKOLB, H. J. 1977. *Earthquake Forces on Tall Structures*. Booklet 2717A, Bethlehem Steel Corp., Bethlehem, Pa.

ECKEL, E. B. 1970. *The Alaska Earthquake March 27, 1964: Lessons and Conclusions*. U.S. Geological Survey Professional Paper 546.

ESPINOSA, A. F., ed. 1976. *The Guatemalan Earthquake of February 4, 1976: A Preliminary Report*. U.S. Geological Survey Professional Paper 1002.

HAYS, W. W. 1980. *Procedures for Estimating Earthquake Ground Motions*. U.S. Geological Survey Professional Paper 1114.

HAYS, W. W., ed. 1981. *Facing Geologic and Hydrologic Hazards: Earth-Science Considerations*. U.S. Geological Survey Professional Paper 1240-B.

JUDSON, S., M. E. KAUFFMAN, and L. D. LEET. 1987. *Physical Geology*, 7th ed. Englewood Cliffs, N.J.: Prentice-Hall, Inc.

NICHOLS, D. R., and J. M. BUCHANAN-BANKS. 1974. *Seismic Hazards and Land-use Planning*. U.S. Geological Survey Circular 690.

OFFICE OF EMERGENCY PREPAREDNESS. 1972. *Disaster Preparedness*. Report to Congress of the United States, vol. 3.

PRESS, F. 1975. Earthquake prediction. *Scientific American* 232(no. 5):14–23.

PRESS, F., and R. SIEVER. 1982. *Earth*, 3d ed. San Francisco: W. H. Freeman and Co.

SEED, H. B., R. U. WHITMAN, H. DEZFULIAN, R. DOBRY, and I. M. IDRISS. 1972. Soil conditions and building damage in 1967 Caracas earthquake. *Journal of the Soil Mechanics and Foundations Division*, American Society of Civil Engineers SM-8:787–806.

U.S. GEOLOGICAL SURVEY AND NATIONAL OCEANIC AND ATMOSPHERIC ADMINISTRATION. 1971. *The San Fernando, California, Earthquake of February 9, 1971*. U.S. Geological Survey Professional Paper 733.

U.S. GEOLOGICAL SURVEY. 1976. *Earthquake Prediction: Opportunity to Avert Disaster*. U.S. Geological Survey Circular 729.

U.S. GEOLOGICAL SURVEY. 1982. *The Imperial Valley, California, Earthquake of October 15, 1979*. U.S. Geological Survey Professional Paper 1253.

WALLACE, R. E. 1974. *Goals, Strategy, and Tasks of the Earthquake Hazard Reduction Program*. U.S. Geological Survey Circular 701.

WALLACE, R. E. 1984. *Faulting Related to the 1915 Earthquakes in Pleasant Valley, Nevada*. U.S. Geological Professional Paper 1274-A.

WIEGEL, R. L., ed. 1970. *Earthquake Engineering*. Englewood Cliffs, N.J.: Prentice-Hall, Inc.

YOUD, T. L., and S. N. HOOSE. 1978. *Historic Ground Failures in Northern California Triggered by Earthquakes*. U.S. Geological Survey Professional Paper 993.

PROBLEMS

1. Why doesn't the elastic rebound theory satisfactorily explain all earthquakes?
2. Describe the instruments used to measure earthquakes.
3. Contrast earthquake magnitude and intensity.
4. Why can't *S* waves travel through the earth's core?
5. How is a tsunami generated?
6. How do the geologic materials beneath a site influence its response to an earthquake?
7. Speculate upon the likelihood and effects of an earthquake on your campus. Consider the location, local geology, and building characteristics (age, size, materials, and so on) in your analysis.
8. What are the major areas of research concerned with earthquake prediction?
9. Summarize the dilatancy model.
10. How might the effects of the San Fernando earthquake have differed if the magnitude were 8.0 instead of 6.6?

CHAPTER
8
HEAT FLOW, GRAVITY, AND MAGNETISM

The interior of the earth is inaccessible to direct human observation, yet the prevailing conditions and processes there exert a great influence upon the surface of the earth. In order to study the internal regions of the earth, we must use indirect methods of investigation. Among these techniques are studies of heat flow, gravity, and magnetism. These topics, along with seismicity, fall into the field of study known as *geophysics*.

THE EARTH'S INTERNAL HEAT

According to most recent hypotheses of the earth's origin, planetary accretion took place at a moderate temperature. During the early part of its history, the earth remained a solid, homogeneous body. Internal temperatures increased, however, as a result of the decay of radioactive elements. A crucial point was reached when the temperatures exceeded the melting point of iron at certain zones within the new planet. As iron melted, it began to separate and sink toward the center of the earth. The movement of molten iron toward the incipient core released gravitational energy in the form of heat, leading to the segregation, or *differentiation*, of the earth into its present layers (Figure 8-1). The assumed present relationship of the temperature profile and melting point curve (Figure 8-1) requires that the earth is now largely solid with the exception of the outer core. Another consequence of differentiation was that the radioactive elements became concentrated in the crust because of their chemical affinity for other elements that accumulated there. The present thermal regime of the earth therefore is dictated by a partially molten core and a radioactive heat source in the crust.

Heat Flow

The unequal distribution of volcanoes, hot springs, geysers, and other indications of high heat flow suggests that the flow of heat from the interior of the earth to the surface is not uniform. We have already mentioned the two major sources of internal heat—the molten core and radioactivity. Heat flows whenever a temperature gradient exists; therefore heat flows from the earth's core to its cooler surface. Three physical processes account for heat transport; these include conduction, convection, and radiation. Conduction is the atomic transfer of energy by vibrating atoms to adjacent atoms vibrating less rapidly. The rate of heat conduction is a function of the temperature gradient and the *thermal conductivity* of the rocks through which the heat is flowing. Rocks are poor heat conductors; fluctuations of temperature due to solar effects at the surface, for example, are felt only in the upper few meters of soil. By itself, conduction is inadequate to explain the distribution of heat flow in the earth. In fact, heat that began to move outward from the center of the earth at the time of its origin would not yet have reached the surface. Although conduction of heat is important in the earth, another process must be occurring. Radiation of heat is unlikely through rocks; therefore, convection must play an important role in heat transport.

Convective heat transfer may be more efficient than conduction because the material being heated moves from areas of higher to lower temperature. A good example of convection is the movement of water in a pan that is heated from below (Figure 8-2). The hottest water in the center of the pan rises to the surface and descends along the cooler sides. The rise of hot water in geysers and hot springs also is partially the result of this mechanism.

Convection in the earth is thought to take place in the mantle. Despite

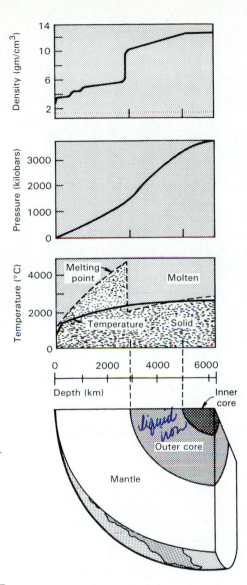

Figure 8-1 Physical conditions in the interior of the earth. The inferred temperature profile of the earth passes above the melting point curve in the region of the earth's outer core. (From E. A. Hay and A. L. McAlester, *Physical Geology: Principles and Perspectives*, 2d ed., copyright © 1984 by Prentice-Hall, Inc., Englewood Cliffs, N.J.)

the rigidity of mantle rocks, as evidenced by transmission of seismic waves, they can flow when thermal or mechanical stresses are applied over long periods of time. Like the water in the pan, plumes of hot mantle rocks slowly rise and spread out laterally beneath the rigid lithosphere (Figure 8-3). Areas above these rising plumes are regions of high heat flow and active volcanism. Circulation is maintained by descending currents and lateral flow at depth

Figure 8-2 Convective heating of water in a pan may be similar to convection cells of mantle material that rise to the base of the lithosphere, spread laterally, and then descend.

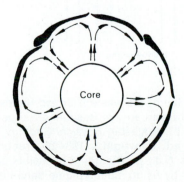

Figure 8-3 Hypothetical configuration of convection cells in the mantle.

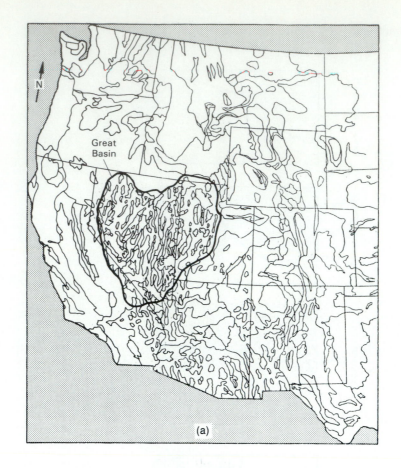

(a)

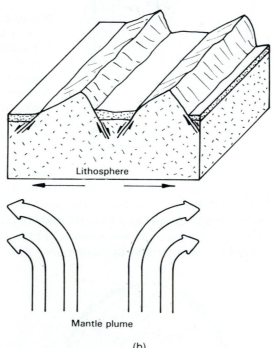

(b)

Figure 8-4 (a) Map of the western United States showing location of Great Basin area of Basin-and-Range Province. (From E. A. Hay and A. L. McAlester, *Physical Geology: Principles and Perspectives*, 2d ed., copyright © 1984 by Prentice-Hall, Inc., Englewood Cliffs, N.J.) (b) Diagram of possible basin-and-range structure caused by rising mantle plume beneath.

toward the rising plumes. The whole mantle may be broken into a series of such *convection cells*. Mantle convection may provide the driving mechanism for plate tectonics. Thus heat flow in the mantle is ultimately responsible for volcanism, mountain building, and earthquakes in the lithosphere.

The second major source of internal heat is radioactive decay. Since differentiation, radioactive elements have been concentrated in the rocks near the earth's surface—in particular, in granitic rocks. Thus radioactive heat production is most significant in the continental crust. Heat flow measured at the surface is the sum of radioactive production from the crust and deep heat flow from the core.

Variations in Heat Flow

Heat-flow measurements have been made both on the continents and in the ocean basins. Continental measurements are made by inserting thermometers into holes drilled into the walls of mines and tunnels. In the oceans, temperature probes are driven into bottom sediments to measure the temperature. In both areas, measurements of temperature changes with depth are used to determine the temperature gradient. When the temperature gradient is multiplied by the measured thermal conductivity of the rock or sediment, the heat flow is obtained.

Heat-flow values throughout the world can be related to the geologic setting of the point of measurement. In the ocean basins, high heat flow is found beneath the midoceanic ridges (Figure 8-3). Midoceanic ridges are thought to be the surface manifestations of rising mantle convection currents. Volcanism in these areas forms new lithosphere, which moves laterally away from the ridges. Heat flow declines with distance from midoceanic ridge crests and reaches minimum values over trenches. Trenches develop over subduction

Figure 8-5 Steam wells at The Geysers geothermal area, California. (Photo courtesy of U.S. Geological Survey.)

zones, where cold lithospheric slabs are forced downward into the mantle. Sub-duction zones constitute the descending plumes of convection current cells; thus heat flow is low in these areas.

On the continents, heat flow also varies, even though the radioactive heat component of total heat is relatively constant. Areas of high heat flow have had recent tectonic and volcanic activity. The Basin-and-Range region of west-ern North America is one such area (Figure 8-4). Volcanism and recent faulting have been common in the Basin-and-Range province, so it is possible that a rising convection current or plume is present beneath this province. The high heat flow in this region is under study for possible production of *geothermal energy*. The Geysers geothermal plant in California (Figure 8-5) already pro-duces 600 MW of electrical power from geothermal energy. Low continental heat flow occurs in geologically old regions, like the Canadian Shield, that have been tectonically inactive for millions of years.

GRAVITY

The force of gravity can be expressed as

$$F = \frac{mM}{r^2}G \qquad \qquad \text{Eq. 8-1}$$

where F is the gravitational attraction between two bodies, m and M are the masses, r is the distance between the bodies, and G is the universal gravi-tational constant. For a body at the surface of the earth, r and M become the radius of the earth and its mass, respectively. If the earth were perfectly spher-ical, nonrotating, and uniform in density, the gravitational acceleration, or the acceleration of a freely falling body, would be constant anywhere upon the earth's surface. Gravitational acceleration varies from place to place, however, and it is these minor variations that make gravity measurements a very useful tool in geology.

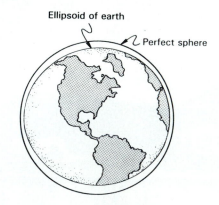

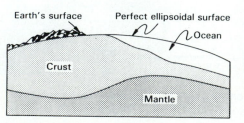

Figure 8-6 Variations in gravitational attraction are caused by the earth's ellipsoidal shape, elevation differences, and differences in density of crustal materials. (From E. A. Hay and A. L. McAlester, *Physical Geology: Principles and Perspectives*, 2d ed., copyright © 1984 by Prentice-Hall, Inc., Englewood Cliffs, N.J.)

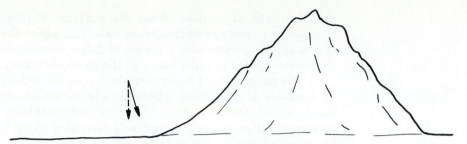

Figure 8-7 Horizontal deflection of a plumb bob by the mass of a prominent mountain range.

Measurement

The unit of measurement used in gravity studies is the Gal, which equals an acceleration of 1 cm/s². The average value of g (gravitational acceleration) at the earth's surface is about 980 Gal, although the measured values vary from place to place. These variations are attributed to (1) rotation of the earth, (2) topography, and (3) variations in density of near-surface rocks. Rotation affects gravity by causing a bulge at the earth's equator that is due to centrifugal force. The radius of the earth is thus greater at the equator than at the poles. The radius also varies from highlands to lowlands or ocean basins. These departures from the ideal ellipsoidal shape of the earth cause minor changes in

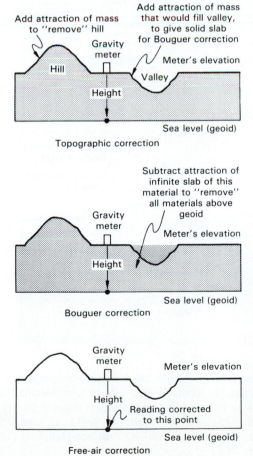

Add attraction of mass to "remove" hill

Add attraction of mass that would fill valley, to give solid slab for Bouguer correction

Gravity meter

Meter's elevation

Hill

Valley

Height

Sea level (geoid)

Topographic correction

Subtract attraction of infinite slab of this material to "remove" all materials above geoid

Gravity meter

Meter's elevation

Height

Sea level (geoid)

Bouguer correction

Gravity meter

Meter's elevation

Height

Reading corrected to this point

Sea level (geoid)

Free-air correction

Figure 8-8 Corrections to measured gravity readings are necessary to discover subsurface geological features. The topographic correction removes the effect of local variations in elevation. The Bouguer correction compensates for a uniform slab of material between the geoid and the point of measurement. The free-air correction results in a reading that would be obtained if the meter were at sea level. (From S. Judson, M. E. Kauffman, and L. D. Leet, *Physical Geology*, 7th ed., copyright © 1987 by Prentice-Hall, Inc., Englewood Cliffs, N.J.)

gravitational attraction across the surface (Figure 8-6). Finally, lateral changes in density of near-surface rocks can affect the value of *g*. The value of *g* would be greater over a region of dense near-surface rocks such as basalt than over lighter granitic rocks at the same elevation.

The magnitude of these variations is on the order of milliGals (0.001 Gal) or fractions of a milliGal. These minute variations are measured by instruments called *gravity meters*. A gravity meter consists of a weight suspended on a spring that expands or contracts according to the local gravitational field. Despite this simple principle, gravity meters are extremely sophisticated, expensive instruments because of the precision they must attain.

After a field measurement is taken, it must be corrected before it will yield geologically useful information. One major correction resolves the effects of local topographic variations. For example, the presence of a nearby mountain range will deflect a pendulum from its vertical position because of the gravitational attraction caused by the mass of the mountains (Figure 8-7). The *Bouguer correction* adjusts the reading to represent a slab of uniform thickness above a hypothetical plane at sea level called the *geoid* (Figure 8-8). Also, a correction is applied to compensate for the material between the geoid and the point of measurement. This is known as the *free-air* correction. The final values are compared to an expected value, which is calculated for an ideal earth with a homogeneous crust. Any differences between the corrected measurement and the expected value are due to local variations in density caused by the composition of near-surface rocks. These differences are called *gravity anomalies*.

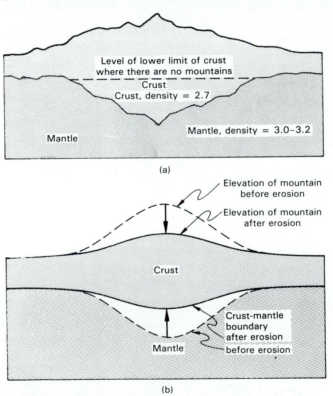

Figure 8-9 Isostatic equilibrium. (a) A mountain range with a low-density crustal root. (b) Isostatic equilibrium is maintained by reduction in size of the root as the mountain mass is slowly eroded. (From S. Judson, M. E. Kauffman, and L. D. Leet, *Physical Geology*, 7th ed., copyright © 1987 by Prentice-Hall, Inc., Englewood Cliffs, N.J.)

The study of gravity anomalies has led to important theories about the earth's crust. For example, young mountain ranges typically yield large negative anomalies. This implies that rocks of low density extend to some depth below the base of the mountain range. Ocean basins have positive anomalies, which indicates that the rocks beneath the ocean basins are denser than average. Thus the density of near-surface rocks is inversely proportional to their topographic position; regions of dense rocks usually lie below sea level, and mountain ranges are composed of lighter rocks. Gravity measurements also have many practical applications. Masses of igneous rock that contain ore bodies often differ in density from the rocks that surround them. In addition, sedimentary rock structures that trap petroleum may cause characteristic gravity anomalies.

Isostasy

The discovery of negative gravity anomalies beneath mountain ranges has important implications concerning the earth's crust and mantle. Negative gravity anomalies beneath young mountain ranges means that mountains must have low-density roots that extend far below the surface (Figure 8-9). In

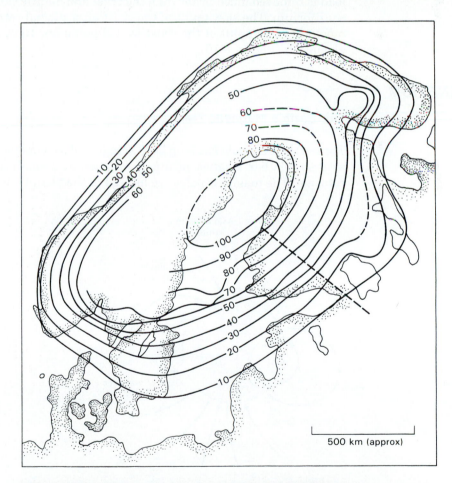

Figure 8-10   Uplift in Scandinavia during the last 5000 years caused by glacial retreat. The contours are the elevations in meters of a shoreline that was at sea level 5000 years ago. (After R. K. McConnell, Jr., 1968, *Journal of Geophysical Research* 73:7090.)

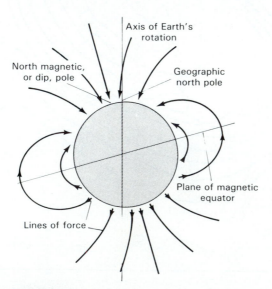

effect, mountain masses are "floating" in the denser mantle rock. This concept
is known as the theory of *isostasy*. Although mantle rocks are rigid solids with
respect to short-term deformations such as the propagation of seismic waves,
they may behave as a viscous fluid over long periods of geologic time. Like an
iceberg in water, the mountain's low-density root lies well below the level of
the plains surrounding it. The mountain mass is supported by buoyant forces
in the mantle similar to the buoyancy that supports any floating object.

Implied in the theory of isostasy is the idea that crustal masses establish
a state of dynamic equilibrium with the mantle. Thus the dense basalts of the
ocean basins lie at a low elevation in comparison with lighter granitic mountain ranges, which float higher in the mantle. Any change in the mass of a
crustal body will be reflected in changes in this *isostatic equilibrium*. For example, as mountains are slowly eroded, mass is removed and the low-density
root is uplifted and decreases in size because less support is required. Thus
the Rocky Mountains have deep roots in comparison with the old, eroded Appalachian Mountains, whose roots have largely disappeared.

Isostatic equilibrium also is illustrated by the advances and retreats of
glaciers. The heavy load of a glacier imposed upon the crust causes the crust
to sink deeper into the mantle. When the ice melts and retreats, however, the
load is removed much faster than the crust and mantle can reestablish isostatic
equilibrium. The slow uplift of the crust after glacial retreat is called *isostatic
rebound*. Slow uplift of the crust is still going on, 10,000 years after the last
glaciation (Figure 8-10).

MAGNETISM

The Earth's Magnetic Field

The earth as a whole has long been recognized as a magnetic body that behaves
as a simple bar magnet, or dipole, aligned at a small angle from its axis of
rotation. The magnetic poles are located in the vicinity of the geographic poles.

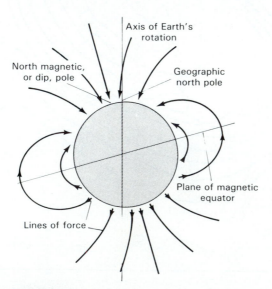

Figure 8-11 Lines of force caused by the earth's magnetic field.
(From S. Judson, M. E. Kauffman, and L. D. Leet, *Physical
Geology*, 7th ed., copyright © 1987 by Prentice-Hall, Inc.,
Englewood Cliffs, N.J.)

Magnetic *lines of force* associated with the earth's magnetic field cause magnetized bodies on the earth to asign themselves parallel to the lines of force (Figure 8-11). The discovery of this principle made possible navigation with the magnetic compass.

The vertical angle between a line of force and the earth's surface is known as magnetic *inclination*. This angle is zero, or horizontal, at the magnetic equator and 90°, or vertical at the magnetic pole. Because the magnetic and geographic poles do not exactly coincide, there is a small angle between the direction of a compass needle pointing to the magnetic north pole and the direction of a line pointing to the geographic north pole. This angle is known as *magnetic declination* (Figure 8-12).

The cause of the earth's magnetism has been and continues to be the object of much speculation. It is now believed that this field is caused by currents moving within the liquid iron core of the earth. These currents would generate electrical currents in the highly conductive iron, which would in turn induce the magnetic field. The core therefore functions as a self-exciting dynamo.

Paleomagnetism

As a magma cools, the atoms in magnetic minerals such as magnetite become aligned in the direction of the earth's magnetic field. This process occurs at a temperature called the *Curie point*, a temperature well below the melting point of the mineral. Once magnetism is acquired by the rock, it is retained unless the rock becomes heated above the Curie point. Sedimentary rocks can also be magnetized during deposition in some sedimentary environments.

Measurement of a rock's magnetic declination and inclination can be made; such measurements are said to indicate the *paleomagnetism* of the rock. Paleomagnetic studies have yielded dramatic results concerning the history of the earth. For example, from declination and inclination values it is possible to determine the position of the magnetic (and therefore geographic) pole at the time of formation of the rock. One of the most fascinating results of paleomagnetic studies is that the magnetic poles have apparently moved great distances through geologic time. For several reasons, movements of the magnetic poles over great distances is not thought to be possible. Therefore, in order to explain the paleomagnetic results, it was proposed that the continents containing the rocks, rather that the magnetic poles, have moved over the

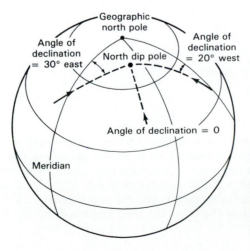

Figure 8-12 Declination is the angle measured at any point on the earth's surface between magnetic north and the geographic north pole. (From S. Judson, M. E. Kauffman, and L. D. Leet, *Physical Geology*, 7th ed., copyright © 1987 by Prentice-Hall, Inc., Englewood Cliffs, N.J.)

earth's surface during geologic time. This was one of the main lines of evidence supporting the theory called *continental drift*.

In the 1950s and 1960s, paleomagnetic studies produced even more spectacular evidence for the tectonic evolution of the earth. This evidence consisted of two types. First, it was discovered that some rocks were magnetized with their magnetic poles exactly reversed from the present pole configuration. In other words, the direction needle of a compass would point to south rather than north for the magnetism in these rocks. As paleomagnetic studies proliferated, it became clear that the earth has suddenly reversed the polarity of its magnetic dipole many times; in fact a magnetic reversal occurs on the average about every half-million years.

The second form of evidence was gathered when *magnetometers* were towed behind research ships crossing the oceans. These instruments were able to measure the magnetism of the rocks beneath the sea floor. When these ships crossed midoceanic ridges, a remarkable magnetic pattern was discovered (Figure 8-13). At the crest of the midoceanic ridge, rocks of normal (that is, in the orientation of the earth's present magnetic field) magnetism were encountered. However, as the ship moved away from the ridge crest, linear bands of rock were crossed that were magnetized in alternately reversed and normal polarities. Even more exciting was the realization that these magnetic bands or stripes were symmetrical on opposite sides of the ridge crests. This was strong evidence that new oceanic crust was formed by volcanism at the ridge crest

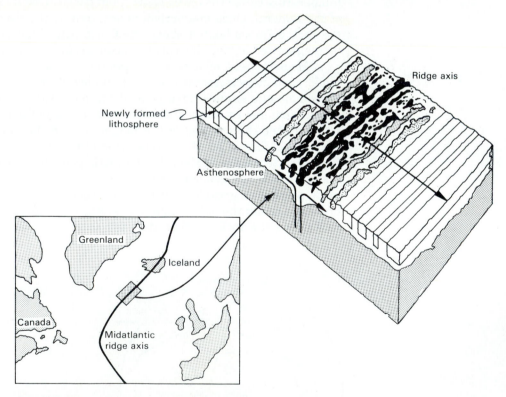

Figure 8-13 Symmetrical magnetic patterns across the mid-Atlantic ridge. Symmetrical bands of similar polarity (same pattern) indicate formation at the ridge and lateral movement away from the ridge in both directions. (From E. A. Hay and A. L. McAlester, *Physical Geology: Principles and Perspectives*, 2d ed., copyright © 1984 by Prentice-Hall, Inc., Englewood Cliffs, N.J.)

in the prevailing polarity of the earth's magnetic field and then split apart, moving in opposite directions away from the ridge. When a magnetic reversal occurred, the rocks forming at the ridge crests would be of that polarity. Thus the magnetic bands were symmetrical about the ridge. This work became nearly irrefutable evidence for the formation of new lithosphere at midoceanic ridges and the movement of lithospheric plates away from these divergent plate margins. This theory became known as *sea-floor spreading*. It is considered to be one of the foundations of plate tectonics.

SUMMARY AND CONCLUSIONS

An introduction to geology, even if its purpose is the practical application of the science, is incomplete without a general understanding of the earth's interior. Geophysicists study this hidden frontier by measuring the heat flow, gravity, and magnetism of the earth. The rewards of this study have been spectacular. In addition to providing support for plate tectonics, the theory that explains the earth's dynamic behavior, geophysical studies have yielded invaluable practical data relating to petroleum and mineral deposits, geothermal power, and earthquakes.

The earth achieved its current internal structure through differentiation; this process separated the iron core, the dense silicate mantle, and the low-density crust. Heat flow from the core generates convection currents in the mantle. Where rising plumes impact the base of the lithosphere, the lithosphere is broken apart and dragged laterally. The point of upwelling frequently coincides with midoceanic ridges, zones of high heat flow where new lithosphere is formed.

Heat flow beneath continents is variable according to the age and tectonic activity of individual geologic provinces. Continental heat flow is the sum of mantle heat and radioactive crustal heat. Tectonically active provinces are centers of high heat flow, whereas stable shield regions yield low heat-flow values.

The study of gravity is mainly used to detect minor variations in the density of near-surface rocks. These variations remain after latitudinal and topographical effects have been corrected from measured gravity values. Interpretation of many gravity readings led to the theory of isostasy. This important concept compares the mantle to a viscous fluid, when considered in terms of geologic time. Crustal masses reach a condition of equilibrium by "floating" in the mantle to a degree determined by their size and density. High mountain ranges are composed of low-density rocks that extend as roots well below the earth's surface. Any rapid change in load causes crustal blocks to rise or sink into the mantle; thus one example of isostatic rebound is the crustal rise of glaciated areas after the last glacial retreat.

The earth's magnetic field can be approximated as a dipole bar magnet oriented near the axis of rotation. It is probable that electrical currents in the iron core induce this magnetic field.

Studies of paleomagnetism have shown that the earth's magnetic field periodically reverses its polarity. This discovery has been most useful in providing evidence for sea-floor spreading. Symmetrical zones of alternating normal- and reversed-polarity rocks adjacent to midoceanic ridges strongly suggest that oceanic lithosphere is produced at midoceanic ridges and then moves laterally away from these spreading centers.

REFERENCES AND SUGGESTIONS
FOR FURTHER READING

HAY, E. A., and A. L. McALESTER. 1984. *Physical Geology: Principles and Perspectives.* Englewood Cliffs, N.J.: Prentice-Hall, Inc.

JUDSON, S., M. E. KAUFFMAN, and L. D. LEET. 1987. *Physical Geology*, 7th ed. Englewood Cliffs, N.J.: Prentice-Hall, Inc.

PRESS, F., and R. SIEVER. 1982. *Earth*, 3d ed. San Francisco: W. H. Freeman and Co.

PROBLEMS

1. What geologic processes are influenced by heat flow from the earth's interior?
2. Why does heat flow vary from one area to another on the earth's surface?
3. What types of geologic evidence would you look for if you were exploring for geothermal energy?
4. What types of corrections are made to surface gravity readings? What are some of the objectives of gravity studies?
5. What effects does isostasy have on the earth's surface topography?
6. What is the origin of the earth's magnetic field?
7. How did studies of the paleomagnetism of the ocean basins support plate tectonics and its preceding theories?

CHAPTER
9
CRUSTAL DEFORMATION AND PLATE TECTONICS

Large-scale convectional movements in the earth's mantle create enormous stresses in the crust. As a result of these stresses, horizontal and vertical movements in the crust have been occurring throughout geologic time. Deformation of crustal rocks takes place during movements of the crust. Brittle deformation, the type that generates earthquakes, is common in the upper crust. At greater depths, plastic or ductile deformation is common. In the geologist's view, the rock structures produced by these deformations are evidence that can be used to unravel the earth's geologic history. From a practical standpoint, studies of geologic structure are used in exploration for petroleum and mineral deposits. Engineers must also understand rock structure because the discontinuities imparted to rocks during deformation frequently govern the engineering behavior of rock masses at the surface or in the shallow subsurface regions of the earth.

MECHANICS OF DEFORMATION

In Chapter 6 the principles of rock deformation were introduced. In that discussion we considered the strain caused by a uniaxial load applied in an unconfined compression test. Although the results of tests such as these are applicable to rock engineering where rock masses are exposed at the surface, natural deformation of rocks occurs from shallow depths downward to very great depths. The conditions at these depths are very different than at the surface. Therefore, it is reasonable to expect different types of deformation.

In a general way, deformation passes from brittle to plastic, or ductile, with increasing depth. The unconfined compression test is no longer representative of the conditions to which rocks are subjected at depth because of the tremendous weight of the overlying column of rocks and soil. This weight translates into a pressure applied from all directions to any given element of rock at depth. This all-around pressure is known as *confining pressure*. It is similar to hydrostatic pressure in that it increases with depth; however, this *lithostatic pressure* is not always equal in all directions. Major principal stress may be applied in vertical or lateral directions.

A laboratory test called the *triaxial test* (Figure 9-1) has been devised to more closely simulate the behavior of rocks or soil at depth. In this test the cylindrical sample is enclosed in a jacket through which a radial confining pressure can be applied. Gases or liquids can be used to apply this all-around confining pressure. An axial load can then be applied, similar to an unconfined compression test, until failure occurs. With this apparatus, tests can be run repeatedly to study the effect of confining pressure upon deformation.

Representative results of triaxial tests are shown in Figure 9-2. Several effects are apparent. First, as the confining pressure is increased, the rock passes through a transition from brittle to ductile behavior. Ductile response becomes dominant at confining pressures above 700 kg/cm^2. This result supports our expectation that rocks whose behavior may be brittle at the earth's surface become ductile under high confining pressures. Second, the tests run at higher confining pressures indicate that the strength of the rock increases with increasing confining pressure.

Confining pressure is not the only change in rock deformation with depth. Because the temperature increases steadily with depth, it is likely that rocks will behave differently at higher temperatures. Triaxial test cells can be constructed to investigate the effects of temperature on rock behavior under stress. Some results are shown in Figure 9-3. A particular rock type was tested at several temperatures under a constant confining pressure. Strength decreased at higher temperatures. In addition, the samples became ductile (reached their

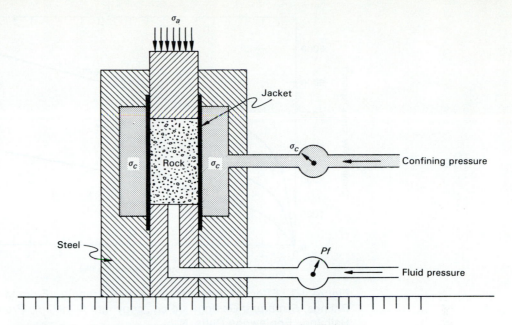

Figure 9-1 Schematic diagram of a triaxial cell. Confining pressure, σ_c, can be applied to the sample outside the jacket. Pore pressure within the rock P_f can also be regulated. σ_a is the axial load that is increased to failure. (From J. Suppe, *Principles of Structural Geology*, copyright © 1985 by Prentice-Hall, Inc., Englewood Cliffs, N.J.)

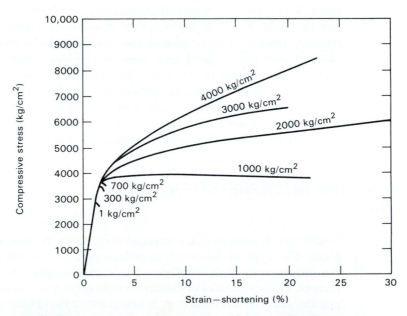

Figure 9-2 Effect of confining pressure on a rock in a triaxial test. Each curve represents a test conducted at a different confining pressure. (From M. P. Billings, *Structural Geology*, 3d ed., copyright © 1972 by Prentice-Hall, Inc., Englewood Cliffs, N.J.)

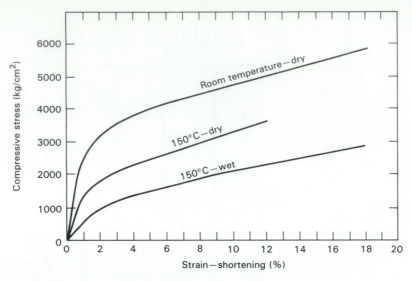

Figure 9-3 Effect of temperature on rock strength. (From M. P. Billings, *Structural Geology*, 3d ed., copyright © 1972 by Prentice-Hall, Inc., Englewood Cliffs, N.J.)

elastic limit) at a lower stress. Thus plastic deformation will become more prevalent with increasing depth because of greater temperatures and pressures.

Another factor that affects rock deformation is time. Pressures are applied to rocks over millions of years in the earth. Although we obviously cannot duplicate such conditions in a lab test, it is possible to vary the rate of strain at which a rock sample approaches failure. In this way the effect of time can be investigated, although extrapolation to geologic time intervals is still difficult. When the strain rate of a triaxial test is decreased, several observations can be made. First, rock strength decreases with decreasing strain rate (increasing time to failure); second, the rocks become more ductile at lower strain rates. These results hold true even at stresses below the elastic limit. The implications, therefore, are that rocks can deform at low levels of stress if the stresses are applied over long periods of time. In fact, the relationship between stress and strain rate can even be compared to that of a viscous material. Rocks, therefore, that are strong and brittle at the earth's surface may deform like a fluid deep within the earth over long periods of time.

MEASUREMENT AND INTERPRETATION OF STRUCTURE

Under the stresses that are imposed upon rocks in the earth, deformation takes place. The type of deformation is dependent upon the material properties of the rock as well as the confining pressure, temperature, strain rate, and other factors. The mechanics of deformation involve various combinations of elastic, plastic, and viscous behavior. When deformation is complete, the rock may be totally changed from its initial state. Among the features of deformed rock are geologic *structures* such as folds, faults, and joints, which can be studied to provide information about the deformational mechanisms. Engineers are more concerned, however, with the properties that structures impart to a rock mass, especially those that impact an engineering project. For example, faults may bring two rock types with vastly different engineering properties into contact.

Joints may lower the strength of a rock mass as well as increase the permeability. Thus a mass of largely intact granite may be a favorable site for a nuclear-waste repository, whereas a jointed granitic rock mass may be totally unsuitable.

Investigation of rocks at any engineering site requires analysis, measurement, and interpretation of rock structures. This process begins with a spatial analysis of the deformed rocks.

Strike and Dip

Most geologic structures are studied by measuring the orientation of planar elements within the rock. The orientation of a plane in space can be specified by measuring its *strike and dip* (Figure 9-4). Strike is the direction of a line

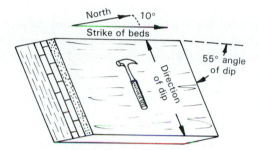

Figure 9-4 Strike is the direction of a horizontal line intersecting a bedding plane or other plane in a body of rock. Dip is the angle of inclination of the plane measured from the horizontal. (After G. D. Robinson et al., U.S. Geological Survey Professional Paper 505.)

Figure 9-5 Measurement of dip with a Brunton compass at the base of a sandstone bed. (J. R. Stacy; photo courtesy of U.S. Geological Survey.)

formed by the intersection of the plane and the horizontal. Dip is the amount of slope of the plane. It is determined by measuring the acute angle between the horizontal and the sloping plane. Dip is always measured in a vertical plane perpendicular to strike in the direction of maximum inclination of the rock plane.

In the field, strike and dip measurements often are made with a Brunton compass (Figure 9-5). The types of planes that are measured in the field vary with the type of rocks and deformational mechanisms. In folded sedimentary rocks, the orientations of bedding planes are measured throughout the area in order to reconstruct the pattern of folding in the region. Fault and joint planes are measured directly so that their distribution throughout a rock mass can be predicted.

Geologic Maps

In a systematic geologic study of an area, many values of strike and dip are measured and then plotted on a base map of the area. Other data gathered in a field study include the distribution of rock types and, in particular, the *con-*

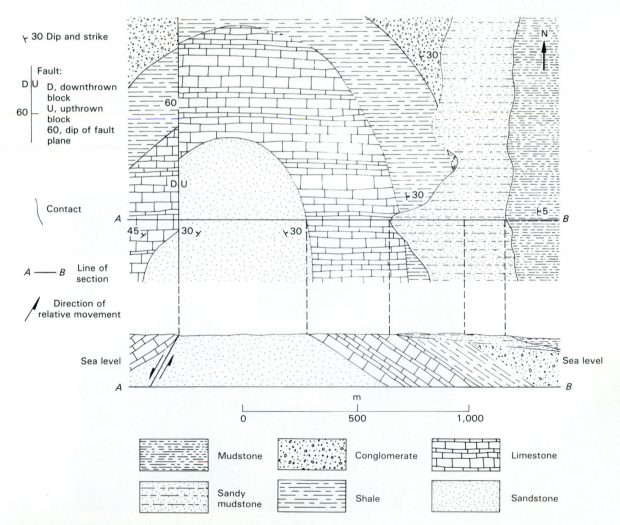

Figure 9-6 Geologic map and cross section. (From S. Judson, M. E. Kauffman, and L. D. Leet, *Physical Geology*, 7th ed., copyright © 1987 by Prentice-Hall, Inc., Englewood Cliffs, N.J.)

tacts, or boundaries between rock units. Aerial photographs are particularly useful for identifying contacts and structures. For this reason, air-photo interpretation is almost always a preliminary phase in geologic studies and engineering site investigations. When field and air-photo data are plotted on the base map along with the structural features and extended throughout the map area, the resulting map is called a *geologic map*. Rock units are shown on a geologic map by colors or patterns (Figure 9-6). Structural measurements are plotted using strike and dip symbols. Various symbols can also be used to show other types of structural data.

Geologic cross sections are commonly constructed from geologic maps (Figure 9-6). Cross sections, which are drawn to show interpreted subsurface geologic conditions, may be based only on lithologic and structural data from the surface. Data from wells or test holes also are used to provide direct subsurface information for cross sections.

Published geologic maps should be examined during the first phase of an engineering site investigation. If existing maps are adequate, the investigation may proceed to surface or subsurface sample collection. If adequate geologic maps of the area are not available, then mapping is often conducted specifically for the engineering project.

FOLDS

Folds are produced by lateral compression of the crust. Under this type of stress, the crustal rocks are deformed into a series of wavelike forms oriented transverse to the direction of maximum stress. Folds are an indication that crustal shortening has occurred.

Folds are most easily visible in sedimentary rock sequences because of the bedding of the rock strata. Bedding planes were initially horizontal prior to deformation (see Principle of Original Horizontality, Chapter 1). These discontinuities facilitate the recognition and interpretation of folds (Figure 9-7).

Two general fold types are recognized. *Anticlines* are formed by the upward bending or buckling of strata, so the fold has the shape of an arch (Figure 9-8). The sides, or *limbs*, of an anticline dip downward and outward from the

Figure 9-7 Folded sedimentary rocks in Hudspeth County, Texas. (C. C. Albritton, Jr.; photo courtesy of U.S. Geological Survey.)

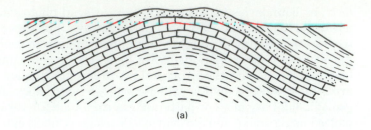

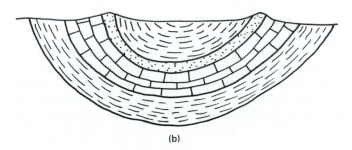

Figure 9-8   Cross sections of (a) an anticline and (b) a syncline.

fold crest. The opposite of an anticline is a *syncline*, in which the central portion is bent downward to form a trough shape (Figure 9-7). The limbs of a syncline dip toward the center of the trough.

Parts of a Fold

The terminology used to describe folds is shown in Figure 9-9. The line connecting the points of maximum curvature is called the *axis*. The *axial plane* is a plane that is passed through the fold and contains the axis.

   The axis of a fold or group of folds may be approximately horizontal or tilted, in which case the structure is classified as a plunging fold (Figure 9-10). The outcrop pattern of plunging folds differs from nonplunging folds as illustrated in Figure 9-11. Erosion of the landscape accentuates these differences. Beds in eroded, plunging folds can be traced through sinuous patterns across the surface (Figure 9-12). The sharp bends in beds, where the strike changes direction, identify the points of intersection between eroded anticlines and the ground surface. Therefore, in a series of plunging folds the direction of plunge is toward the bend, or nose, of an anticline (Figure 9-12). For a

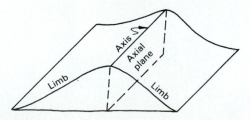

Figure 9-9   Parts of a fold. The axial plane divides the fold into two equal portions, and the axis is a line connecting the points of maximum curvature.

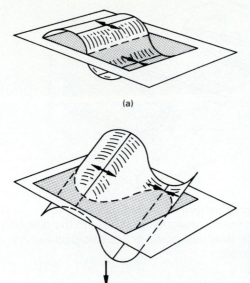

(a)

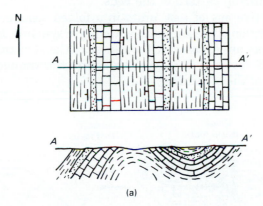

(b)

Figure 9-10 Comparison of (a) nonplunging and (b) plunging folds. (From E. A. Hay and A. L. McAlester, *Physical Geology: Principles and Perspectives*, 2d ed., copyright © 1984 by Prentice-Hall, Inc., Englewood Cliffs, N.J.)

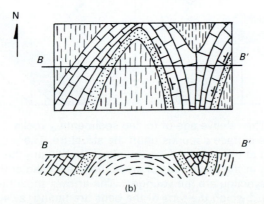

Figure 9-11 (a) Map and cross section of nonplunging folds. (b) Map and cross section of folds plunging north.

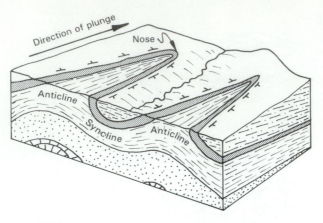

Figure 9-12 Outcrop pattern of eroded, plunging folds. Beds in a plunging anticline form a sharp bend or nose in the direction of plunge. (After G. D. Robinson et al., U.S. Geological Survey Professional Paper 505.)

syncline the direction of plunge is toward the opposite direction from the sharp bend in the outcrop pattern of the beds.

The relative age of rock units in a folded sequence can be determined after identification of the fold as an anticline or syncline. In an eroded anticline, the oldest rocks are exposed at the center of the structure (Figure 9-13). The relationship is reversed in a syncline, where the youngest rocks are exposed at the center of the structure.

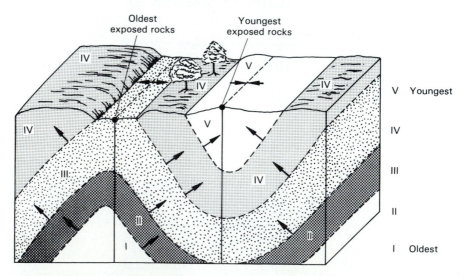

Figure 9-13 The relative age of folded sedimentary rocks exposed at the surface depends upon the structure. The rocks exposed at the center of the anticline, Unit III, are the oldest exposed rocks in the sequence, whereas those exposed at the center of the syncline are the youngest. The arrows show the direction toward which the tops of the beds are facing: away from the axial plane of the anticline and toward the axial plane of the syncline. (From E. A. Hay and A. L. McAlester, *Physical Geology: Principles and Perspectives*, 2d ed., copyright © 1984 by Prentice-Hall, Inc., Englewood Cliffs, N.J.)

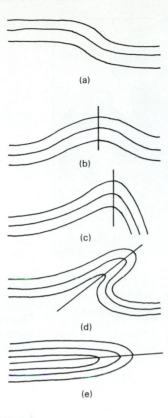

Figure 9-14 Cross sections of fold types as defined by geometry and position of axial plane: (a) monocline; (b) symmetrical anticline; (c) asymmetrical anticline; (d) overturned anticline; and (e) recumbent anticline.

Types of Folds

Folds can vary in size from structures visible on satellite images to examples that must be observed through a microscope. In addition to this size variation, there are also many types of folds. Several of the basic forms are shown in Figure 9-14. Fold types are differentiated by the geometry of the limbs and the orientation of the axial planes. In *symmetrical* folds, the limbs dip in opposite directions in about the same amount. Dips of opposing limbs are not equivalent in *asymmetrical* folds. If both limbs dip in the same direction, the folds are said to be *overturned*. This condition can be caused by an intense stress applied in one direction. In regions of more severe deformation, axial planes may be nearly horizontal. These types of folds are termed *recumbent*.

Not all folds consist of two limbs. A steplike bend in strata without two well-defined limbs is known as a *monocline* (Figure 9-14). Finally, large-scale warps in relatively stable continental crustal regions resemble folds. These structures are commonly circular or oval in plan view. Uplifts of this type are called *domes*, and downwarps are known as *basins* (Figure 9-15). Basins are particularly important for their petroleum resource potential. During the slow subsidence of the crust to form basins, thick sections of sedimentary rocks are deposited. Under the proper conditions, petroleum is formed and trapped in these sequences.

FRACTURES

The brittle failure of rock produces fractures. These curved or planar discontinuities can be subdivided into *joints* and *faults*. Both types are extremely important features to identify and evaluate at the site of an engineering project.

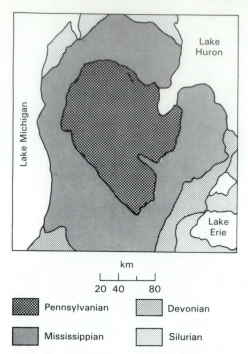

Figure 9-15 Map of the Michigan Basin showing progressively younger rock units toward the center of the structure.

Joints

Joints (Figure 9-16) are fractures in which there has been no movement parallel to the failure plane. Although joints are produced by unloading (Chapter 10) and cooling of lava (Chapter 3), our main concern here is for joints caused by tectonic stresses. These fractures commonly occur in sets consisting of numerous planes with the same strike and dip distributed throughout a rock

Figure 9-16 Closely spaced joints in an igneous rock body. (M. H. Staatz; photo courtesy of U.S. Geological Survey.)

mass at regular intervals. One or more joint sets, each having a distinctive orientation, are often present.

Joints can be produced by tensional, shearing, and compressional stresses. If structures other than joints are found in an area, the joints may be related to the stresses that are responsible for those structures. For example, Figure 9-17 shows two possible orientations of intersecting joint sets associated with folds. Joint sets that intersect at high angles are known as *conjugate joint sets*.

The engineering properties of rock masses are established by the type, orientation, and spacing of joints and other discontinuities (Chapter 6). Strength is controlled by these structures because failure usually occurs along planes of weakness rather than within intact rock. The orientation of joints may impart a degree of anisotropy to the rock mass; in other words, strength may vary with the direction from which the stress is applied because of the spatial distribution of joints.

Rock-slope stability is also controlled by the spacing and orientation of joints. These properties control the size and direction of failure of blocks that affect the engineering of highway cuts, open-pit mine slopes, and other excavations.

Another rock-mass property that is influenced by jointing is permeability. In many rock types, most of the movement of groundwater, petroleum, and other fluids is through fractures rather than through pores in the intact rock. In rocks of low permeability, exploration for groundwater involves identifying major joint traces on the ground surface. The point of intersection of two or more joints is a prime location for high well yields in such areas. Planning for waste-disposal facilities must also consider fracture patterns. Contaminated groundwater can migrate rapidly along joints and faults; these fractures must therefore be identified and monitored.

Faults

Faults are fractures along which movement of rock masses has occurred parallel to the fault plane. Faults are classified according to the type of movement, or slip, that has taken place. The two categories established on this basis include dip-slip faults and strike-slip faults.

Dip-slip faults, in which the relative movement of rock masses is in the direction of dip of the fault plane, include *normal* faults, *reverse* faults, and

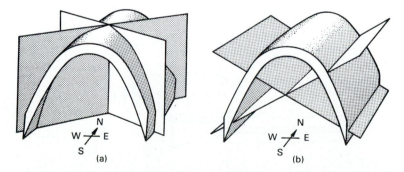

Figure 9-17 Conjugate joint sets caused by stress that produced folding: (a) fold with vertical diagonal joints; (b) fold with strike joints dipping about 30°. (From M. P. Billings, *Structural Geology*, 3d ed., copyright © 1972 by Prentice-Hall, Inc., Englewood Cliffs, N.J.)

thrust faults (Figure 9-18). Normal and reverse faults can be differentiated if the relative displacement of a bed or other lithologic feature can be determined. The blocks on opposite sides of the fault plane are defined as the *hanging wall* and the *foot wall*. The hanging wall is always the block above the fault plane, and the foot wall is the block below the plane. Therefore, by definition, if the hanging wall has moved downward with respect to the foot wall, the fault is a normal fault (Figure 9-19), and relative movement in the opposite sense is characteristic of reverse faults. Thrust faults are very similar to reverse faults except that the angle of the fault plane with respect to horizontal is quite low. Dip-slip faults sometimes occur in pairs (Figure 9-20). A down-dropped block bounded by two normal faults is called a *graben*. The opposite case, in which the center block is up-thrown, forms a *horst*.

   Strike-slip faults are characterized by the relative, lateral movement of fault blocks in the direction of strike of the fault plane (Figure 9-21). These faults are either *right-lateral* or *left-lateral* types depending on the relative movement of fault blocks. The two varieties can be distinguished by facing the fault from either side and noting the direction, either left or right, toward the continuation across the fault of a linear feature such as a stream, fence,

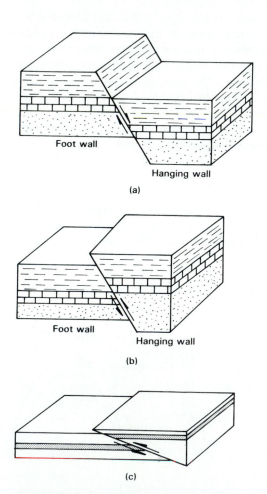

(a)

(b)

(c)

Figure 9-18 Types of dip-slip faults: (a) normal fault; hanging wall is displaced downward with respect to foot wall; (b) reverse fault; hanging wall is displaced upward with respect to foot wall; (c) thrust fault; similar to reverse fault except that fault plane is inclined at lower angle.

Figure 9-19 A normal fault cutting cross-bedded sandstone. (T. L. Finnell; photo courtesy of U.S. Geological Survey.)

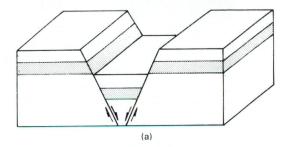

(a)

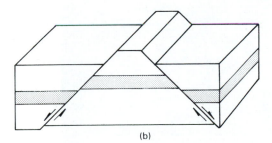

(b)

Figure 9-20 Pairs of normal faults forming (a) a graben and (b) a horst.

or road that formerly passed straight across the fault prior to its movement. The San Andreas fault is an excellent example of a right-lateral, strike-slip fault (Figure 9-22).

Dip-slip and strike-slip faults are merely end members of fault types that may display any combination of dip-slip and strike-slip displacement. Faults that contain components of both types of movements are often called *oblique-slip* faults. The actual type of movement that takes place along any individual fault is the result of the applied stresses. As shown in Figure 9-23, lateral stresses of compressional and tensional nature are commonly associated with

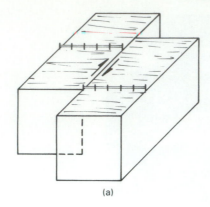

(a)

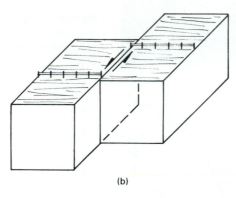

(b)

Figure 9-21 (a) Right-lateral and (b) left-lateral strike-slip faults. Fence on surface can be used to determine amount of displacement.

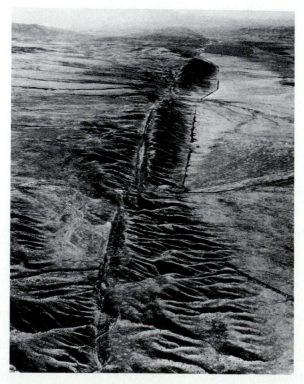

Figure 9-22 The San Andreas fault in the Carrizo Plains of California. (R. E. Wallace; photo courtesy of U.S. Geological Survey.)

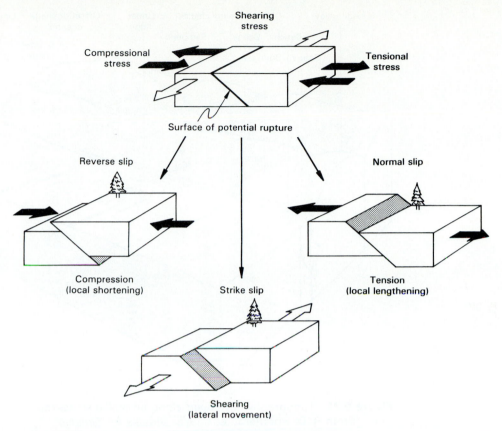

Figure 9-23 Relationship of stress orientation to the style of faulting. (From E. A. Hay and A. L. McAlester, *Physical Geology: Principles and Perspectives*, 2d ed., copyright © 1984 by Prentice-Hall, Inc., Englewood Cliffs, N.J.)

reverse and normal faults, respectively. Strike-slip faults are the result of shearing stresses acting as a couple on a crustal block.

When fault movement takes place, the type and amount of displacement is readily observable. Fault scarps (Figure 7-17) can be used to determine the amount of displacement of a dip-slip fault. These surface ruptures displace any surface features that are present. An active strike-slip fault can often be recognized by a characteristic set of topographic features (Figure 9-24). These include bends in stream courses or offsets of other linear topographic features as well as depressions or *sag ponds* along the trend of the fault.

Fault scarps and other types of topography associated with active faults generally are short lived because of erosion. There is often no surface indication of inactive faults. These must be located by careful geologic mapping. Clues may be provided by linear stream courses or other topographic features. Stream courses are readily established along fault zones because of the low resistance to erosion caused by the crushing and grinding of rock along the fault plane.

The identification of active or potentially active faults is critical to several types of projects. Nuclear power plants, schools, hospitals, and other buildings should be sited so as to avoid active faults. Inactive faults are also relevant to engineering design because they may bring rocks of different engineering properties into contact or control the permeability and groundwater movement through a rock mass.

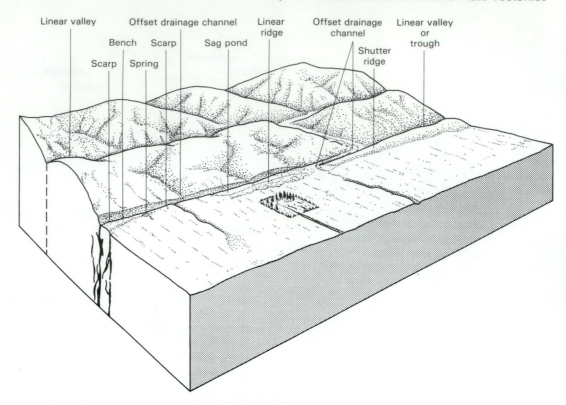

Figure 9-24 Topography developed along an active strike-slip fault. (From R. D. Borcherdt, ed., 1975, *Studies for Seismic Zonation in the San Francisco Bay Region*, U.S. Geological Survey Professional Paper 941-A.)

PLATE TECTONICS

In several sections of this book the theory of plate tectonics has been mentioned. This theory is important because it provides a unifying model for changes that have affected the earth's crust through geologic time. In addition, it provides an explanation for the distribution of volcanoes, earthquakes, and many other geologic phenomena. In this section, we will briefly consider some of the main points of evidence for this theory.

Continental Drift

In 1912 Alfred Wegener, a German meteorologist, proposed that the continents had moved great distances laterally across the earth's surface through geologic time. The shape of the coastlines of Africa and South America, which would fit together like the pieces in a jigsaw puzzle if the continents were closer, was one of the main lines of evidence for this idea. This theory, which became known as continental drift, proposed that all the continents were once assembled into a huge continental mass called Pangaea (Figure 9-25) and then drifted apart to their present positions.

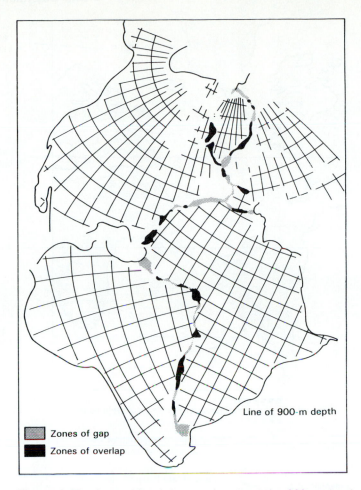

Zones of gap
Zones of overlap

Line of 900-m depth

Figure 9-25 Assembly of the continents at the 900-m depth of the ocean bottom to form Pangaea. (From E. Bullard et al., 1965, *Phil. Trans. Royal Soc.* 1088:41–51.)

In addition to noting the fit of continents, Wegener summarized geologic evidence suggesting that the continents were once joined together (Figure 9-26). This evidence included several types. First, mountain belts, structural trends, and rock types found on different continents would be continuous if the continents were assembled as in Pangaea. Second, the distribution of numerous fossil species found on several continents could be explained if the continents were originally joined. These species included land plants and animals that could not have crossed an ocean to colonize widely separated continents. Third, rock types that form under specific climatic conditions were encountered in regions whose present climate is very different. For example, rocks containing coal and tropical plants were discovered in polar regions, and rocks formed by glaciation were mapped near the present equator. The locations of these rocks would be difficult to explain unless the land mass had moved into different climatic zones.

Wegener's hypothesis was largely discounted by scientists of the day because of the implausible mechanism to which he attributed the drift of continents. Wegener suggested that the granitic rock of the continents had moved laterally through the stronger, denser mantle rock. Other scientists quickly pointed out the mechanical impossibility of this process, and Wegener's ideas were ignored by most geologists for several decades.

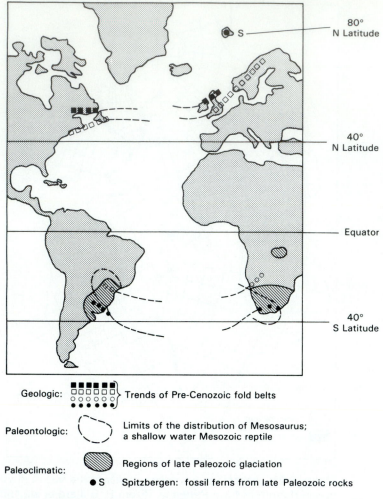

Geologic:	▪▪▪▪▪ □□□□□□ ●●●●●●	Trends of Pre-Cenozoic fold belts
Paleontologic:	(⌒)	Limits of the distribution of Mesosaurus; a shallow water Mesozoic reptile
Paleoclimatic:	▨	Regions of late Paleozoic glaciation
	● S	Spitzbergen: fossil ferns from late Paleozoic rocks

Figure 9-26 Evidence for continental drift as proposed by Alfred Wegener. (From E. A. Hay and A. L. McAlester, *Physical Geology: Principles and Perspectives*, 2d ed., copyright © 1984 by Prentice-Hall, Inc., Englewood Cliffs, N.J.)

The Theory of Sea-Floor Spreading

In the first several decades after World War II, tremendous advances in exploration of the oceans were realized. Topographic surveys of the ocean floor using sounding devices discovered the immense midoceanic ridges positioned near the center of the ocean basins (Figure 9-27). Rock samples from the ridge crests were composed of fresh basaltic lava covered by very little sediment. The thickness of bottom sediment above bedrock increases laterally away from the ridge crests, implying an increase in age of the oceanic crust with distance from the midoceanic ridges.

The most dramatic evidence pertaining to the origin of the ocean basins was obtained from paleomagnetic studies of the oceanic crust (Chapter 8). The symmetrical bands of alternately magnetized volcanic crustal rock provided important evidence for the *sea-floor spreading* theory, which proposed that new crustal material was formed by volcanic eruptions at crests of the midoceanic ridges and that slow lateral movement of the crust away from the ridges was occurring. Other geophysical evidence suggested that the layer in

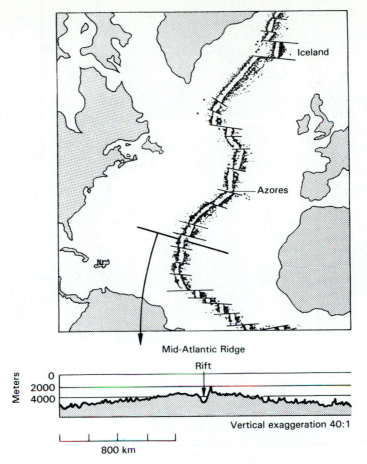

Iceland

Azores

Mid-Atlantic Ridge

Rift

Meters

0
2000
4000

Vertical exaggeration 40:1

800 km

Figure 9-27 Discovery of midoceanic ridges was an important step in the development of the sea-floor spreading hypothesis. Dots show positions of earthquakes along the crest of the Mid-Altantic Ridge. (From B. C. Heezen, The Rift in the Ocean Floor, *Scientific American*, Oct. 1960.)

motion consisted of the crust and upper mantle, a layer named the lithosphere. This rigid slab could move laterally above a hot plastic layer called the asthenosphere. Thus Wegener's basic premise was vindicated: Continents do move laterally but as part of a thicker, rigid slab that slides along above the weak asthenosphere.

Plates and Plate Margins

The sea-floor spreading theory quickly became integrated into a more comprehensive model called plate tectonics, which considers the rigid lithosphere to be divided into 12 major slabs, or plates (Figure 9-28). Most of the dynamic crustal processes such as volcanic eruptions and earthquakes take place at the boundaries of the plates. Three major types of plate boundaries are recognized (Figure 3-7). At divergent boundaries, new lithosphere is produced by volcanic eruptions. Midoceanic ridges fall into this category. Divergent boundaries also develop beneath continents. The Red Sea, separating Africa and the Arabian Penninsula, is considered to be a spreading center beginning to split apart a continental mass. Volcanic activity and shallow- to moderate-focus earthquakes are characteristic of divergent plate boundaries.

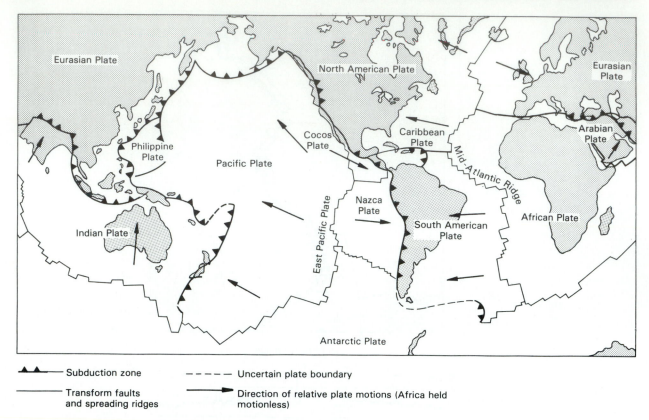

Subduction zone — — — — Uncertain plate boundary

Transform faults ——▶ Direction of relative plate motions (Africa held
and spreading ridges motionless)

Figure 9-28   The major lithospheric plates, with directions of
movement and types of boundaries shown. (From E. A. Hay and
A. L. McAlester, *Physical Geology: Principles and Perspectives*, 2d
ed., copyright © 1984 by Prentice-Hall, Inc., Englewood Cliffs,
N.J.)

Convergent plate boundaries mark the zone of contact or collison where
plates move toward each other (Figure 3-7). Because of the enormous crustal
compression localized in these zones, one of the plates is thrust downward
below the other one in a *subduction zone*. Earthquake foci, including those of
deep earthquakes, define the subducted lithospheric slab. Features of convergent boundaries include oceanic trenches that lie above subduction zones, and
mountain belts that are formed at the edges of continents. These mountain
belts are composed of andesitic composite volcanoes (Chapter 3) and/or deformed crustal material that has been folded, faulted, and uplifted by the lateral compression generated by the colliding plates.

The third major type of plate boundary is the shear, or *transform fault*,
boundary. Lithospheric plates are sliding past each other along these boundaries. Major strike-slip faults, such as the San Andreas, mark the location of
shear boundaries. Earthquakes are very common as the plates grind slowly
past each other, but volcanic activity is not as common as along divergent and
convergent boundaries. Numerous transform faults offset segments of midoceanic ridges (Figure 9-27).

The theory of plate tectonics is one of the most important advances in
the earth sciences in the twentieth century. It has provided a model that explains the distribution of volcanoes, earthquakes, mountain belts, and other
geologic phenomena. Despite the usefulness of this model, many problems remain unsolved. We can expect many refinements of the hypothesis in the forthcoming years.

CASE STUDY 9-1
EFFECTS OF ROCK STRUCTURE ON DAM CONSTRUCTION

Careful geologic mapping was necessary prior to construction of the Arbuckle Dam near Ardmore, Oklahoma, which was completed in 1966 by the U.S. Bureau of Reclamation. The complex geologic setting of this dam consists of a series of tightly folded and faulted Paleozoic sedimentary rocks and beautifully illustrates the role of geology in a major engineering project (Jackson, 1969).

The Arbuckle Dam is located in a region of rolling hills known as the Arbuckle Mountains. Rocks of the Arbuckle Mountains consist primarily of a sequence of shales and limestones deposited in a Paleozoic sea (Table 9-1). In late Paleozoic time, the rocks were deformed by compression from the southwest into a series of broad anticlines and synclines. Some of the folds were overturned and some rocks in the anticlines were thrust over younger rocks in the adjacent synclines. A later sequence of events included uplift, high-angle normal faulting, and erosion. Sediments eroded from the rising mountain mass were deposited around the flank of the uplift. During Mesozoic time, marine rocks were deposited over the Paleozoic sequence and then removed by erosion. Continued erosion during the Cenozoic era has produced the present landscape. Major structural elements of the region are shown in Figure 9-29.

Geologic conditions at the dam site were defined by careful geologic mapping and test drilling (Figure 9-30). The stream valley is established in a sequence of steeply dipping, faulted shales, and limestones. The center of the valley is a graben formed by several normal faults, including the highly deformed Reagan fault zone at the base of the left abutment. Rocks are

TABLE 9-1

Generalized Stratigraphic Section of Arbuckle Dam and Reservoir

	Formation	Thickness (feet)
Quaternary	ALLUVIUM	0–30
Pennsylvanian	VANOSS FM.—Poorly to moderately cemented conglomerate, shale, and sandstone of outwash debris on north flank of mountains.	650–900
	DEESE FM.—Varicolored shale; well-cemented conglomerate; thin beds of hard limestone and sandstone	up to 2000
	SPRINGER FM., GODDARD MBR.—Black shale, noncemented but very compact, bituminous, appears graphitic.	1000 to 3000
Mississippian	CANEY FM.—Dark-colored shale, moderately hard limy, fissile.	150
	SYCAMORE FM.—Blue to weathered limestone, hard, tough.	0–50
	WOODFORD FM.—Dark-colored shale, siliceous and cherty, platy and brittle.	400–500
Devonian-Silurian	HUNTON GR.—Thick to medium bedded limestone, marlstone	250
Cambrian-Ordovician	SYLVAN FM.—Gray-green fissile shale.	200
	VIOLA FM.—Bituminous limestone.	450
	ARBUCKLE GR.—Limestone, dolomite, shale, sandstone.	5000 to 8000

Source: From J. L. Jackson, 1969, Geologic studies at Arbuckle Dam, Murray County, Oklahoma, *Bulletin of the Association of Engineering Geologists*, 5:79–100.

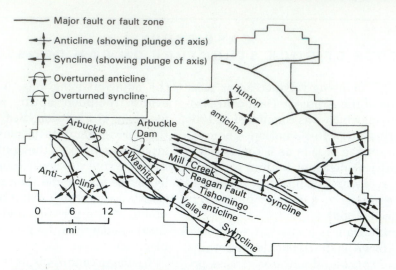

Figure 9-29 Geologic setting of the Arbuckle Mountains, showing major faults and folds. (From J. L. Jackson, *Bulletin of the Association of Engineering Geologists* 5:79–100, copyright © 1969 by the Association of Engineering Geologists.)

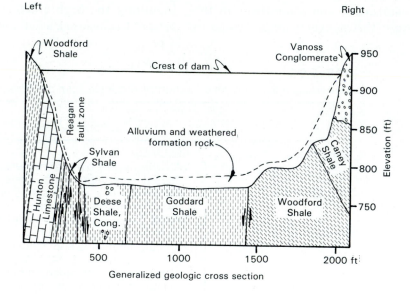

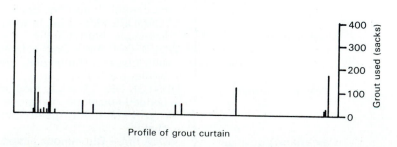

Figure 9-30 Cross section of the Arbuckle Dam site and profile of the grout curtain emplaced beneath the dam. (From J. L. Jackson, *Bulletin of the Association of Engineering Geologists* 5:79–100, copyright © 1969 by the Association of Engineering Geologists.)

fractured, crushed, and altered along the fault and shear zones.

One of the major concerns in dam construction is to prevent excessive leakage from the reservoir through the foundation rocks beneath the base of the dam. Leakage can result in loss of reservoir storage as well as the possibility of failure and collapse of the dam. A common procedure to prevent such leakage involves sealing dam foundations with a grout curtain. The grout curtain is produced by injecting grout (cement) under high pressure into holes drilled in a line across the valley floor. Liquid grout is forced into pores, cavities, and fractures within the rock and then hardens to form a low permeability seal for the dam. At the Arbuckle Dam,

135 grout holes were used, ranging in spacing from 25 to 110 ft. The profile of the grout curtain shown in Figure 9-30 illustrates the amount of grout needed to seal the foundation. The most permeable area was the Reagan fault zone at the base of the left abutment. Relatively small amounts of grout were needed across the valley, with the exception of a fault and unconformity in the vicinity of the right abutment.

Construction of the Arbuckle Dam required a thorough understanding of the geology of the site. At this project the distribution of folds, faults, and other structures was critical to the design of the dam.

SUMMARY AND CONCLUSIONS

The study of the mechanisms and effects of rock deformation is one of the principal subdivisions of geology. In order to explain rock structures, we must take into account the high confining pressures and temperatures at the depths where deformation occurs, as well as the long periods of time available for the application of stress.

The interpretation of aerial photographs and construction of geologic maps is the first step in characterizing the structure and geologic history of an area. These maps are made by plotting the distribution of rock types and the spatial orientation of structural elements on base maps. The strikes and dips of planes associated with bedding, joints, and faults are measured in order to reconstruct the structures of the area prior to erosion. Cross sections utilizing these data can be derived from geologic maps. Geologic maps and cross sections are essential tools for engineering-site investigations.

Folding encompasses a range of mechanical processes in which horizontal sequences of rocks are deformed into undulating, wavelike forms. Anticlines and synclines are the main types of folds, although folds can be described as symmetric, asymmetric, overturned, or recumbent, depending upon the orientations of the axial plane and the limbs. Domes and basins are large-scale analogues of anticlines and synclines. Plunging folds develop complex outcrop patterns after erosion.

Fractures can be classified as either joints or faults. Both types control the strength and permeability of rock masses. Fault terminology is based on the type of slip displayed and the relative movement of opposing blocks. The types of faults that occur are governed by the type and orientation of stresses in the deforming rock mass. Thrust and reverse faults develop under compression, whereas tensional stresses often are responsible for normal faults. Strike-slip faults are found where shear stresses dominate between crustal blocks.

During the past several decades geologists have formulated a unifying model to explain many of the geologic processes observed in the crust. Predecessors of this model were the theories of continental drift and sea-floor spreading. Continental drift was discounted because of the lack of a mechanism to explain the lateral movements of the continents. Sea-floor spreading provided an explanation for the lateral movement of the rigid lithosphere above the plastic asthenosphere, which rafts the granitic continental masses along in the process.

The currently accepted plate tectonics model includes about a dozen major lithospheric plates that move within the earth's uppermost layer. Tectonic and volcanic processes are concentrated at plate boundaries, which may be divergent, convergent, or shear types.

REFERENCES AND SUGGESTIONS FOR FURTHER READING

BILLINGS, M. P. 1972. *Structural Geology*, 3d ed. Englewood Cliffs, N.J.: Prentice-Hall, Inc.

BULLARD, E., J. E. EVERETT, and A. G. SMITH. 1965. The fit of the continents around the Atlantic. *Philosophical Transactions of the Royal Society of London* 1088:41–51.

HAY, E. A., and A. L. McALESTER. 1984. *Physical Geology: Principles and Perspectives*, 2d ed. Englewood Cliffs, N.J.: Prentice-Hall, Inc.

JACKSON, J. L. 1969. Geologic studies at Arbuckle Dam, Murray County, Oklahoma. *Bulletin of the Association of Engineering Geologists* 5:79–100.

JUDSON J., M. E. KAUFFMAN, and L. D. LEET. 1987. *Physical Geology*, 7th ed. Englewood Cliffs, N.J.: Prentice-Hall, Inc.

PRESS, F., and R. SIEVER. 1982. *Earth*, 3d ed. San Francisco: W. H. Freeman and Co.

ROBINSON, G. D., A. A. WANEK, S. H. HAYS, and M. E. McCALLUM. 1964. *Philmont Country: The Rocks and Landscape of a Famous New Mexico Ranch*. U.S. Geological Survey Professional Paper 505.

SUPPE, JOHN. 1985. *Principles of Structural Geology*. Englewood Cliffs, N.J.: Prentice-Hall, Inc.

WESSON, R. L., E. J. HELLEY, K. R. LAJOIE, and C. M. WENTWORTH. 1975. "Faults and future earthquakes." In *Studies for Seismic Zonation in the San Francisco Bay Region*, edited by R. D. Borcherdt. U.S. Geological Survey Professional Paper 941-A, pp. A5–A30.

PROBLEMS

1. How can the deformation of rocks deep within the earth's crust be simulated in laboratory tests of rock samples?
2. What evidence leads us to believe that rocks exposed at the surface were deformed at great depths?
3. What field measurements are made in the study of rock structures? What do they indicate about the structures?
4. In what type of geologic setting or province would you be likely to find highly folded rocks?
5. What is the difference between faults and joints?
6. Describe the relationship between fault type and stress orientation.
7. Summarize the ideas that led up to the theory of plate tectonics.
8. What are the characteristics of each type of plate margin?
9. List as many ways as you can that folds and faults would influence the construction of tunnels and dams.
10. Describe the major structural features of your state.

CHAPTER
10
WEATHERING AND EROSION

We have followed the various igneous, metamorphic, and sedimentary processes that lead to the formation of rocks below the surface of the earth's crust. We have then seen how tectonic forces driven by the earth's internal energy elevate these deeply formed rocks to the earth's surface. Progression through the geologic cycle now brings us to the processes acting upon, or just below, the earth's surface associated with the action of water, ice, wind, and gravity. These processes, fueled by external energy derived from the sun, are components of the hydrologic cycle. Together, they lead to the decay, disintegration, and transportation of continental rocks to the ocean basins, where sedimentation precedes the formation of future continental rocks.

Weathering is the mechanical and chemical breakdown of rocks exposed at the earth's surface to smaller particles that may differ in composition from the original substance. Although the mechanical and chemical aspects of weathering will be discussed separately, they occur simultaneously and are related in many ways.

Weathering processes are of interest and importance to a broad range of scientists and engineers. Soil scientists study weathering processes in relation to the genesis of agriculturally productive soils from rocks and unconsolidated deposits. Hydrogeologists investigate the chemical constituents that are released by weathering reactions in the upper few meters of the soil zone and are then carried downward to influence the chemical composition of groundwater. Geologists and engineers have demonstrated the role of weathering in landslides and other types of slope failure. In the construction industry the effect of weathering on building stone is of great concern, an effect obvious to anyone who has observed the faded and indistinct inscriptions on old tombstones. The deterioration of building stone, road aggregate, and other construction materials is a serious and expensive consequence of weathering phenomena. Perhaps the most important influence of weathering is its relevance

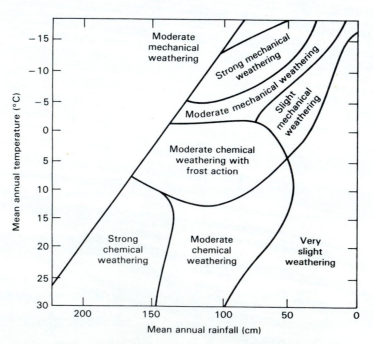

Figure 10-1 Climatic influences on types of weathering processes. (From Peltier, 1950. Reproduced by permission from the *Annals of the Association of American Geographers* 40:219, Fig. 3, 1950.)

to the design and performance of structures built upon weathered rock. Weathering gradually weakens rock; in some climates the end result is a deep mantle of residual soil with vastly different engineering properties from the parent rock. The effects of weathering must be assessed at every foundation site where design requires the presence of predominantly unaltered rock.

The types and intensity of weathering processes that occur in a particular area are primarily the result of climate. Temperature controls the mechanical processes caused by freezing and thawing of water as well as influencing the rates of chemical weathering reactions. The other main climatic variable is precipitation, because water plays a part in both physical and chemical processes. The dominant type of weathering that can be expected in a particular climatic region is indicated in Figure 10-1, which utilizes mean annual rainfall and mean annual temperature to characterize climate.

MECHANICAL WEATHERING

Mechanical weathering includes processes that fragment rocks into smaller particles by exerting forces that exceed the strength of the rock. Usually, these forces act within rock masses to overcome the tensile strength of the rock, which is much lower than its compressive strength. The principal mechanical and chemical weathering types are shown in Table 10-1.

Frost action is one of the most effective mechanical weathering processes. When water freezes within a rock mass, its tendency to expand exerts pressure

TABLE 10-1

Mechanical and Chemical Weathering Processes

Mechanical
   Frost action
   Salt weathering
   Temperature changes
   Moisture changes
   Unloading
   Biogenic processes

Chemical
Solution:

$$CaCO_3 + H_2CO_3 \rightarrow Ca^{2+} + 2HCO_3^-$$
(Calcite) (Carbonic   (Calcium   (Bicarbonate
               acid)   ion)   ion)

Hydrolysis:

$$2KAlSi_3O_8 + 2H^+ + 9H_2O \rightarrow Al_2Si_2O_5(OH)_4 + 4H_4SiO_4 + 2K^+$$
(Orthoclase)   (Kaolinite)   (Dissolved   (Dissolved
                                                          silica)   potassium)

Hydration:

$$CaSO_4 + 2H_2O \rightarrow CaSO_4 \cdot 2H_2O$$
(Anhydrite)   (Gypsum)

Oxidation:

$$4FeSiO_3 + O_2 + H_2O \rightarrow 4FeO(OH) + 4SiO_2$$
(Pyroxene)   (Limonite)   (Dissolved
                                                 silica)

greater than the tensile strength of most rocks. Often, water penetrates into existing cracks and joints in the rock. When it freezes, rock walls on opposite sides of the joint are forced apart. The size and shape of the angular rock fragments produced depends upon the joint spacing and orientation in the rock (Figure 10-2). On slopes, rock fragments produced in this manner may fall, slide, or roll to the base of the slope, where they accumulate in piles called *talus.* Water within the pores of a rock may also fracture the rock during freezing, but this is a complex and poorly understood process.

2. Crystallization of salt within a rock is somewhat similar to freezing of water. The expansion and weakening of the rock caused by salt crystallization is called *salt weathering* and it presents one of the most serious threats to the durability and appearance of stone buildings. The salt that crystallizes from solution in the rock includes not only sodium chloride but also a number of chloride, sulfate, and carbonate salts that can enter the rock in several ways. Sedimentary rocks may retain salts from their original deposition. Other dissolved constituents may be contributed by chemical weathering processes. The salt may also be introduced after the rock is quarried. Polluted urban air and rainfall, for example, are excellent sources of salts.

Salt can crystallize from an aqueous solution as a crust on the rock surface or within rock pores and voids. Often, a concentration of salt forms in a thin layer just below the rock surface. In this case, the surficial layer gradually deteriorates until it splits, or spalls, off from the rock face, at which time the process begins again. Inside the rock, salt crystals can weaken the rock in at least three ways. One is the pressure and expansion caused by the initial growth of the salt crystal. Thermal expansion of the salt crystal is another mechanism. Upon heating, the salt crystals expand more than the surrounding minerals of the host rock, so that pressure is created. Finally, salt crystals tend to absorb water into their structure to form hydrated crystals. This entails a volume expansion and an increase in pressure upon the adjacent grains, which can reach extremely high values (Figure 10-3). These hydration pres-

Figure 10-2 Angular blocks of rock produced from an outcrop by frost action.

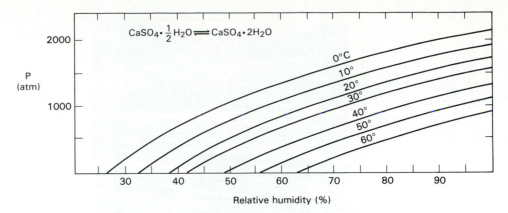

Figure 10-3 Hydration pressures produced by hydration of $CaSO_4 \cdot \frac{1}{2}H_2O$. (From E. M. Winkler and E. J. Wilhelm, 1970, *Geological Society of America Bulletin*, 81:567–572.)

sures can at times exceed the tensional strength of the rock, causing spalling of thin, exterior layers.

③ *Temperature changes* have long been suspected as a weathering mechanism because of thermal expansion and contraction of the minerals within a rock. Although lab experiments tended to dispute the effectiveness of heating and cooling in rock weathering, more recent work suggests that the experiments did not fully duplicate natural conditions and that the mechanism may be significant.

④ *Moisture changes*, in the form of alternate wetting and drying, have experimentally been shown to cause expansion and contraction that may lead to weakening of the rock. This mechanism may be even more effective when combined with temperature changes.

Rock, like any other material, deforms under the application of stress. Deeply buried rock experiences both elastic and plastic compression under the weight of the overlying rocks. Upon removal of the overburden, which could occur during stream erosion of a valley or by deep excavation for a construction project, the elastic component of the deformation is recovered and the rock expands. The expansion caused by *unloading* is often sufficient to fracture the rock along planes parallel to the surface on which stress has been released. ⑤ This type of weathering, also called *exfoliation*, produces thin slabs of rock bounded by joints oriented parallel to the ground surface as the overlying material is removed and stress is released (Figure 10-4). The jointed rock slabs generated by exfoliation are susceptible to further weathering and erosional processes. As the rock slabs are gradually removed from the slope, the unloading action continues and new joints and slabs form. A process that produces ⑥ a somewhat similar result is called *spheroidal weathering*, in which thin concentric layers form and gradually split off the outside of weathering boulders. Unlike exfoliation, the forces that cause the curved slabs to deteriorate and break away from the central core are due to chemical weathering reactions. Exposures of deeply weathered granitic rock often contain spheroidal cores of unweathered rock surrounded by weathered rock (Figure 10-5). The cores are gradually decreased in size by spheroidal weathering.

⑦ Organisms participate in both mechanical and chemical weathering processes. The mechanical effects are dominated by plant roots that exploit thin joints or cracks in rock as they grow. The pressures exerted by the growing

Figure 10-4 Exfoliation results in the separation of thin slabs of rock from a rock mass along planes parallel to the surface. (N. K. Huber; photo courtesy of U.S. Geological Survey.)

Figure 10-5 Rounded boulders exposed in a road cut produced by spheroidal weathering. The spheroidal boulders on the land surface were formed by the same mechanism.

roots can wedge apart blocks of rock, leading to acceleration of other weathering processes in the larger openings.

CHEMICAL WEATHERING

Most rocks orginally formed under conditions very different from those that exist at the earth's surface. In particular, the igneous and metamorphic rocks crystallized at very high temperatures and pressures. When these rocks are exposed to the lower temperatures and pressures present at the surface, they are unstable and tend to chemically react with components of the atmosphere to form new minerals that are more stable under those conditions. The most important atmospheric reactants are oxygen, carbon dioxide, and water. In polluted air, however, other reactants are available.

Two important controls on chemical weathering reactions are temperature and pH. Weathering is more intense in warm climates. The pH, or the negative log of the hydrogen-ion concentration, is a measure of the acidity of the weathering reactions. Water, as it reaches the ground in the form of precipitation, is slightly acidic (pH ~ 5.9) because of reactions with carbon dioxide and other constituents of the atmosphere. Even more acidic precipitation, known as acid rain, is produced by industrial emissions into the atmosphere. Rainfall therefore has the potential to initiate weathering processes.

Figure 10-6 illustrates conditions in the zone of plant growth—the upper meter or so of the soil zone—as rainfall or snowmelt percolates downward from the surface. Decaying organic matter in this zone, under aerobic conditions, generates carbon dioxide. For example, the reaction

$$O_2 + CH_2O \rightleftharpoons CO_2 + H_2O \qquad \text{Eq. 10-1}$$

indicates that a carbohydrate (CH_2O) will react with oxygen to produce carbon dioxide. Carbon dioxide in the soil zone can reach levels of several orders of magnitude greater than its concentration in the atmosphere, and then react with infiltrating rainfall or snowmelt by the reaction

$$H_2O + CO_2 \rightleftharpoons H_2CO_3 \qquad \text{Eq. 10-2}$$

to produce carbonic acid. Production of carbonic acid lowers the pH by partial dissociation according to the reaction

$$H_2CO_3 \rightleftharpoons H^+ + HCO_3^- \qquad \text{Eq. 10-3}$$

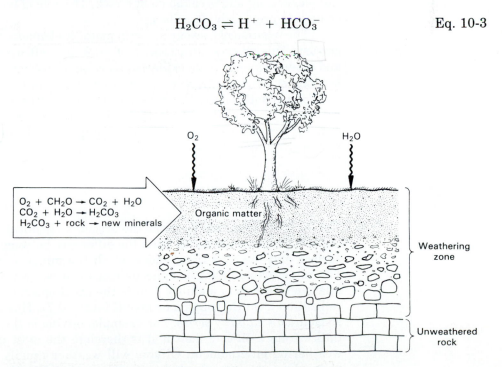

Figure 10-6 Production of acidic weathering solutions by formation of CO_2 in the soil zone.

The resulting increase in hydrogen-ion concentration (lower pH) causes a more intense attack on minerals by the solution.

The specific types of weathering reactions that have been recognized are listed in Table 10-1 along with example chemical reactions. When a mineral completely dissolves during weathering, the reaction is known as *solution*. The tendency of a mineral to dissolve in weathering solutions is described as its *solubility*. Evaporite minerals dissolve readily in water, while carbonate minerals are somewhat less soluble. Even so, the dissolution of limestone can lead to the formation of huge limestone caves beneath the surface. Some silicate minerals weather by solution, but their solubilities are very low. Quartz, for example, usually dissolves appreciably only under the intense weathering conditions in the tropics.

Hydrolysis is the reaction between acidic weathering solutions and many of the silicate minerals, including the feldspars. The reaction illustrated in Table 10-1 indicates that in hydrolysis a feldspar mineral reacts with hydrogen ions to form several dissolved products as well as a solid product, the clay mineral kaolinite. Other clay minerals are produced by similar weathering reactions. This reaction is an example of the breakdown of a mineral that is stable at high temperature and pressure to form a new mineral that is stable under conditions near the earth's surface. Notice that hydrolysis produces dissolved silica and potassium. These constituents are carried downward out of the weathering zone in the water in which they are dissolved toward the water table, where they become part of the groundwater flow system. Hydrolysis reactions are responsible for producing deposits of clay that are mined for use in many industrial processes.

The absorption of water into the lattice structure of minerals is called *hydration*. The formation of gypsum from anhydrite by hydration is illustrated in Table 10-1. Clay minerals are also susceptible to hydration. The volume expansion that accompanies hydration is an important contributor to the physical weakening and breakdown of a rock. The role of hydration in salt weathering was previously described.

The reaction of free oxygen with metallic elements is familiar to everyone as rust. This process, an example of *oxidation*, affects rocks containing iron and other elements. In an oxidation reaction, iron atoms contained in minerals lose one or more electrons each and then precipitate as different minerals or amorphous substances. For example, in the weathering of pyroxene by oxidation, as illustrated in Table 10-1, iron is oxidized and hydrated to form the mineral limonite. The presence of limonite in rocks or soils is indicated by brownish or reddish staining.

STABILITY

The stability of minerals under the action of chemical weathering agents at the earth's surface depends upon the difference between the conditions at the surface and the conditions under which the mineral originally crystallized. The minerals occurring in igneous rocks can be grouped into a sequence of relative weathering stability that is the exact opposite of their order of crystallization from the magma or lava (Figure 10-7). This relationship is known as *Goldich's stability series*. For example, olivine is the first mineral to crystallize from a silicate melt and is therefore the most unstable mineral in a weathering environment. Olivine will weather rapidly when exposed to the atmosphere. Quartz, the last mineral to crystallize from a magma, is very stable in most weathering environments. It dissolves so slowly that, for all practical purposes, it is considered to be insoluble except under humid, tropical

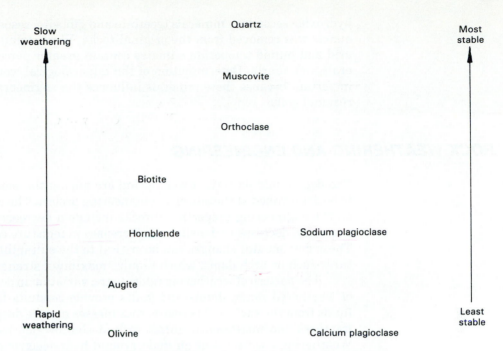

Figure 10-7 Goldich's stability series, which shows the relative stability under weathering conditions of primary minerals in igneous rocks.

climatic conditions. The stability of quartz explains why it is so common in sedimentary rocks such as sandstone. Feldspars and ferromagnesian minerals from the original rocks weather relatively rapidly to clay minerals, leaving quartz to be transported and deposited in depositional environments. The clay minerals that form by the weathering of feldspars and other minerals are stable weathering products that are deposited as muds in low-energy depositional environments and later are altered to shales.

The weathering products of rock-forming minerals vary as a function of climate and other factors. Table 10-2 shows the secondary mineral composition of residual soils weathered from rocks of initially similar composition in three different types of climate in Australia. The intense leaching of the humid tropical climate left a weathering residue dominated by iron and aluminum-

TABLE 10-2

Mineral Composition of Weathered Rocks as a Function of Climate

Climate	Hot arid	Humid tropical	Humid temperature
Mineral		Percentages	
Quartz	28	8	50
Illite	32	—	24
Chlorite	15	—	—
Montmorillonite	12	—	7
Kaolinite	—	12	14
Gibbsite	—	68	—
Geothite	5	10	—
Gypsum	5	—	—
Calcite	2	—	—
Others	1	2	5

SOURCE: From F. C. Beavis, *Engineering Geology*, copyright © 1985 by Blackwell Scientific Publications, Inc., Melbourne.

hydroxide secondary minerals (goethite and gibbsite, respectively). Most of the quartz was removed from the original rocks. Weathering products in the hot arid and humid temperate climates contain greater percentages of clay minerals and quartz. Determination of the mineralogical weathering products is important because these minerals influence the engineering properties of the residual soils.

ROCK WEATHERING AND ENGINEERING

The degree and pattern of weathering are among the most important factors to be determined at the site of an engineering project. The effects of weathering on the engineering properties of rocks include a decrease in strength, loss of elasticity, decrease in density, and increases in moisture content and porosity. These detrimental changes can be critical to the suitability of a site for structures such as arch dams, which require maximum strength and elasticity.

The pattern of weathering refers to the variation in depth and distribution of weathered zones. Joints and faults provide conduits for the movement of fluids from the surface. Therefore, rock masses may be deeply weathered along fractures and unaltered in intact rock at shallow depths between fractures. Weathering conditions at an underground hydroelectric power plant in Australia are shown in Figure 10-8. Weathering zones are classified as fresh (F), slightly weathered (SW), highly weathered (HW), and completely weathered (CW). Tunnel and underground excavations for the project were sited in order to avoid faults and associated deep-weathering zones.

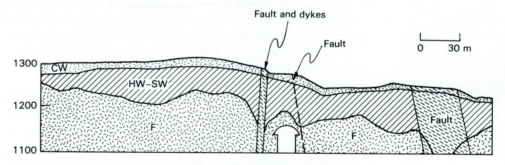

Figure 10-8   Degree of rock weathering at the site of an underground power station in Australia. Weathering zones include fresh rock (F), slightly weathered rock (SW), highly weathered rock (HW), and completely weathered rock (CW). (From F. C. Beavis, *Engineering Geology*, copyright © 1985 by Blackwell Scientific Publications, Inc., Melbourne.)

EROSION

The chemical and physical breakdown of rocks by weathering processes leads to the formation of materials that can easily be transported by processes operating at the earth's surface. *Erosion* describes the removal and transportation of surficial materials under the action of one or more of the forces and agents listed in Table 10-3. The materials that are susceptible to erosion include not only weathered residual products developed from rocks but any unconsolidated surficial deposit. We will refer to these materials as soil in the

TABLE 10-3

Erosional Agents

Gravity (mass wasting)
Water (fluvial erosion)
Wind (eolian erosion)
Glacial ice
Waves

engineering sense, that is, any unconsolidated material regardless of thickness or origin.

Erosion by Water

Throughout the world, erosion by water, or fluvial erosion, is the most important type of erosion in terms of the amount of sediment removed from the land surface. Unfortunately, agricultural lands are the source of much of this sediment. Erosion of land altered by human activity is much more rapid than erosion of land in its natural state. The loss of soil productivity caused by the erosion of topsoil is one of the most serious problems facing the human population.

Erosion in stream and river channels is also a significant problem. Bridge piers, levees, and other structures can be damaged or destroyed by river erosion.

Processes.   Soil erosion by water occurs in a variety of ways. Initiation of erosion often includes detachment of soil particles from the surface by *raindrop erosion.* The energy transferred to the surface by the impact of raindrops is sufficient to dislodge individual soil particles from particle aggregates at the soil surface. Downslope movement of particles also begins by the splash of the raindrop.

In the early stages of a rainstorm, most of the precipitation infiltrates into the ground. If the duration and intensity of the precipitation are sufficient, a saturated zone develops just below the soil surface and water begins to pond on the surface. We will discuss these relationships in more detail in Chapter 12. When ponded water reaches a depth greater than the height of surface irregularities, water flows downslope over a broad area as *overland flow.* This shallow flow of water, which transports detached soil particles, is given the name *sheet erosion*.

Sheet erosion is a rather inefficient erosional process and it is soon replaced by the formation of small channels, or *rills*, on the slope and the transport of particles by *rill erosion* (Figure 10-9). The channelized flow of water in rills has a higher velocity and a greater ability to transport particles downslope. With time, rills enlarge into gullies (Figure 10-10), which may render the land unfit for agriculture.

Erosional Rates.   Attempts have been made to predict the amount of erosion that will occur within a given amount of time for various types of land settings. These methods yield only rough estimates because of the large number of variables in the soil erosion process. Some of the main variables are listed in Table 10-4. Some of these factors have been combined into the *Universal Soil Loss Equation*, which is stated

$$A = RKLSCP$$ Eq. 10-4

Figure 10-9 Severe rill erosion on a steeply sloping crop field. Notice sediment deposition at base of slope. (Photo courtesy of USDA Soil Conservation Service.)

Figure 10-10 Progressive loss of agricultural or grazing land is caused by development and enlargement of gullies. (Photo courtesy of USDA Soil Conservation Service.)

TABLE 10-4
Variables Affecting Soil Erosion

Rainfall
    Intensity
    Duration
Soil Characteristics
    Porosity
    Permeability
    Moisture content
    Grain size and shape
Topography
    Orientation of slope
    Slope angle
    Length of slope
Vegetation
    Type and distribution on slope

where

A = average annual soil loss (tons/acre)

R = rainfall factor

K = soil-erodibility factor

LS = slope length–steepness factor

C = cropping factor

P = conservation practice

The rainfall factor in this equation is obtained from the product of the kinetic energy of raindrops and the maximum 30-min intensity of the storm. The factor K, relating to the characteristics of the soil, is a function of the grain-size distribution. Soils composed of silt and fine sand are the most erodible types. The topography of the area is evaluated by the product of L (slope length) and S (slope angle expressed as a percent). The cropping factor ranges from high values (severe erosion) for barren land and row crops, to low values for natural grasses and woodlands. Crop-management techniques are incorporated into the conservation factor. Erosion-control measures such as contour plowing and terracing reduce the value assigned to this factor from its maximum, in the case where no erosion control methods are implemented. Tables and nomographs are available for obtaining numerical values for the factors in the Universal Soil Loss Equation.

    The Universal Soil Loss Equation was developed mainly for agricultural lands. Other methods of estimating erosion rates include measuring the amount of sediment carried by rivers downstream from the eroded area and measuring the amount of sediment trapped in reservoirs constructed along river courses. These methods indicate that erosion is greatly accelerated by certain changes in land use. An excellent example of these changes is illustrated in Figure 10-11 (Wolman, 1967). When Europeans arrived in the United States, they initiated a series of land-use changes. One of the first changes to occur in arable areas was the conversion of land to crop production. This process involved clearing forested land or breaking sod in prairie regions. Both activities caused a decrease in infiltration, an increase in runoff, and an increase in soil erosion. Marginal agricultural areas often reverted back to forest or

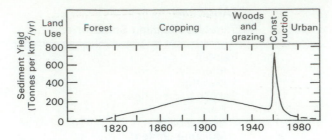

Figure 10-11   Sediment yield from lands undergoing various land-use changes. (From M. G. Wolman, 1967, *Geografiska Annaler* 385. Used by permission of the Swedish Society for Anthropology and Geography.)

grazing as more productive farm land was developed. The erosion was correspondingly decreased in those areas.

The most drastic changes in erosion and sediment yield to streams accompany urbanization of portions of a drainage basin. This activity produces an immense increase in soil erosion. Erosion is most severe during and immediately after construction (Figure 10-12). After the construction phase, erosion decreases because much of the land surface is covered by buildings and paved areas. In addition, runoff is collected and conveyed to stream channels through an artificial network of drains and storm sewers.

Erosion Control.   The use of erosion-control methods must be increased throughout the world in an attempt to combat excessive soil erosion. In agricultural lands, methods include *contour farming, strip cropping*, and *terracing*. The use of these methods becomes more important as the slope of the land increases. Contour farming is the orientation of crop rows parallel to elevation contours of the land. This practice disrupts overland flow of water and thereby inhibits sheet and rill erosion. Strip cropping is the alternation of strips of

Figure 10-12   Soil erosion from cleared, unprotected slopes during housing construction. (Photo courtesy of USDA Soil Conservation Service.)

erosion-inhibiting crops such as grasses with more erodible row crops. Grasses trap sediment eroded from crop rows and also increase moisture infiltration. The optimum width of the strips is determined by the amount of land slope. Terracing is a more intensive method of erosion control requiring construction of alternating embankments and terraces on the slope.

Although the area converted to urban use is small in comparison to the amount of farmland, the sediment yield from urbanizing areas can be greater by a factor of 100 or more than natural or agricultural areas nearby. The problem of controlling erosion in land undergoing urbanization is directly related to the problem of sedimentation, the deposition of the eroded soil in undesirable locations. Excessive sedimentation can clog highways, ditches, and drains, causing expensive cleanup operations (Figure 10-13). Increased sediment yield to stream systems kills fish, increases flooding, and decreases the capacity of reservoirs. Chemical pollutants, including pesticides and herbicides, are often adsorbed on soil particles and carried by erosion to streams and lakes.

In urban areas, new construction is likely to pose the most serious erosion and sedimentation problems. Sediment yields from areas that annually produce from 70 to 280 tonnes/km^2 can increase to as much as 28,000 tonnes/km^2 during certain types of construction. Guy (1976) has documented many examples of effective and ineffective erosion and sediment-control measures and has also suggested methods to reduce erosion by applying sound hydrologic and geomorphic principles to the design of these systems. Soil erosion can often be drastically reduced by planning construction during periods of low expected rainfall, dividing the construction site so that only small areas are cleared of vegetation at any one particular time, and utilizing temporary vegetation cover on soil stockpiles and other exposed areas whenever possible.

When additional erosion control measures are necessary, *diversions, bench terraces,* and *detention basins* can be constructed. Diversions consist of

Figure 10-13 Newly constructed street (notice curb and partially buried drain in foreground) filled with sediment eroded from adjacent housing lots. (Photo courtesy of USDA Soil Conservation Service.)

channels and ridges designed to collect overland flow and carry it away from areas that must be protected from erosion or sedimentation. Bench terraces reduce the length of the slope and, as implied in Eq. 10-4, reduce erosion. Diversions are often routed to detention basins (Figure 10-14), which are designed to trap sediment carried by runoff from the construction site. Temporary sediment-control measures are often installed until vegetation or paving reduces the area of unprotected ground at a site. Straw or hay bales, for example, form effective sediment traps around storm-drain inlets or culverts, when flow velocities are not extremely high. These temporary measures often fail, however, when the amount and velocity of overland flow overwhelm the straw-bale structure (Figure 10-15).

Figure 10-14 A sediment detention basin used to trap sediment in an urban area. (Photo courtesy of USDA Soil Conservation Service.)

Figure 10-15 Failure of a straw-bale filter designed to trap sediment in an unvegetated ditch along a newly constructed highway. Flow in the ditch overtopped the straw-bale dam and eroded a new channel adjacent to the straw bales.

Erosion by Wind

The impact of wind erosion has never been more clearly demonstrated than in the Great Plains of the United States during the 1930s (Figure 10-16). Dry conditions coupled with steadily increasing destruction of the natural grass cover devastated the agricultural economy of the region.

The erosion of soil by wind is similar to erosion by water. Wind, however, is not as effective an erosional agent as water because of its lower density. Variables in the wind-erosion system can be divided into characteristics of the wind and characteristics of the soil and land surface. Wind variables include velocity, duration, and length of open area without obstacles over which the wind blows. Important properties of the soil and land surface include particle size and size distribution, moisture content, and vegetation. Particles of typical density within the size range of 0.1 to 0.15 mm in diameter (very fine to fine sand) are most susceptible to wind erosion. The distribution of grains upon the surface is also very important. Soils containing large particles gradually become protected as wind erosion proceeds. As the fine particles are removed, coarser grains become concentrated at the surface, preventing further erosion of the soil.

Very fine particles in the soil, including silt and clay, promote the formation of soil aggregates. These clumps composed of smaller particles are held together by the cohesion of the silt and clay. In order for soil aggregates to be eroded, they must first be broken down by raindrop impact, abrasion by wind-transported sediment, and other processes.

Moisture is one of the most important soil properties. Moist soil is cohesive and resistant to erosion. The cohesive forces provided by the soil water are lost when the soil dries. This explains why droughts and wind erosion go hand in hand.

Vegetation is the final soil and land-surface variable. Vegetation provides physical protection for the soil as well as holding moisture, increasing the roughness of the surface, and adding organic binding agents to the soil. The removal of vegetation during plowing and other activities is one of the primary factors in increased wind erosion.

Figure 10-16 A dust storm in Colorado during the 1930s. Notice the drifts of sand around fences and buildings. (Photo courtesy of USDA Soil Conservation Service.)

Figure 10-17 A shelterbelt planted to minimize erosion of soil by wind. (Photo courtesy of USDA Soil Conservation Service.)

Wind erosion can be greatly reduced by proper farming and land-use practices. *Shelterbelts* are one of the most common attempts to control wind erosion (Figure 10-17). These linear bands of trees or other plants decrease wind velocity and the unobstructed distance over which the wind blows. Farming methods for wind-erosion control include strip cropping and planting of temporary cover crops rather than allowing the soil to remain bare between periods of crop production.

CASE STUDY 10-1
EROSION, SEDIMENTATION, AND RESERVOIR CAPACITY

The interrelationship between erosion and deposition is well illustrated by the accompanying photos of Lake Ballinger Dam in Texas.

Figure 10-18a shows the dam and reservoir prior to its abandonment in 1952. Lake Ballinger Dam was built in 1920 to provide a munic-

(a)

Figure 10-18 (a) Lake Ballinger Dam, Texas, before abandonment. (b) The dam after it was abandoned in 1952 because of excessive sedimentation. (Photos courtesy of USDA Soil Conservation Service.)

(b)

Figure 10-18   (*continued*)

ipal water supply. Sedimentation gradually reduced the maximum depth of the lake from about 11 m to slightly more than 1 m by 1952. The excessive sedimentation was caused by runoff from cultivated fields in the small drainage basin of the reservoir. Figure 10-18b, taken after the final drainage of the reservoir, shows the greatly decreased capacity caused by excessive sedimentation. Better soil conservation practices in the drainage basin could have alleviated the problem. Thus the problem of erosion in one area is compounded by undesirable sedimentation in another area.

SUMMARY AND CONCLUSIONS

Rocks in contact with the atmosphere at the earth's surface are subjected to a wide variety of physical and chemical weathering processes that lead to their gradual decay and disintegration. Expansive pressures within rocks are caused by frost action, salt weathering, moisture-content changes, temperature changes, and removal of overburden load. These physical processes are complemented by chemical reactions between atmospheric constituents and unstable minerals to form new compounds that are more stable under the surface conditions of low temperature and pressure. Chemical weathering reactions include complete dissolution of the original mineral (solution), formation of new minerals (hydrolysis), addition of water to the mineral structure (hydration), and reaction with oxygen to form new substances (oxidation). Minerals derived from igneous rocks weather in reverse order from their order of crystallization. The resistance of quartz to weathering explains its abundance in sedimentary rocks produced by the weathering and erosion of previously existing rocks.

Erosion by water and wind is most severe when land is altered from its natural state. Prediction of rates of fluvial erosion by such methods as the Universal Soil Loss Equation is difficult because of the large number of variables involved in the process. Erosion damage is often initiated by raindrop impact and sheet erosion. Sheet erosion progresses to rill and gully erosion during the gradual destruction of arable land by fluvial action. Land cleared for construction presents the most vulnerable condition for soil erosion. Various farming practices can be adopted to minimize erosion. Similarly, erosion of lands under development can be greatly limited by construction planning and utilization of erosion-control structures designed with an understanding of the hydrologic and geomorphic processes involved.

Wind erosion poses its greatest threat to agricultural lands. Appropriate farming practices constitute the most effective approach to minimizing crop and soil losses.

REFERENCES AND SUGGESTIONS FOR FURTHER READING

BEAVIS, F. C. 1985. *Engineering Geology*. Melbourne: Blackwell Scientific Publications, Inc.

COOKE, R. V., and J. C. DOORNKAMP. 1974. *Geomorphology in Environmental Management*. Oxford: Clarendon Press.

COSTA, J. E., and V. R. BAKER. 1981. *Surficial Geology: Building with the Earth*. New York: John Wiley.

GUY, H. P. 1976. "Sediment-control methods in urban development: some examples and implications." In *Urban Geomorphology*, edited by D. R. Coates. Geological Society of America Special Paper 174, pp. 21–35.

PELTIER, L. 1950. The geographical cycle in periglacial regions as it is related to climatic geomorphology. *Annals of the Association of American Geographers* 40:214–236.

WINKLER, E. M., and E. J. WILHELM. 1970. Salt burst by hydration pressures in architectural stone in urban atmosphere. *Geological Society of America Bulletin* 81:567–572.

WOLMAN, M. G. 1967. A cycle of sedimentation and erosion in urban river channels. *Geografiska Annaler* 49-A:385–395.

PROBLEMS

1. How does salt weathering cause deterioration of building stone?
2. What are the conditions under which exfoliation becomes an important weathering process?
3. What gas is produced by the aerobic decay of organic matter in the soil zone?
4. Describe the primary chemical process by which the feldspar minerals weather.
5. Why do minerals fit into their respective positions in Goldich's stability series?
6. What engineering properties of rock are affected by weathering, and what changes take place?
7. What variables affect soil erosion on slopes?
8. What methods are used to minimize erosion during construction?
9. What are the consequences of excessive soil erosion? Which one(s) would you expect to be most severe in your area?

Chapter
11

SOILS,
SOIL HAZARDS,
AND LAND
SUBSIDENCE

Perhaps no term causes more confusion in communication between various specialized groups of earth scientists and engineers than the word *soil.* The problem arises in the reasons for which different groups of professionals study soils. Soil scientists, or *pedologists*, comprise a group interested in soils as a medium for plant growth. For this reason, pedologists focus most of their attention on the organic-rich, weathered zone that supports plant growth—the upper meter or so beneath the land surface—and refer to the rocks or sediments below the weathering zone as *parent material*. Soil scientists have developed a complex system of classification for soils that is based on the physical, chemical, and biological properties that can be observed and measured in the soil.

Soils engineers, the corresponding group of engineering soil specialists, take an entirely different approach to the study of soil. To engineers, the word soil connotes any material that can be excavated with a shovel. This definition places no limitations on depth, origin, or ability to support plant growth. The engineering classification of soil (Chapter 6) is based on the particle size, particle-size distribution, and plasticity of the material. These characteristics relate closely to the behavior of soil under the application of load.

Most geologists fall somewhere between pedologists and soils engineers in their approach to soils. Geologists are interested in the weathered soils of the pedologist as indicators of past climatic conditions, as environments where rock is converted to sediment, and, in some areas, as ore deposits. The pedological soil classification is too complicated for nonspecialists and, therefore, older soil classifications often are utilized by geologists. The loose material below the weathering zone usually is called sediment or unconsolidated material by geologists. The term "unconsolidated" may be confusing to engineers because the process of consolidation has very specific implications to a soils engineer.

An important aspect of soil with respect to engineering applications is whether the material can be classified as *transported* or *residual*. Transported soils are deposits of rivers, glaciers, and other surficial processes. Residual soils are those that developed in place by the weathering processes discussed in Chapter 10. This distinction is important because the origin of the material is relevant to its thickness, continuity or discontinuity over a landscape, and the nature of the lower contact with unweathered bedrock.

THE SOIL PROFILE

In the first part of this chapter we shall investigate soil from the viewpoint of soil scientists. This discussion will, therefore, be limited to the zone in which *pedogenic*, or soil-forming, processes occur. Pedogenic processes include the chemical weathering processes that we discussed in Chapter 10, as well as certain additional physical, chemical, and biological processes. The result of this activity is a sequence of recognizable layers, or *horizons*, that constitute the *soil profile*.

A number of possible horizons can be present in a soil profile observed in a road cut or pit (Figure 11-1). A schematic example of the major horizons is given in Figure 11-2. The basic divisions of the profile are the A, B, and C horizons. A surficial horizon, the O, is used to designate fresh or partly decomposed organic matter. The A horizon, recognized by its dark color, is a zone with a high content of decomposed and altered organic matter called *humus*. Subdivisions of the A horizon include the A_1, a zone of mixed organic and mineral material, the A_2, a light-colored horizon produced by leaching of iron, aluminum and clay, and the A_3, a transitional area. The B horizon consists

Figure 11-1 A soil profile exposed in a pit dug for soil classification. (Photo courtesy of USDA Soil Conservation Service.)

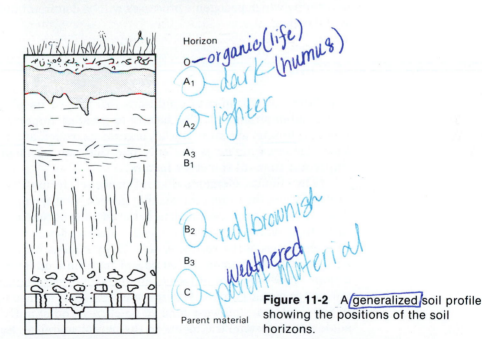

Horizon

O — *organic (life)*

A₁ — *dark (humus)*

A₂ — *lighter*

A₃
B₁

B₂ — *red/brownish*

B₃

C — *weathered parent material*

Parent material

Figure 11-2 A *generalized* soil profile showing the positions of the soil horizons.

B-takes leached iron & aluminum

of the transitional B₁ and B₃ horizons above and below the B₂, which is generally reddish or brownish in color and contains accumulations of iron and aluminum oxides and clay. The C horizon is a zone of partly altered parent material, in which the original characteristics of the soil or rock are still recognizable. In residual soils the transition from the C horizon to unweathered bedrock is gradual and highly variable in depth. Along fractures and other discontinuities, weathering may extend to much greater depths than the average depth to the base of the C horizon.

Many pedogenic processes are involved in the formation of soil horizons. For example, in humid climates soluble salts such as calcium carbonate are leached out of the profile and carried downward to the water table. In arid climates, however, calcium carbonate and other salts are leached from the upper horizons and redeposited in the C horizon. Calcium carbonate can accumulate to the extent that an extremely hard layer, or *caliche,* may develop. This zone may be so hard that blasting is required for excavation. Other pedogenic processes are involved in the transfer of materials from one soil horizon to another. The movement of clay and metal oxide from the A to the B horizon includes several pedogenic processes. Dissolved organics are very important in the removal of iron and aluminum from the A horizon. Organic molecules bond to normally insoluble metal ions in a process called *chelation.* In the B horizon, the organic molecules are destroyed and iron and aluminum oxides precipitate from solution.

In the tropics, warm, moist conditions combine to produce thick, deeply weathered residual soils. Intense leaching removes even quartz, which is practically insoluble in arid and temperate climates. Tropical soils, therefore, are mainly residual accumulations of iron and aluminum oxides (Table 10-2). The concentrations of these metals are so high in some places that the soil is mined. Aluminum-rich soils called *bauxites* provide most of the world's aluminum ore. The iron-rich soil of tropical areas is called *laterite.* Laterites are reddish brown in color because of the abundance of iron oxide.

The number and type of soil horizons present are not constant from one area to another. In the next section we will examine the major factors that determine which pedogenic processes will be dominant and which soil horizons will develop.

SOIL-FORMING FACTORS

Soil forms in response to conditions existing in its physical, chemical, and biological environment. The factors of greatest importance in soil formation include climate, parent material, organisms, relief, and time (Figure 11-3). Of these factors, climate is the only one that can be identified as being more important than all the other factors.

The climatic elements that influence soil formation are temperature and precipitation, the same controls that govern weathering processes. In the early days of soil science, pedologists believed that parent material—rock or sediment from which the soil developed—was most important in determining the characteristics of the resulting soil. Many studies have shown, however, that similar soils occur upon varied parent materials under constant climatic conditions. The influence of parent material is greatest in young soils that developed in a mild climate. For example, soils that formed from glacial sediment in the American midwest tend to be clayey and slightly alkaline because of shale and limestone fragments in the parent material. Soils occurring on glacial sediments in New England, on the other hand, are sandier and acidic, reflecting the composition of the predominantly siliceous igneous rocks from which the glacial sediment was derived.

Organisms include both plants and animals that inhabit a particular soil. If all other soil-forming factors were held constant, soils developed under different vegetative communities would exhibit numerous differences. A good example is the difference in soil properties observed across the transition from prairie to forest vegetation. As shown in Figure 11-4, prairie soils have thicker A horizons containing more organic matter. Forest soils have a lower pH than prairie soils, have highly leached zones within the A horizon, and display greater accumulation of clay and iron oxide in the B horizon.

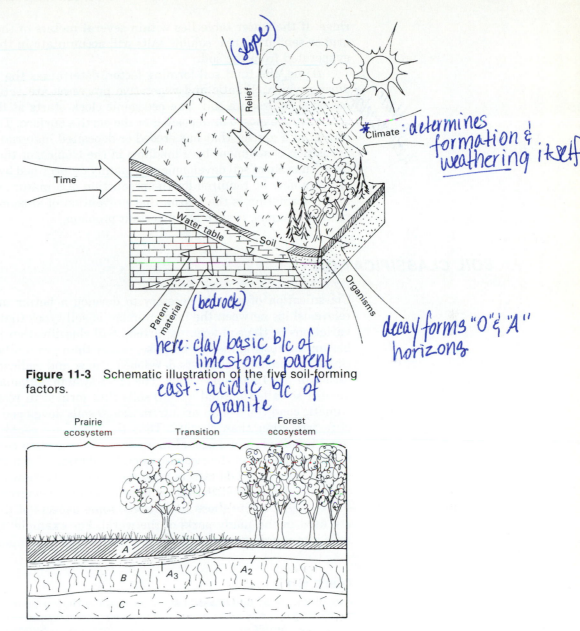

(handwritten annotations)
(slope)

Climate: determines formation & weathering itself

here: clay basic b/c of limestone parent

east: acidic b/c of granite

(bedrock)

decay forms "O" & "A" horizons

Figure 11-3 Schematic illustration of the five soil-forming factors.

Prairie ecosystem   Transition   Forest ecosystem

Figure 11-4 Changes in soil from a forest to prairie area. (Reprinted by permission from S. W. Buol, F. D. Hole, and R. J. McCracken, *Soil Genesis and Classification,* 2d ed., copyright © 1980 by the Iowa State University Press, Ames, Iowa.)

(handwritten annotation)
steep = no depth or development

Relief, the factor that relates to the position of the soil in a particular landscape, can cause drastic changes in soil characteristics within a short distance. The aspects of relief that influence soil development include slope and topographic position. Slope is important because of erosional and depositional processes. Soil on a steep hillslope is likely to be thin because of more intense erosion in comparison to soils at the base of the slope, where accumulation of the soil eroded from the hillslope occurs (Eq. 10-4). Topographic position refers to the location of the soil within the local landscape. The water table, for example, may be much closer to the surface in a valley than at the top of a ridge. If the water table lies within several meters of the land surface in an arid or semiarid area, soluble salts will accumulate in the soil as moisture is evaporated from the soil.

Time, the final soil-forming factor, determines the length of the period during which climatic and vegetative processes are acting upon the parent material to produce soil. The pedogenic clock starts at the instant a rock or sediment is exposed or deposited at the earth's surface. This could include the retreat of a glacier that has eroded or deposited material on a landscape, the deposition of sediment by a mudflow in the tropics, or the extrusion of a lava flow. The rate of soil development is mainly determined by climate. Thousands of years may be required for the production of a mature soil. The time factor in soil formation is the reason that prevention of erosion, particularly of agricultural lands, is such an important problem.

[handwritten: lg time → well-develop]

SOIL CLASSIFICATION

Classification of soils is necessry to develop a better understanding of the relationships between the large number of soil types that can be produced by variations in the soil-forming factors. Soil classification in the United States has an interesting history because, rather than gradually modifying an older classification, soil scientists developed a new and radically different classification in the 1950s and 1960s. The older system, summarized in Table 11-1, divided so-called normal (zonal) soils that formed in response to the typical climatic conditions of an area from azonal soils developed under the influence of factors other than climate. Thus the same soil could be considered to be zonal in one area and azonal in another. This led to a certain amount of confusion in the use of the classification. In addition, the highest divisions of the system were based on factors presumed to be important in soil genesis rather than soil characteristics that could be precisely measured and compared from soil to soil. Despite these limitations, some aspects of the classification still are used in the many parts of the world. For example, the division of zonal soils into *pedalfers* and *pedocals* is a useful way of comparing soils over large

TABLE 11-1

Part of the Older Soil Classification Used in the United States

Order		Suborder
Zonal soils	Pedocals	Soils of the cold zone 1. Light-colored soils of arid regions 2. Dark-colored soils of the semiarid, arid, sub-humid, and humid grasslands
	Pedalfers	3. Soils of forest-grassland transition 4. Light-colored podzolized soils of the timbered regions 5. Lateritic soils of forested warm-temperature and tropical regions.
Intrazonal soils		1. Halomorphic (saline and alkali) soils of imperfectly drained arid regions and littoral deposits 2. Hydromorphic soils of marshes, swamps, seep areas, and flats 3. Calomorphic

[handwritten: (specific zone (climate)]
[handwritten: temp. climate good rain]

Source: From 1938 USDA Yearbook of Agriculture.

areas. Pedalfers, with the abbreviations for aluminum (al) and iron (fe) incorporated into the name, are soils that accumulate iron and aluminum oxides within the profile. These soils are most common in moist, forested regions. Pedocals, which contain calcium carbonate (cal) within the profile, occur primarily in arid and semiarid areas. The distribution of pedalfers and pedocals is shown in Figure 11-5.

The new classification system was developed by the U.S. Soil Conservation Service to rectify problems and confusion caused by use of the older system. It is based on measurable properties of soil profiles. The divisions of the classification are shown in Table 11-2. Ten soil orders comprise the first category of the classification. The characteristics of the orders are illustrated in Table 11-3. At the other end of the classification, about 10,000 soil series have been identified in the United States. Although this classification is accurate and precise, its use by geologists and engineers has been limited because of its complexity.

SOIL HAZARDS

In Chapter 6 the basic mechanical principles of rock-particle systems, or soils, were explained. The behavior of soil under imposed engineering loads, however, is often much more complex. In order to identify and predict potentially hazardous soil processes, the engineer must determine the relationships and interactions between the soil and rock units, the groundwater flow system, the climate, and the vegetation of an area. All these factors can influence soil processes that can damage existing structures or require special design for new projects.

Expansive Soils

Surprisingly, damage to structures caused by soils that expand by absorbing water is the most costly natural hazard in the United States on an average annual basis. The annual bill for repair and replacement of highways and buildings affected by expansive, or swelling, soils runs into billions of dollars. The portions of the United States underlain by expansive soils are shown in Figure 11-6.

The tendency for soils to swell can be explained by the characteristics of clay particles that we have already considered. The unbalanced charges on clay-particle surfaces draw water molecules into the area between silicate sheets, thus forcing them apart. The cations attracted to the clay surfaces provide another factor in swelling behavior. Because of the attraction of the negatively charged clay-particle surfaces for cations, small spaces within or between clay particles may contain a higher concentration of cations than larger pores within the soil. These conditions (Figure 11-7) create an *osmotic potential* between the pore fluids and the clay-mineral surfaces. Normally, cations diffuse from a higher concentration to a lower concentration in order to evenly distribute the ions throughout the solution. In expansive soils, because ions are held by the clay particles, water moves from areas of low ionic concentration (high concentration of water) to areas of high ionic concentration (lower concentration of water) within clay particles or aggregates. This influx of water exerts pressure, which causes the clay to swell.

If a clay soil is subjected to drying conditions, for example, when evaporation is removing water from the soil, a suction effect is exerted on the soil that causes water molecules that are not held tightly to clay particles to be

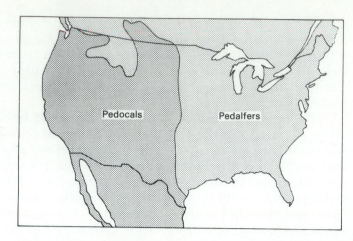

Figure 11-5 Generalized distribution of pedalfers and pedocals in the United States. (From S. Judson, M. E. Kauffman, and L. D. Leet, *Physical Geology,* 7th ed., copyright © 1987 by Prentice-Hall, Inc., Englewood Cliffs, N.J.)

TABLE 11-2

Subdivisions of the United States Comprehensive Soil Classification System

Category	Number of taxa
Order	10
Suborder	47
Great group	Approximately 206
Subgroup	
Family	
Series	Approximately 10,000 in United States

TABLE 11-3

Simplified Definitions of Soil Orders in United States Comprehensive Soil Classification System

Alfisol	Soil with gray to brown surface horizon, medium to high base supply, and a subsurface horizon of clay accumulation.
Aridisol	Soil with pedogenic horizons, low in organic matter, usually dry.
Entisol	Soil without pedogenic horizons.
Histosol	Organic (peat and muck) soil.
Inceptisol	Soil with weakly differentiated horizons showing alteration of parent materials.
Mollisol	Soil with a nearly black, organic-rich surface horizon and high base supply.
Oxisol	Soil that is a mixture principally of kaolin, hydrated oxides, and quartz.
Spodosol	Soil that has an accumulation of amorphous materials in the subsurface horizons.
Ultisol	Soil with a horizon of clay accumulation and low base supply.
Vertisol	Cracking clay soil.

SOURCE: From A. L. Bloom, *Geomorphology,* copyright © 1978 by Prentice-Hall, Inc., Englewood Cliffs, N.J.
NOTE: Base supply refers to amount of exchangeable base cations such as calcium, magnesium, sodium, and potassium that remain in the soil.

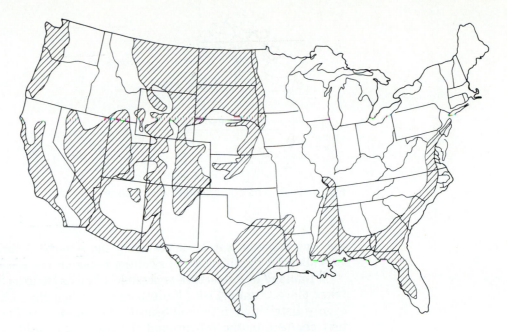

Figure 11-6 Distribution of expansive soils in the United States. (From H. A. Tourtelot, 1974, *Bulletin of the Association of Engineering Geologists* 11. Used by permission of the Association of Engineering Geologists.)

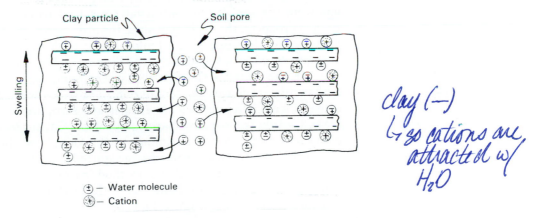

⊕ — Water molecule
⊕ — Cation

clay (—)
↳ so cations are
attracted w/
H_2O

Figure 11-7 Swelling of clay-rich soils. Water molecules diffuse into clay particles in order to equalize the ionic concentration between the particle and the soil pore.

when clay dries
H_2O is sucked up
to surface =
SHRINKAGE

drawn out into the larger pores of the soil and move toward the land surface to replace the evaporated water. This loss of water from the clay leads to shrinkage, the reversal of the swelling process.

Swelling and shrinkage of a soil can occur only with changes in water content. Therefore, the potential for damage from expansive soil is limited to the upper zone of the soil in which seasonal changes in moisture content take place. This zone is called the *soil-moisture active zone* (Figure 11-8). Below this zone, even though the soil may have the potential to shrink and swell, volume changes will not take place because the water content of the soil is constant.

Prediction of swelling behavior can be accomplished by lab tests designed to measure the pressure generated by soil being wetted. A quick estimate of

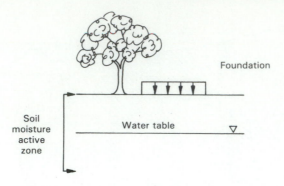

Figure 11-8 The soil-moisture active zone.

swelling behavior can be made using the Atterberg limits of the soil. Table 11-4 shows a classification of swelling based on the plasticity index.

Damage to a structure is possible when as little as 3% volume expansion takes place (Figure 11-9). Failure results when the volume changes are unevenly distributed in the soil beneath the foundation. For example, water-content changes in the soil around the edge of a building can cause swelling pressure beneath the perimeter of the building, while the water content of the soil beneath the center remains constant (Mathewson et al., 1975). The type of failure resulting from this situation is called *end lift* (Figure 11-10). The factors that could cause end lift include excessive lawn or shrub watering beside the building and insufficiently long drain spouts that do not carry drainage from the roof far enough away from the structure. If a house lot is not graded properly, the land may slope toward the house. This would concentrate drainage at the edge of the building and high infiltration could occur. The opposite of edge lift is *center lift* (Figure 11-10), where swelling is focused beneath the center of the structure or shrinkage takes place under the edges. Either process causes the foundation to bulge beneath the center. A possible cause of soil shrinkage near the edges of a structure is the close proximity of trees. Certain trees remove large quantities of water from the soil by the process of *transpiration*; soil shrinkage in the vicinity of the roots is the result.

If highly expansive soils are recognized before construction, special foundation designs can be used. One method involves support of the building by a grade beam resting upon belled piers that extend below the base of the soil-moisture active zone. (Figure 11-11).

TABLE 11-4

Correlation of Swelling Behavior with Plasticity Index

Degree of expansion	Plasticity index
Very high	>35
High	25–41
Medium	15–28
Low	<18

SOURCE: From R. D. Holtz and W. D. Kovacs, *An Introduction to Geotechnical Engineering*, copyright © 1981 by Prentice-Hall, Inc., Englewood Cliffs, N.J.

Figure 11-9 Structural damage to a group of apartment buildings in San Antonio, Texas, caused by shrink-swell behavior of expansive soils. (Photo courtesy of USDA Soil Conservation Service.)

End lift

→ watering
downspouts
poorly graded

Hydrocompaction

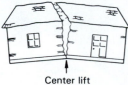

Center lift

Figure 11-10 Two types of structural failure caused by swelling soils. (From C. C. Mathewson et al., 1975, *Bulletin of the Association of Engineering Geologists* 12:275–302. Used by permission of the Association of Engineering Geologists.)

A different type of soil problem is presented by certain soils in arid regions. When water and load are applied to these soils, their structure collapses to a more dense state. This type of behavior, called *hydrocompaction*, is the result of the origin and history of the soils. Sediments deposited by wind or water usually develop a very loose, open structure. Soils susceptible to hydrocom-

→ trees take H₂O

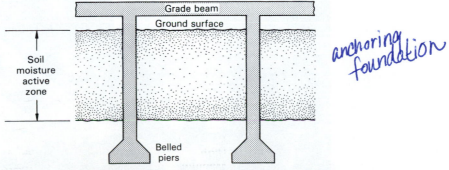

anchoring
foundation

Figure 11-11 Belled pier and grade beam foundation utilized to prevent damage from expansive soils. (From B. M. Das, *Principles of Foundation Engineering*, copyright © 1984 by Wadsworth, Inc., Belmont, Calif.)

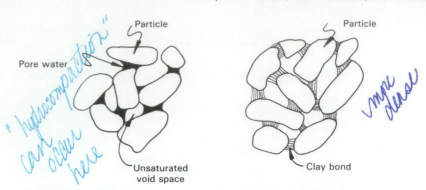

[handwritten annotations: "hydrocompaction" can occur here; more dense]

Figure 11-12 Loose, unsaturated soil structure maintained by pore water and clay bonds.

[handwritten annotation: → susceptible to hydrocompaction]

paction, which are dominantly composed of silt and sand grains rather than clay particles, exist in an unsaturated state. The low-density structure is maintained by weak bonds formed by water, clay, or soluble precipitants that bridge pores between particles. (Figure 11-12). When the pores are only partially filled with water, the water exerts tensional forces that tend to hold adjacent particles together. In addition, small amounts of clay or soluble precipitants (such as gypsum) between particles act as a temporary cementing agent. When water saturates the soil, both the water and clay bonds are destroyed. Pore water loses its tensional effect when the pores become saturated, and clay bonds are suspended in the solution upon wetting. The soil will then collapse to a denser structure under its own weight or the weight of a structure built upon the soil. Saturation is often caused by lawn watering around a foundation and leakage

[handwritten annotation: clay bonds are destroyed]

Figure 11-13 Rotation of buildings due to liquefaction of saturated sand during the 1964 Niigata, Japan, earthquake. (Photo courtesy of U.S. Geological Survey.)

of water lines, storm sewers, and canals. When the collapse of the soil occurs, settlement of foundations or rupture of utility lines is usually the result.

Liquefaction

Several types of soils can begin to flow as a liquid or plastic material under certain circumstances. The term liquefaction is commonly applied to the conversion of saturated sand to a liquid state under rapid or cyclic stresses of the type that might be caused by earthquakes, vibrations, or explosions. Normally, a sand would respond to increased stress or loading by the expulsion of pore water associated with a decrease in void ratio. Water can be driven out readily because of the high hydraulic conductivity of sand. Under rapid or cyclic loading, however, the increased stress is transferred to the pore water as the particles are forced closer together. If the pore-water pressure is great enough to suspend the particles within the pore fluid, total loss of shear strength and fluid behavior results. An excellent example of liquefaction occurred in the Niigata, Japan, earthquake of 1964 (Figure 11-13). Liquefaction of a saturated sand beneath the foundations of the apartment buildings was responsible for a temporary loss of shear strength in the soils supporting the buildings. When this happened, the buildings rotated or sank into the fluidized soil. Little structural damage resulted from these movements, and the buildings were later jacked upright and new foundations were constructed.

LAND SUBSIDENCE

Subsidence describes several related localized or areally widespread phenonema associated with sinking of the land surface. The vertical movements involved range from sudden collapse to slow, gradual declines in surface elevation. In a general way, subsidence can be subdivided into types caused by (1) removal of subsurface fluids, (2) drainage or oxidation of organic soil, and (3) surface collapse into natural or excavated subsurface cavities.

Subsidence attributed to withdrawal of fluids is a widely distributed occurrence in the United States and other countries. Although it is most commonly heavy pumping of groundwater that leads to this type of subsidence, depletion of oil and gas reservoirs has been responsible for subsidence in some areas. A prominent example is the Wilmington oil field in Long Beach, California, in which maximum elevation decreases of about 9 m were attributed to the withdrawal of petroleum. Direct damage to structures and indirect damage from increased flooding proved very costly to the city.

Groundwater withdrawals have accounted for subsidence over thousands of square kilometers of land in the western United States and other areas. The problem occurs in agricultural areas like the San Joaquin Valley in California (Figure 11-14), where the water is used for irrigation, as well as in urban areas including Las Vegas, Houston, and Mexico City, where the water is withdrawn for municipal supply.

Accompanying subsidence in many locations are ruptures of the land surface. These ground failures take the form of fissures as much as several kilometers in length, in which the surface cracks or separates laterally (Figure 11-15), or faults, in which vertical displacement across the rupture occurs. Faults of this type in the Houston area have caused millions of dollars of damage.

The subsurface mechanism causing surface subsidence from fluid withdrawal is the compaction of compressible beds of rock or sediment. It is likely

Figure 11-14 Subsidence in the San Joaquin Valley, California, caused by withdrawal of groundwater. (Photo courtesy of Thomas Holzer.)

Figure 11-15 Fissures formed during land subsidence. (Photo courtesy of Thomas Holzer.)

that most of the compaction occurs within beds of silt and clay in the sedimentary sequence rather than within the more permeable and less compressible beds of sand from which the fluids are extracted. This process is another example of consolidation, in which clay beds compact as water moves slowly out of them into the adjacent sand beds.

Subsidence by a different mechanism occurs when highly organic soils are drained during land-use conversion to agriculture or construction. Soils of this type, frequently termed peat or muck, are formed in marshes, bogs, and other poorly drained settings with water tables near the land surface. The high water table inhibits oxidation of organic matter, so the soils accumulate high percentages of organics. Other characteristics of the soils include dark color, low density, and high compressibility. When drained, organic soils are exposed to oxidizing conditions as the water table drops. The resulting decomposition of the organic content leads to gradual subsidence of the land surface. Damage to buildings can be caused by settlement of the structure itself, or if the building is constructed on foundations that bear upon stronger materials below the organic soils, problems are caused by subsidence of the soil around the structure (Figure 11-16).

Figure 11-16 Subsidence of peat and muck soils around a house built on piling. Land surface around house has subsided about 0.6 m in 2 years. Porch and steps have fallen off. (Photo courtesy of USDA Soil Conservation Service.)

Figure 11-17 Sinkholes formed by subsidence over abandoned coal mines. (Photo courtesy of North Dakota Public Services Commission.)

(3) The third major type of subsidence is associated with the collapse of overlying materials into large underground cavities. Although the processes that cause this type of subsidence generally occur in bedrock below the soil zone, collapse of the surface soils into the voids below is the usual manifestation. The cavities may be excavated or natural. Underground coal mines constitute

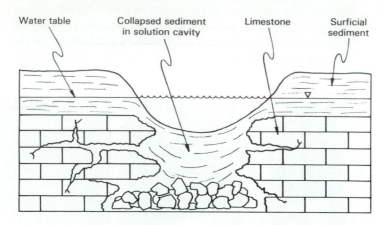

Water table   Collapsed sediment   Limestone   Surficial
                in solution cavity   sediment

Figure 11-18 Natural sinkholes form when surficial sediment collapses into cavities formed by solution in soluble rocks.

Figure 11-19 The Winter Park, Florida, sinkhole as it appeared just after formation in 1981. (Photo courtesy of Jammal & Associates.)

the most common example of subsidence over excavated cavities. This process has affected about 8000 km² in the United States, mostly in the eastern part of the country. The effects of subsidence depend upon the method of mining and the thickness of the overburden. When the coal is entirely removed from a seam during mining, surface subsidence is usually contemporaneous with mining. A basin-shaped depression commonly develops over the mined-out area if the coal seam is deeper than about 30 m. Alternatively, columns or *pillars* of coal are left in place to provide support for the mine roof. Subsidence above mines of this type may take place many years later after the mine is abandoned. Unfortunately, buildings, or even towns, have been built unknowingly over abandoned underground mines. Subsidence over abandoned mines results in either *sinkholes* or *troughs*. Sinkholes (Figure 11-17) are circular pits formed by collapse of the surface soil into the mine voids when the coal seam is fairly shallow. Troughs are larger subsidence features that develop over deeper mines.

Natural cavities form in rock primarily by dissolution of the rock by circulating groundwater. Limestone caves are the most common example of solution cavities, although other soluble rock types behave similarly. Surface collapse, leading to the development of sinkholes, often occurs in dry periods when water tables are declining. Groundwater below the water table provides a measure of support for the sediment above a limestone cavern by the hydrostatic pressure exerted by the fluid. When the water table drops, this support is lost and surface collapse is initiated (Figure 11-18). The Winter Park, Florida, sinkhole (Figure 11-19), which formed in 1981, provides a spectacular example of urban disruption by subsidence over a cavity in soluble rock. The collapse occurred over a period of several days with no previous warning.

SUMMARY AND CONCLUSIONS

The term soil has different meanings to various groups of earth scientists and engineers. Soil scientists are concerned with material produced by pedogenic processes that can sustain plant growth. Soils engineers consider soil to be any transported sediment or residual material that can be excavated with hand tools.

Pedogenic processes acting upon rock or sediment produce recognizable horizons. The main horizons include the A, the zone of high organic-matter content and intensive leaching of mineral constituents, the B, the zone of accumulation of clay and metal oxides, and the C, a transition between highly altered and nonaltered parent material.

The development of horizons in the soil profile is controlled by the five soil-forming factors. Parent material and topographic position (relief) set the stage for the action of climate and organisms over a period of time. Of these factors, climate is dominant. The possible variations and interactions between the soil-forming factors can produce a vast range of soil types. As a result, pedogenic classification of soils has been difficult. A modern classification system based entirely on measurable soil properties has replaced an older, inadequate system. Because of the complexity of the new system, some terms and concepts from the older system are still in use.

Engineering projects must often be built in areas where hazardous soil processes are active. The tendency of clay soils to absorb water leads to volume expansion that can be very damaging to structures. Within the soil-moisture active zone, changes in water content of the soil will lead to swelling or shrinking. Land-use changes are often responsible for this behavior. Construction and landscaping practices that promote swelling include improper lot grading and excessive lawn watering. If highly expansive soils are present, special foundation designs may be necessary.

Hydrocompaction is a type of soil collapse that takes place when arid-region soils are saturated and loaded. Saturation destroys soil moisture or clay bonds that impart temporary strength to the soils. When these bonds are removed, the soil collapses to a more dense state.

Subsidence causes numerous engineering problems due to the collapse or sinking of the land surface. These phenomena originate by withdrawal of fluids, oxidation, or drainage of organic soils, and collapse of surface material into subsurface cavities.

REFERENCES AND SUGGESTIONS FOR FURTHER READING

BUOL, S. W., F. E. HOLE, and R. J. McCRACKEN. 1973. *Soil Genesis and Classification.* Ames, Ia: The Iowa State University Press.

BLOOM, A. L. 1978. *Geomorphology: A Systematic Analysis of Late Cenozoic Landforms.* Englewood Cliffs, N.J.: Prentice-Hall, Inc.

DAS, B. M. 1984. *Principles of Foundation Engineering.* Monterey, Calif.: Brooks/Cole Engineering Division.

HOLTZ, R. D., and KOVACS, W. D. 1981. *Introduction to Geotechnical Engineering.* Englewood Cliffs, N.J.: Prentice-Hall, Inc.

HOLZER, THOMAS L., ed. 1984. Man-induced land subsidence. *Geological Society of America Reviews in Engineering Geology,* vol. 6.

JUDSON, S., M. E. KAUFFMAN, and L. D. LEET. 1987. *Physical Geology,* 7th ed. Englewood Cliffs, N.J.: Prentice-Hall, Inc.

MATHEWSON, C. C., J. P. CASTLEBERRY, and R. L. LYTTON. 1975. Analysis and modeling of the performance of home foundations on expansive soils in central Texas. *Bulletin of the Association of Engineering Geologists* 12:275–302.

McCARTHY, D. F. 1977. *Essentials of Soil Mechanics and Foundations.* Reston, Va.: Reston Publishing Co., Inc.

TOURTELOT, H. A. 1974. Geologic origin and distribution of swelling clays. *Bulletin of the Association of Engineering Geologists* 11:259–275.

PROBLEMS

1. Why is there some confusion or disagreement in the use of the term "soil"?
2. Summarize the processes that lead to the formation of soil horizons.
3. Under what conditions do soils become mineral deposits?
4. Why is climate the most important soil-forming factor?
5. What are the characteristics of pedocals and pedalfers, and what factors control their formation?
6. Explain the process of soil swelling.
7. What construction techniques can be used to minimize swelling?
8. Why are noncohesive soils susceptible to hydrocompaction?
9. List all the mechanisms you can that cause land subsidence.

CHAPTER
12
GROUNDWATER

The importance of groundwater to our society has never been greater. There is no more fundamental natural resource than fresh water; we depend on it daily for drinking, sanitation, agriculture, industry, and recreation. In the past, most of our needs could be met by such easily accessible surface-water sources as rivers, lakes, and reservoirs. Those days have passed, however; the potential for expansion of surface-water sources is limited. Groundwater is our only alternative for obtaining large amounts of fresh water at a reasonable cost. In fact, by the year 2000 about half of the total water used in the United States will be groundwater.

The steadily increasing use of groundwater has raised important concerns about this resource. First, we must learn to evaluate and manage groundwater reservoirs so that they are not mined to exhaustion like ore deposits or other nonrenewable resources. Second, we have realized that groundwater is not a pristine substance that exists in total isolation from activities taking place at land surface. Instead, it is now known that past waste-disposal practices have gradually contaminated groundwater supplies in many locations. An extensive governmental and private effort is now underway to clean up the most severe contaminant sites and protect the vast majority of unaffected groundwater reservoirs.

The occurrence and movement of groundwater were once thought to be mysterious and unpredictable. Even today, misconceptions about groundwater are held by the general public, the media, and even some geologists. A thorough understanding of groundwater is critical for engineers because subsurface fluids are of major importance in civil, environmental, mining, petroleum, and other branches of engineering.

GROUNDWATER FLOW

Darcy's Law

A nineteenth-century French engineer named Henri Darcy laid the groundwork for the modern study of groundwater with experiments involving the flow of water through a column filled with sand. These experiments established that the volumetric flow rate of water through saturated sand was proportional to the energy gradient, or loss of energy per unit length of flow path. A schematic of Darcy's apparatus is illustrated in Figure 12-1. To express *Darcy's Law* in a commonly used manner, we will define the *specific discharge, v,* as the volumetric flow rate Q, measured in cubic meters per second or similar units, divided by the cross-sectional area of the flow tube, A. Thus

$$v = \frac{Q}{A}$$

Eq. 12-1

The form of energy involved in the flow process is the mechanical energy possessed by the fluid at each point in its flow path. This energy is the sum of three components: elevation, or position (potential energy); movement (kinetic energy); and pressure. When dealing with groundwater, we can neglect the kinetic energy contribution because of the very slow velocities of groundwater flow. The remaining components, elevation and pressure, need not be evaluated individually but, instead, can be measured together by the level to which water at any point in the flow system will rise above an arbitrary datum. This parameter, as indicated in Figure 12-1, is called *hydraulic head*. The thin tubes through which water rises to measure the head are called *manometers*.

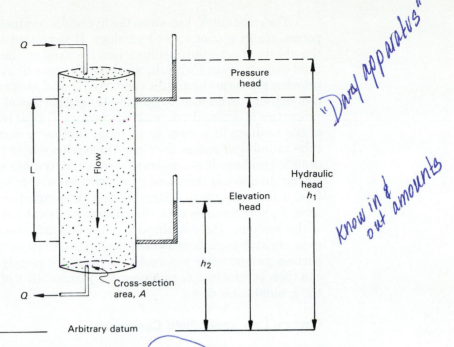

"Darcy apparatus"

Know in & out amounts

Figure 12-1 Schematic diagram of a lab apparatus that illustrates the parameters involved in Darcy's Law.

Also shown in the diagram are the two components of hydraulic head. Elevation head is the distance above the datum to the level of the manometer intake, and pressure head is the height of rise of water in the tube above the intake. Notice that the elevation above the datum, or head, in the manometers decreases in the direction of flow. This is a consequence of the loss of mechanical energy along the flow path as it is converted to heat through friction between the fluid and the sand grains and also because of friction between water molecules in the fluid. Darcy's Law requires evaluation of the *hydraulic* (energy) *gradient* and this is expressed as

$$I = \frac{h_1 - h_2}{L} \qquad\qquad \text{Eq. 12-2}$$

where I is the hydraulic gradient, h_1 and h_2 are head values at the points where the manometers are inserted into the flow system, and L is the distance between the manometers measured in the flow direction.

A final parameter is required for a complete expression of Darcy's Law, and that is the constant of proportionality between the specific discharge and the hydraulic gradient. With this term, K, Darcy's Law can be stated as

$$\frac{Q}{A} = v = -K\frac{h_2 - h_1}{L} \qquad\qquad \text{Eq. 12-3}$$

or, in differential form,

$$v = -K\frac{dh}{dl} \qquad\qquad \text{Eq. 12-4}$$

The minus sign simply indicates that groundwater flow is a mechanical process with an irreversible loss of mechanical energy, or head, in the direction of flow.

The constant K, known as the hydraulic conductivity, is a very important parameter in groundwater hydrology. It is related to intrinsic permeability, the ability of a porous medium, like the sand in the flow tube, to transmit a fluid under a given hydraulic gradient (Chapter 6). A distinction is often made between the terms hydraulic conductivity and permeability because hydraulic conductivity is defined to include the properties of the fluid as well as the properties of the medium, whereas permeability is restricted to the properties of the medium. It is easy to visualize that more water would flow through a tube containing gravel than one filled with silt or clay, under the same hydraulic gradient. It is also true, however, that more water would flow through a sand than would molasses, again with a constant hydraulic gradient. Therefore, it is important to remember that the proportionality constant in Darcy's Law, K, encompasses both the characteristics of the fluid and the properties of the medium. The combination of the two into one constant is convenient in groundwater work because the density and viscosity of water in most near-surface groundwater reservoirs do not vary greatly. Hydraulic conductivity can then be assumed to represent the permeability of the units through which the groundwater flows.

Darcy's Law under Field Conditions

One of the reasons Darcy's Law is so useful is that it can be directly applied to field situations. The device used to measure head in the field is called a *piezometer,* (Figure 12-2) and, as in a manometer, the head measurement depends on the level to which water in a tube rises above a datum. In this case, the datum used is sea level and the tube is a pipe inserted into a hole drilled into the ground, much like a normal water well. A piezometer differs from a well in several respects, however. A well usually contains a water intake section, or *screen*, that is as long as possible so that the yield of the well can be maximized. A piezometer, on the other hand, has a very short screened interval in order to obtain a head measurement at one specific point in a flow system. Also, piezometers are usually smaller in diameter than wells and should be carefully sealed with cement or clay above the screen to isolate the point of measurement from other parts of the groundwater flow system.

With the head measurements obtained from several piezometers, it is possible to calculate the hydraulic gradient. Head measurements in a piezometer are made by determining the depth to water from the top of the pipe with a tape and then subtracting the depth from the elevation of the top of the pipe. Figure 12-3 illustrates the similarity between groundwater flow through a subsurface rock unit and flow through the Darcy apparatus (Figure 12-1). The hydraulic gradient indicates the direction of groundwater flow. In many places, groundwater has a vertical as well as a horizontal component of flow. In these areas it is necessary to install a network of piezometers at various depths to determine the three-dimensional distribution of head in the flow system.

The Water Table

During drilling into the ground for the installation of a well or a piezometer, water is encountered under different physical conditions. The uppermost zone contains water in the pores of rock and soil along with a gas phase. The incomplete filling of void spaces by water gives rise to the name *unsaturated zone* for this part of the subsurface (Figure 12-4a). Below the unsaturated zone, at a depth determined by many factors, including climate, topography, and

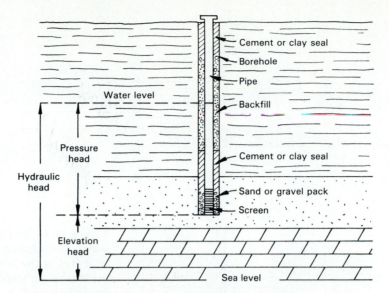

Figure 12-2 Design of a piezometer, the instrument used for measuring hydraulic head in the field.

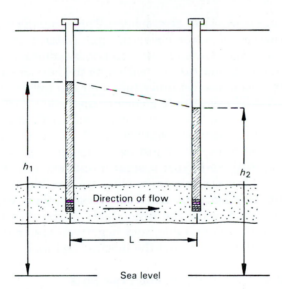

Figure 12-3 Groundwater flow through a bed of rock or sediment. The flow direction is indicated by a decrease in head from left to right. Hydraulic gradient can be determined in the same way as the Darcy experiment.

geologic setting, void spaces in the material are filled with water; this region is called the *saturated zone*. The saturated zone can be subdivided according to the value of fluid pressure in the pore water at a particular point. Just as fluid pressure in a lake increases from zero at the surface (atmospheric) to greater values with depth, the fluid pressure in the saturated zone increases with depth from the surface where it is equal to atmospheric pressure, a surface that is defined as the *water table*. The zone below the water table, which usually lies just beneath the top of the saturated zone, is known as the *phreatic zone*. The water table can be located by measuring the water-level elevation in a shallow well that just penetrates the water table.

The saturated material above the water table contains fluid in which the

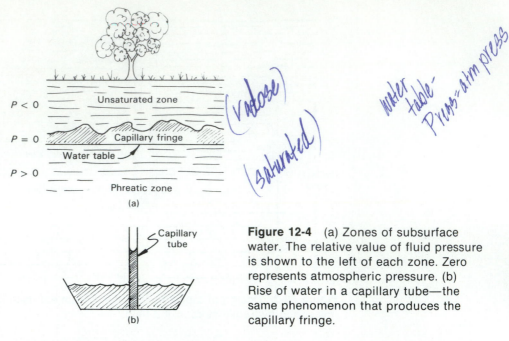

Figure 12-4 (a) Zones of subsurface water. The relative value of fluid pressure is shown to the left of each zone. Zero represents atmospheric pressure. (b) Rise of water in a capillary tube—the same phenomenon that produces the capillary fringe.

fluid pressure is less than atmospheric. This zone, the *capillary fringe*, is analogous to the rise of water above the free-water surface in a capillary tube (Figure 12-4b). The capillary rise is caused by adhesion between the water and the capillary tube and *surface tension* at the curved surface, or meniscus, of water at the top of the capillary tube. Surface tension is an upward-directed force that results from the tendency of the water surface to assume a shape of minimum area when in contact with another fluid (air) with which it does not mix. If the pores in the soil are small, as in clay and silt, capillary rise above the water table occurs just as in a single capillary tube. The capillary effect varies with soil type; a small-to-nonexistent rise occurs in gravel, whereas rises of several meters or more are possible in fine-grained soils with high silt and clay contents. The fluid pressure, then, will be positive (greater than atmospheric) below the water table in the phreatic zone, equal to atmospheric at the water table, and negative (less than atmospheric) in the saturated capillary fringe and the unsaturated zone.

Groundwater Flow Systems and Flow Nets

Figure 12-5a shows a cross section through an area of hilly topography. The position of the water table, high beneath the hills and near the surface in the valley, gives a good indication of the distribution of head. Groundwater will flow from the areas where head is highest, called *recharge areas*, to areas where head is lowest, called *discharge areas*. The discharge area in Figure 12-5a is a river channel. Groundwater emerges as springs and seeps along the sides and bottoms of the channel, thereby sustaining the flow of the stream during dry periods. The component of stream flow derived from groundwater is the *base flow*. Rivers that flow all year because of groundwater discharge are *perennial streams*, while streams that flow only during periods of rainfall that generate surface runoff are *ephemeral streams*.

Aside from recognizing that groundwater will somehow flow from the recharge areas to the discharge area in Figure 12-5a, we would find it useful to know exactly what paths the groundwater will follow in this flow system.

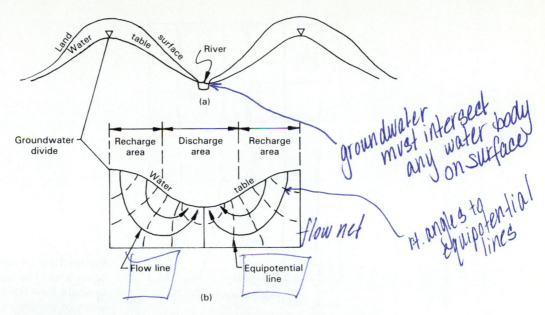

(handwritten annotations:) groundwater must intersect any water body on surface

flow net

rt. angles to equipotential lines

Figure 12-5 (a) A cross section showing the relationship of the water table to topography. (b) A flow net showing the position of flow lines and equipotential lines in the flow system.

If we make some simplifying assumptions, the problem can be solved mathematically and graphically. A solution requires specified boundary conditions for the region of flow. For this flow system we can specify *no-flow* boundaries, that is, boundaries normal to which there is no flow, at the bottom and sides of the flow region in Figure 12-5b. The left and right no-flow boundaries are considered to be imaginary because if the hills are symmetrical, a *groundwater divide* will develop near the crest of the hill and serve as a plane of symmetry. Groundwater will tend to flow away from this vertical plane as it moves toward discharge points in valleys on opposite sides of the hill. The lower no-flow boundary could represent an impermeable bed of rock at this depth in the flow system.

With the boundary conditions that we have established, we can determine the head at any point by developing and solving equations that describe flow in this system, using numerical methods that require computers to solve for the distribution of head, or by using graphical techniques. Once we know the distribution of head, we can draw contour lines called *equipotential lines* on the diagram as in Figure 12-5b. Equipotential lines connect points of equal head. The configuration of the equipotential lines will allow us to draw one other group of lines, *flow lines*. These lines indicate the paths that groundwater will follow in this flow system, the objective of this exercise. Thus the resulting *flow net* indicates that under these conditions, groundwater travels along long, curving paths from a recharge point to a discharge point. Flow nets are very important in predicting flow through earth dams and other permeable earth structures.

When groundwater flow in an area is predominantly horizontal, a flow net can also be drawn on a map, or plan view. The contours of head used in this case are simply contours of the water-table elevation (Figure 12-6). The flow lines, which show the direction of groundwater movement, are very useful in predicting the movement of pollutants that may be introduced into the flow system by a landfill or other waste-disposal site.

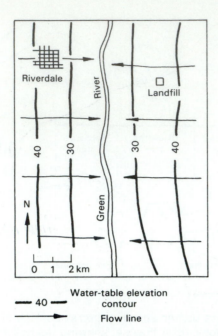

Figure 12-6 A flow net drawn on a map of an area. The water-table contours indicate flow from the east and west sides of the area toward a discharge area in the river valley.

Recharge and Discharge Processes

The input to the dynamic, circulating groundwater flow systems below the earth's surface is recharge in the form of rainfall and snowmelt that percolates downward from the surface to the water table. Without recharge, the water table would steadily drop, wells would go dry, and the great groundwater reservoirs of the earth would be depleted.

When rainfall hits the ground surface or snow melts, several possible pathways are available for water movement (Figure 12-7). If the water sinks

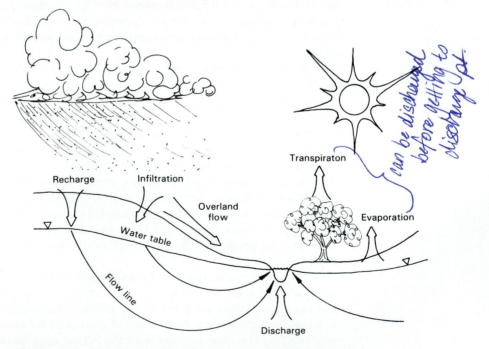

Figure 12-7 The interaction between atmospheric water and groundwater.

into the soil, *infiltration* has occurred. Soils differ greatly in the rate at which water is absorbed, which is called the *infiltration capacity*. The infiltration capacity is also influenced by such other factors as slope, type of vegetation, and the existing soil-moisture condition. As infiltration continues, the infiltration capacity decreases in a manner shown in Figure 12-8. The limiting value of the infiltration capacity is the hydraulic conductivity of the soil. After the infiltration capacity drops below the rainfall rate, water ponds on the surface and soon after begins to flow across the surface to lower elevations if the land is sloping. This process, *overland flow*, may supply sufficient water to stream channels to initiate a flood.

The infiltration of water into the ground does not necessarily mean that groundwater recharge will result, because there are two processes that can recycle water back to the atmosphere. Direct evaporation of the water from the soil is one mechanism. The other, *transpiration* (Figure 12-7), is the utilization of soil water and release of water vapor by plants. Some plants, including certain species of trees, have the ability to take up large amounts of water through their roots and then return a high percentage of it to the atmosphere through their leaves. Evaporation and transpiration are so difficult to separate quantitatively that they are frequently discussed together under the term *evapotranspiration*.

The amount of water available for groundwater recharge is, therefore, the fraction not lost by overland flow or evapotranspiration. Even part of the remaining water may be held in the unsaturated zone and not actually reach the water table. Some water must move downward to the water table for recharge to occur. In many areas recharge may take place only for short periods of the year. The rise in the water table during the spring and early summer is the most common indication of groundwater recharge. In some arid and semiarid areas this may be the only period of recharge in a typical year, aside from occasional heavy thunderstorms.

Groundwater discharge is the opposite of recharge; in discharge areas water moves from the saturated zone into the unsaturated zone and perhaps to land surface. If the near-surface sediments are homogeneous, discharge areas will occur in topographic lows where the water table intersects land surface, as in lakes and streams. The elevation of the stream or lake actually represents the position of the water table at that point. A surficial body of water is not required, however; sometimes discharge areas can be recognized by indications of a high water table and abundant evapotranspiration. Evidence for this type of discharge area includes persistent swampy conditions during all or most of the year, plants like willow and cottonwood that extend their roots below the water table to obtain moisture (*Phreatophytes*), and saline or alkaline soil conditions. Saline and alkaline soil conditions indicate evaporation of groundwater from the soil, with the dissolved salts carried by groundwater left behind in the soil. In some areas, particularly where the water table

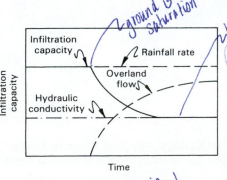

Figure 12-8 The decrease in infiltration capacity that occurs after a zone of saturation develops near the soil surface. The minimum value of infiltration capacity is hydraulic conductivity. (From R. A. Freeze and J. A. Cherry, *Groundwater*, copyright © 1979 by Prentice-Hall, Inc., Englewood Cliffs, N.J.)

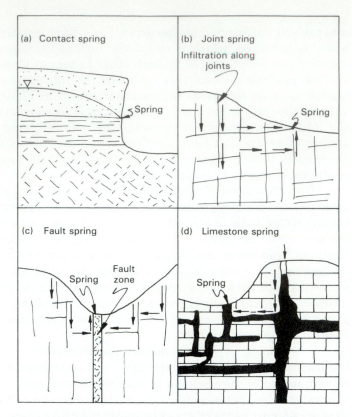

Figure 12-9 Some geologic settings in which springs occur.

Figure 12-10 Comal Springs, near San Antonio, Texas, discharges over 8000 L/s of water from a limestone aquifer system.

is deep, groundwater recharge, rather than discharge, occurs from topographic low points. In these situations most of the recharge is derived from ephemeral stream channels or lakes.

Springs, which are localized discharge points, represent a more obvious type of discharge. Springs can vary greatly in flow rate, ranging from barely a trickle to more than 10 m³/s. Some general types of springs are shown in Figure 12-9. In stratigraphic sequences with alternating permeable and non-permeable beds, contact springs can develop. Joints, faults, and fractures often provide conduits for the upward movement of groundwater to the surface. The high temperature and highly mineralized chemical composition of some springs attest to the depths to which some of these conduits extend. Some of the most spectacular springs are located in areas of soluble bedrock like limestone. Groundwater chemically reacts with these rocks and, over thousands of years, can dissolve large amounts of rock, forming subsurface caves and passages. Tremendous amounts of water may issue from springs in limestone terrains (Figure 12-10).

GROUNDWATER RESOURCES

The most obvious reason for studying groundwater is its value as a resource. In some parts of the United States groundwater constitutes the only water supply. Surface water, the alternative source in other areas, must be extensively treated before drinking and is quite often limited in quantity. Increases in demand for water in the United States will have to be met predominantly by expanding utilization of groundwater.

Aquifers

Economically significant quantities of groundwater can be obtained from bodies of sediment or rock called *aquifers*. The most important property of an aquifer is its hydraulic conductivity, for it is the ability of an aquifer to transmit water to a pumping well for replacement of the pumped water that mainly determines the yield of the aquifer. Geologic materials that constitute productive aquifers include sand and gravel, highly fractured rocks, and soluble rocks containing large subsurface openings for groundwater movement and storage.

Materials that do not transmit economically significant quantities of groundwater are known as *aquitards*. Aquitards are usually composed of dense rock units or sedimentary beds of silt and clay. Sedimentary rock terrains often consist of alternating aquifers and aquitards. Exploration for groundwater in these areas involves location of the permeable rock horizons.

Aquifer type is determined by the presence or absence of an overlying aquitard. Aquifers that lack overlying aquitards and have the water table as their upper boundary are *unconfined aquifers*. These often occur in surficial deposits of sand and gravel. Because the water table is the upper boundary, contour lines of water-table elevation drawn on a map indicate the direction of flow of groundwater in an unconfined aquifer. As shown in Figure 12-11, the slope of the water table is a close approximation of the hydraulic gradient, providing the flow in the aquifer is nearly horizontal. Similarly, the elevation of the water table is closely equivalent to the hydraulic head in an unconfined aquifer.

Confined aquifers are bounded above and below by aquitards. As a result of this confinement, there are important differences between the flow of

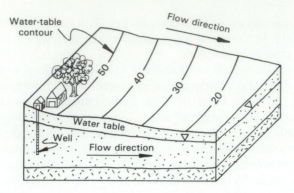

Figure 12-11   An unconfined aquifer showing water-table contours, which are elevations above sea level of the water table at any point.

groundwater in confined aquifers and the flow in unconfined aquifers. The water level, or head, in a well in a confined aquifer may be above or below the water table; in fact, there is no specified relationship between the head in a confined aquifer and the water table. This distinction between the two aquifer types is shown in Figure 12-12.

   If the heads in a confined aquifer are contoured on a map, as previously shown for the water table in unconfined aquifers, the contours will define an imaginary surface called the *potentiometric surface* (Figure 12-12). The potentiometric-surface map is similar to the water-table contour map in that it indicates the distribution of head and therefore the direction of water movement in a confined aquifer. The major difference between the potentiometric surface and the water table is that the potentiometric surface is not an actual

Piezometric

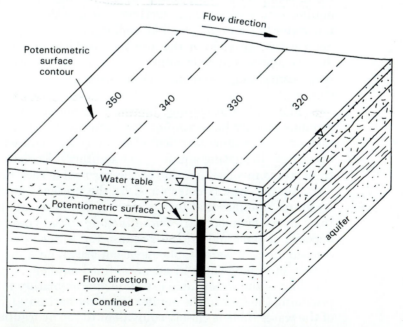

Figure 12-12   A confined aquifer, in which the water level in a well rises to the potentiometric surface. Contours of the potentiometric surface are shown on the land surface.

surface in the ground analagous to the water table; it is a hypothetical level defined by the elevations of water levels in wells in confined aquifers.

When the head in a confined aquifer rises above the top of the aquifer, it is known as an *artesian aquifer*. The fluid pressure in some artesian aquifers may be so high that the head, or potentiometric surface, is above land surface. A well completed in this type of aquifer will flow without the aid of a pump; these wells are known as *flowing artesian wells*. Flowing wells are not unlike the plumbing system in a house in that the water will flow out under its own pressure as long as the faucet is left open.

Geologic and topographic conditions combine to produce confined aquifers. In Figure 12-13 a classic example of a confined aquifer is illustrated. The aquifer in this case is exposed at land surface at the left end of the cross section, where recharge occurs. To the right, the aquifer dips downward beneath a thick aquitard. The potentiometric surface shown above the aquifer indicates that just to the right of the recharge area, the aquifer develops confined conditions. The surface topography of the area determines where flowing artesian conditions will occur; as land surface slopes to the right more rapidly than the potentiometric surface, flowing conditions are common. Figure 12-13 is actually a simplified version of the regional flow system in the Great Plains of the United States. The confined aquifer system, the Dakota Sandstone of Cretaceous age, crops out in the Black Hills region of western South Dakota. When wells were first drilled into the Dakota Sandstone during settlement of the eastern Dakotas, heads in the aquifer were as much as 100 m above land surface in places. Heads declined over the years as flowing wells discharged water from the aquifer.

A final aquifer type is the *perched aquifer*. Beds of material with low hydraulic conductivity in the unsaturated zone lead to establishment of perched conditions (Figure 12-14). The downward flow of water through the unsaturated zone toward the regional water table is retarded by the clay unit. A localized water table and a perched aquifer then develop above the clay. Perched aquifers cannot usually sustain high well yields because of their limited areal extent.

Production of Water from Aquifers

Unlike piezometers, wells are designed to maximize the yield of aquifers. Larger diameters and greater lengths of screen are therefore utilized. If a well penetrating an unconfined aquifer is limited to that aquifer, and if vertical

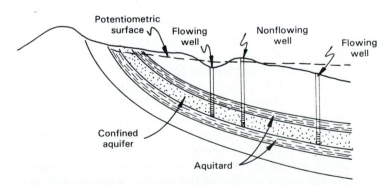

Figure 12-13 A geologic and topographic configuration that produces flowing wells wherever the potentiometric surface is above the land surface.

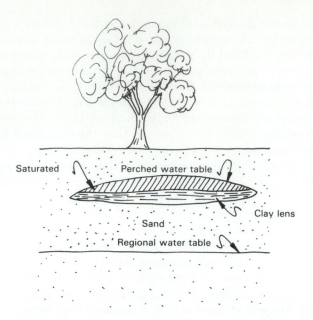

Figure 12-14 Development of a perched aquifer above a low-permeability unit that occurs within the unsaturated zone of a unit of higher permeability.

components of flow in the aquifer are not significant, the *static*, or unpumped water level in the well may be a close approximation to the water table. Similarly, if a well in a confined aquifer is isolated from groundwater in overlying and underlying rock units, the water level in the well may be a good indication of the potentiometric surface in the aquifer. But if the well is pumped changes in the hydraulic regime of the aquifer occur.

When an unconfined aquifer is pumped, the response to this stress is a decline of the water table in the vicinity of the well. These effects are usually assumed to be equal in all directions, so the area of influence of pumping is a cone-shaped region with greatest decline in water table adjacent to the well (Figure 12-15). The affected region changes from a saturated to an unsaturated state during the drop in the water table. Because of its shape it is known as a *cone of depression*. As pumping continues, the cone expands radially outward from the well. The actual shape and dimensions of the cone depend on the hydraulic conductivity, thickness, and storage properties of the aquifer. If the pumping rate is not excessive, the cone of depression will eventually reach an equilibrium position. If pumping is excessive, the water table will continue to drop, causing increased pumping costs. The situation in which groundwater withdrawals exceed recharge is referred to as "groundwater mining." The implications of this situation are discussed in Case History 12-1.

In a confined aquifer the effects of pumping are somewhat different. Here the cone of depression develops in the potentiometric surface without any dewatering of the aquifer (Figure 12-16). To explain the response of the aquifer to the lowering of the potentiometric surface we must invoke the *Principle of Effective Stress*, a concept that has many applications in geology. Referring to Figure 12-17, consider the distribution of stress in a confined aquifer. The total weight of rock, soil, and water above the aquifer exerted per unit area on the upper surface of the aquifer is designated as the *total stress* (σ_T). This stress is opposed in the aquifer by two stresses, the *fluid pressure* (P) and a stress transmitted through the solid grains of the aquifer, the *effective stress* (σ_e). The equilibrium relationship defined by this group of stresses can be

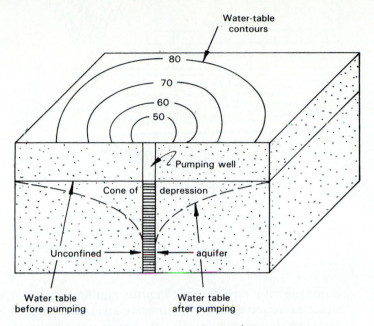

Figure 12-15 Development of a cone of depression in an unconfined aquifer by lowering of the water table.

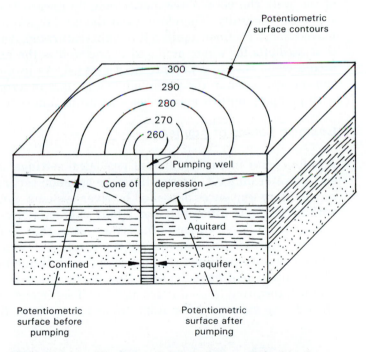

Figure 12-16 The cone of depression produced in the potentiometric surface by pumping a confined aquifer.

expressed as the effective stress equation,

$$\sigma_T = \sigma_e + P \qquad \text{Eq. 12-5}$$

Despite its simplicity, this equation is of utmost importance in many areas of geology and engineering. When a confined aquifer is pumped and a cone of depression develops in the potentiometric surface, the fluid pressure is reduced.

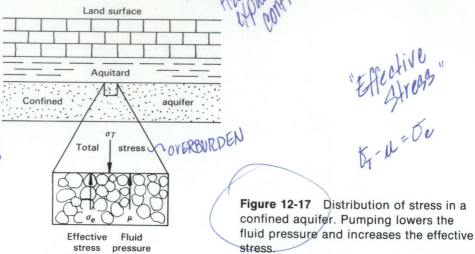

*(compressing)
dec. press.
"inc. σe*

*Remove/Add
H2O =
expand or
contract*

*Moisture content =
unstable
slopes*

*"Effective
stress"

σT - u = σe*

Figure 12-17 Distribution of stress in a confined aquifer. Pumping lowers the fluid pressure and increases the effective stress.

Since the total stress exerted on the aquifer is unchanged, the effective stress equation requires that the effective stress increase. When the grain-to-grain pressure in a material increases, the material tends to compact due to closer packing of the particles. This reduction in porosity is not possible unless water within the pores is expelled. Since the head has been lowered in the vicinity of the well, the pore water moves radially toward the well and the aquifer compacts vertically. Therefore, even though large amounts of water are removed from a confined aquifer, it remains saturated. An additional mechanism that accounts for water produced in the well is the expansion of water that occurs when the fluid pressure is decreased. As in unconfined aquifers, the cone of depression in the potentiometric surface in a confined aquifer expands with pumping until it approaches an equilibrium condition.

Geologic Setting of Aquifers

Aquifers occur in vastly different geological settings, making exploration for groundwater difficult. Among the more common settings for aquifers are the sedimentary sequences deposited by river systems. Such *alluvial aquifers* lie within valleys of different types. Many characteristics of the aquifer can be explained by classification of the river system that deposited the aquifer sediments as *braided* or *meandering* (Figure 12-18). Braided segments flow in numerous interconnected channel segments within a broad, shallow channel. The channel sediment deposited by these rivers is predominantly coarse grained, resulting in productive aquifers if sufficient thicknesses are present. *Meandering* streams differ from braided streams in that they deposit large

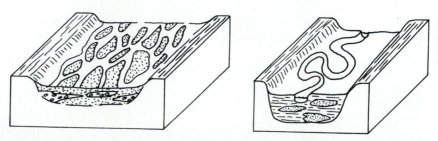

Figure 12-18 Aquifers produced by braided (left) and meandering (right) river systems.

quantities of *overbank* or *flood-plain sediment* on flat surfaces (flood plains) adjacent to the sinuous channels. Unlike channel sediment, overbank sediment does not have good aquifer potential because it is fine grained. Thus an aquifer in alluvial sediments of a meandering river system will consist of narrow, sinuous bodies of channel sediment surrounded by less permeable overbank sediment; however, braided-river deposits may underlie the meandering-river sediment in some places.

In regions where structural deformation has produced alternating mountain ranges and valleys by movement along faults, alluvium of *tectonic-valley aquifers* is present. The alluvium is deposited by streams eroding the rising mountain blocks. A generalized tectonic-valley setting, similar to those in the Basin-and-Range province of the western United States, is illustrated in Figure 12-19. In arid regions tectonic-valley aquifers may provide an excellent supply of groundwater, although development of the aquifers must be carefully managed because of the low rate of recharge.

Although predominantly of alluvial origin, *glacial aquifers* also include shoreline and deltaic deposits of glacial lakes. Rivers carrying meltwater from glaciers are generally braided, high-discharge streams that flow on top of and within glaciers as well as away from glacier margins. Sorted alluvial sediment deposited by streams flowing away from glacial margins, called *outwash*, may be laid down over broad, gently sloping plains known as *outwash plains* or confined to valleys that lead away from the glaciers called *valley trains* (Figure 12-20).

Some glacial aquifers consist of valleys filled with alluvium buried by glacial sediment (till) deposited during a later glacial advance. These *buried valleys*, as shown in Figure 12-21, form some of the most important aquifers in glaciated regions.

Other types of aquifers existing in unconsolidated sediment include *coastal-plain* aquifers (Figure 12-22), which are often composed of permeable beach sediment deposited when sea level was higher at some time in the geologic past. Coastal-plain aquifers are characterized by a zone of interface between fresh groundwater and saline groundwater derived from the ocean. When the aquifers are developed by pumping, the lowering of the water table allows landward encroachment of the saline groundwater. Increasing salinity

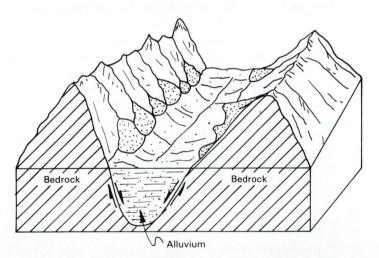

Figure 12-19 A tectonic-valley aquifer formed by deposition of alluvium in a down-faulted basin bounded by uplifted mountain blocks.

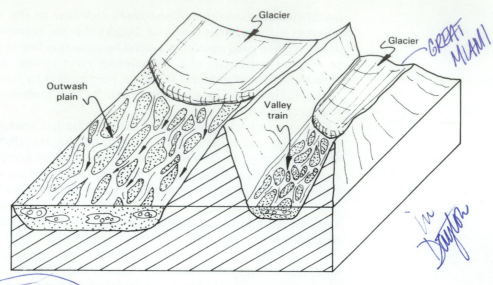

Great Miami

Dayton

Figure 12-20 Deposits of glacial-meltwater streams that form productive aquifers.

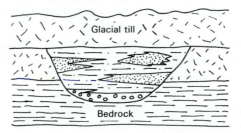

Great Little Miami

Figure 12-21 A buried-valley aquifer formed where a valley filled with alluvium is covered by glacial sediment (till) from a later glacial advance.

of drinking water can therefore be a serious problem for coastal cities that depend upon coastal-plain aquifers for their water supply.

Aquifers also develop in many types of rock. Sedimentary rock formations are among the most important rock aquifers in many parts of the world. Sandstone and conglomerate rock units usually have high hydraulic conductivities and therefore constitute productive aquifers unless the rock is highly cemented, so that its potential as an aquifer is decreased. We have already mentioned the great confined sandstone aquifer system that underlies the Great Plains.

Carbonate rocks—limestone and dolomite—also contain high-yielding

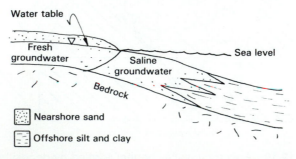

Figure 12-22 A coastal-plain aquifer showing the boundary between fresh and saline groundwater.

aquifers in many places. The suitability of these rocks as aquifers is determined by the amount of dissolution that has taken place during groundwater circulation. Cavernous limestones can yield phenomenal amounts of water if they occur below the water table.

Many rocks either have a low initial hydraulic conductivity because of their fine grain size or have a greatly reduced porosity because of cementing. These rocks may also function as aquifers, however, if they are fractured. Even shales can yield high quantities of groundwater if the network of joints, faults, and other fractures is dense, closely spaced, and interconnected. Coals sometimes constitute aquifers through development of such *fracture permeability*. Fractures are also important in carbonate rocks, because it is along these cracks that water circulation and, therefore, dissolution of the soluble calcium carbonate proceeds. A fractured rock aquifer is shown in Figure 12-23. One important aspect of groundwater production from fractured rocks is the problem of intersecting the fractures with a well screen. Well yields vary greatly in fractured rocks, depending on the number, size, and orientation of fractures in contact with the screen.

Igneous and metamorphic rocks, with several exceptions, are dense, interlocking aggregates of crystals with low initial porosity and permeability. Fracturing is the only mechanism that allows significant production from wells. Among the exceptions to this generalization are lava flows with vesicular texture; these rocks may be very porous and highly permeable. Because vesicular zones often occur near the tops of lava flows (Figure 12-24), groundwater exploration is commonly focused on contacts between flows. A sequence

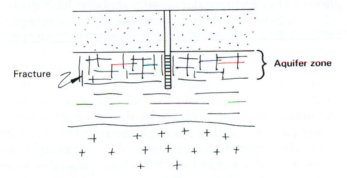

Figure 12-23   A fractured-rock aquifer yielding water to wells from a fractured zone in a shale rock unit.

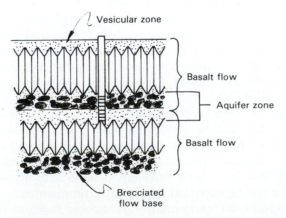

Figure 12-24   An aquifer present at the contact between two basalt flows.

of lava flows may therefore contain aquifers at the flow contacts separated by aquitards formed by the main body of the flows.

GROUNDWATER CONTAMINATION

Natural Groundwater Quality

Quality, or chemical composition, of groundwater is as important as quantity in terms of groundwater supply. The most productive aquifer in the world will be useless if the water is highly mineralized or contains certain chemical components that make the water unsafe for drinking.

Groundwater almost always contains a higher dissolved mineral content than surface water. The dissolved mineral content of groundwater, reported as *total dissolved solids* (TDS), is derived from chemical reactions between the water and soil and rock along its flow path from the point of infiltration to the point of sampling. All minerals and amorphous solids are soluble to some extent in water, despite great variations in the degree of solubility and the rates of dissolution. Minerals that originally precipitated from water, such as halite, gypsum, calcite, and dolomite, are most easily dissolved by circulating groundwater. The cementing agents calcium cabonate and silica that originally precipitated from groundwater can return to solution as conditions change through geologic time. A generally observed relationship is that TDS increases with residence time in the groundwater flow system. Therefore, deep, slow-moving groundwater usually contains high dissolved-solids content.

We have already considered one of the most important components of the groundwater chemical system, the weathering reactions that characterize the soil zone (Chapter 10). There, the CO_2-charged infiltrating water dissolves some minerals and reacts with others to produce clays. In the process, silica, bicarbonate, sulfate, and cations may reach the water table during recharge. In the aquifer additional chemical processes may modify the composition of the initial recharge. Ultimately, the chemical concentration of groundwater at any point in the flow system is the result of the type of solid materials that it has encountered along the way. A water sample taken from an aquifer in limestone and dolomite rocks will contain high concentrations of calcium, magnesium, and bicarbonate. Waters in contact with gypsum will be rich in sulfate. Sedimentary basins containing highly soluble evaporite rocks will yield groundwater with very high concentrations of sodium and chloride.

The quality of groundwater from a particular aquifer can be assessed by comparing the sample to *water quality standards*. Chemical components of groundwater can be divided into major, minor, and trace constituents by concentration as shown in Table 12-1. Standards for some of these parameters and other organic, radioactive, and biological components have been established for various purposes, for example, use as drinking water. These levels are reported in Table 12-2, where they are divided into *recommended limits* and *maximum permissible concentrations*. The recommended limits are assumed to cause no severe health effects, even if exceeded; the maximum permissible concentrations, on the other hand, are established for concentrations of more hazardous constituents. The maximum permissible concentrations are the levels that have been determined to be safe by various health agencies and should not be exceeded under any circumstances.

An additional water-quality parameter familiar to many well owners is *hardness*. The concentration of polyvalent metallic cations—calcium and magnesium for the most part—determines the hardness of water. Hard water,

TABLE 12-1
Classification of Dissolved Inorganic Constituents in Groundwater

Major constituents (>5 mg/L)	
Bicarbonate	Silica
Calcium	Sodium
Chloride	Strontium
Magnesium	
Minor constituents (0.01–10.0 mg/L)	
Boron	Nitrate
Carbonate	Potassium
Fluoride	Strontium
Iron	
Trace constituents (<0.1 mg/L)	
Aluminum	Nickel
Antimony	Niobium
Arsenic	Phosphate
Barium	Platinum
Beryllium	Radium
Bismuth	Rubidium
Bromide	Ruthenium
Cadmium	Scandium
Cobalt	Selenium
Copper	Silver
Gallium	Thallium
Germanium	Thorium
Gold	Tin
Indium	Titanium
Iodide	Tungsten
Lanthanum	Uranium
Lead	Vanadium
Lithium	Ytterbium
Manganese	Zinc
Molybdenum	Zirconium

*Natural make-up of groundwater

Source: From R. A. Freeze and J. A. Cherry, *Groundwater*, copyright © 1979 by Prentice-Hall, Inc., Englewood Cliffs, N.J.

whereas it is not particularly harmful to drink, produces several undesirable effects, such as scale on plumbing fixtures and boilers, and lack of soap suds. Iron is another parameter of concern for reasons other than health. Iron concentrations above 0.3 mg/L will produce stains and solid precipitates on plumbing and clothing.

Generation and Movement of Contaminants

In recent years we have become increasingly aware of changes in groundwater quality resulting from human activities including disposal of wastes, agricultural activities, spills, and leaks. Among the most serious of the problems discovered was the Love Canal incident at Niagara Falls, New York. Migration of toxic chemicals buried in an old canal eventually forced the relocation of 238 families from houses built over the old dump site. In one respect, groundwater contamination is more troublesome than surface-water contamination because of the very slow rate of movement of groundwater. Groundwater velocities, which can range from less than 1 m per year to several tens of meters per year, indicate that contaminants may remain in groundwater flow systems for hundreds or thousands of years before reaching discharge points. Efforts

TABLE 12-2
Selected Drinking Water Standards

Constituent	Recommended concentration limit (mg/L)
Inorganic	
Total dissolved solids	500
Chloride (Cl$^-$)	250
Sulfate (SO$_4^{2-}$)	250
Nitrate (NO$_3^-$)	10 (as N)
Iron (Fe)	0.3
Manganese (Mn)	0.05
	Maximum permissible concentration
Arsenic (As)	0.05
Barium (Ba)	1.0
Cadmium (Cd)	0.01
Chromium (Cr^{6+})	0.05
Selenium (Se)	0.01
Lead (Pb)	0.05
Mercury (Hg)	0.002
Fluoride (F)	1.4–2.4
Organic	
Cyanide	0.06
Endrin	0.0002
2,4-D	0.1
2,4,5-TP silvex	0.01
Phenols	0.001
Synthetic detergents	0.5

SOURCE: From U.S. Environmental Protection Agency, 1975, Water Programs: national interim primary drinking water regulations, *Federal Register* 40 (no. 240); World Health Organization, European Standards, 1970.

to decontaminate aquifers are in their infancy and are proving to be very expensive.

Movement of contaminants in the subsurface can be separated into processes operating in the unsaturated zone and those occurring in the saturated zone. In the unsaturated zone, soluble chemicals are carried by water percolating downward toward the water table. Soluble constituents dissolved or leached from waste materials on or below the ground surface form a degraded fluid called *leachate*. The amount of time required for leachate to reach the water table depends on the depth to the water table and the rate of movement in the unsaturated zone. Figure 12-25 shows the downward movement of leachate formed by water percolating through the waste materials of a *sanitary landfill*.

Once contaminants reach the water table, they travel in the direction of groundwater flow as a coherent zone of degraded groundwater known as a *leachate plume*. The direction of movement may be horizontal, vertical, or inclined, depending on the gradient of the flow system. A classic example of leachate movement is shown in Case Study 12-2.

Contaminants move in aquifers by two predominant mechanisms—*advection* and *hydrodynamic dispersion*. Advection is the transport of contaminants by the bulk movement of the flowing water. The rate of movement of contaminants can be estimated by determining the *average linear velocity* of the groundwater. This velocity is described as average because of the nature of flow in porous media. Variation in size of the microscopic flow channels followed by individual particles of water leads to a difference in velocity, de-

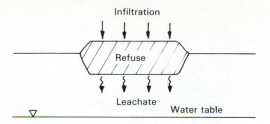

Figure 12-25 Generation of leachate and downward movement to the water table.

pending on which particular path a particle of water follows. Since it is not possible to determine the infinite variety of possible flow paths, an average velocity is used. When an average linear velocity is reported, it must be kept in mind that some groundwater, and therefore contaminants, will be traveling along certain flow paths at higher velocities. The average linear velocity can be calculated by the formula

$$\bar{v} = \frac{k}{n} \frac{dh}{dl}$$

Eq. 12-6

where $\bar{v}$ is average linear velocity, K is hydraulic conductivity, n is the porosity of the aquifer material, and dh/dl is the hydraulic gradient.

Hydrodynamic dispersion results in the spreading of the zone of contamination along the flow path. The random microscopic flow channels in the aquifer material cause some contaminant particles to move away from the center of mass of the contaminant plume both laterally and in the direction of flow, thus enlarging the volume of aquifer containing some portion of the contaminant. Despite the spreading effect of hydrodynamic dispersion shown in Figure 12-26, the influence is much less than the mixing caused by turbulent flow in a river. It is correct to think of leachate masses moving in compact units through aquifers, modified only by minor spreading caused by hydrodynamic dispersion.

Because dispersion spreads a fixed mass of contaminant over a larger volume of aquifer, a reduction in contaminant concentration occurs. If the particular contaminant is very toxic at low concentrations, the dispersive tendency will have an undesirable result because more area is affected even though the concentration is decreased.

The ultimate path of contaminants in groundwater can be predicted only by knowing the nature of the flow system. Water-table contour maps and potentiometric-surface contour maps will give an indication of the horizontal component of movement. The importance of vertical components is illustrated in Figure 12-27, where a contaminant source is added to the flow system shown in Figure 12-5. Well A will not be contaminated because it is too shallow, even though it is in the direction of flow of groundwater from the waste-disposal site.

Figure 12-26 The expansion of a contaminant in an aquifer by hydrodynamic dispersion. (From R. A. Freeze and J. A. Cherry, *Groundwater*, copyright © 1979 by Prentice-Hall, Inc., Englewood Cliffs, N.J.)

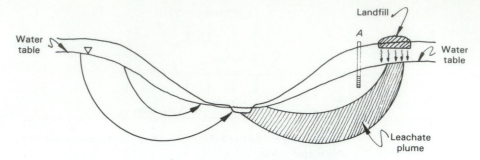

Figure 12-27 Movement of a leachate plume in a regional aquifer. A well at *A* would not be contaminated because of the vertical component of groundwater movement in the flow system.

Contamination of a well can occur even when the natural gradient suggests that the well will be safe. A pumping well located upgradient of a disposal site, like the one shown in Figure 12-28, can become contaminated because the cone of depression it creates reverses the natural gradient and induces movement of the contaminated water toward the well.

Attenuation Mechanisms

As dissolved contaminants move through the unsaturated and saturated zones, a number of physical, chemical, and biological processes may alter the concentration of the solute. Chemical reactions, including acid-base, and oxidation-reduction reactions can change the chemical species of the contaminant and, in doing so, alter its mobility.

Some important processes that decrease the concentration of contaminants in the subsurface, or *attenuation mechanisms*, are listed in Table 12-3. One of these, *adsorption*, is particularly important. Adsorption, the attraction of an ion to the surface of a solid particle in the aquifer, is really a form of *ion exchange*. In order to maintain an overall neutral charge, the particles of clay and other substances that attract ions must release ions of the same charge. The most common reaction involves the exchange of cations. Contaminant cations are drawn to negatively charged particle surfaces in exchange for other cations released from the particle surface to the groundwater solution. Clays have the strongest ability to effect this exchange reaction, although other

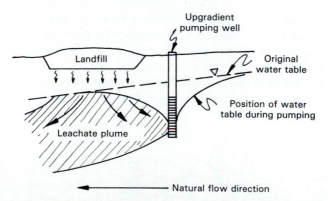

Figure 12-28 Contamination of an upgradient pumping well due to the gradient reversal caused by pumping.

TABLE 12-3

Attenuation Mechanisms

Hydrodynamic dispersion
Adsorption-desorption reactions
Acid-base reactions
Solution-precipitation reactions
Oxidation-reduction reactions
Microbial cell synthesis
Filtering, die-off, and adsorption of microorganisms
Radioactive decay

material, including iron oxide coatings or particles, may also participate in exchange processes. Contaminants may also be removed from solution by chemical precipitation, radioactive decay, and die-off of microorganisms. Hydrodynamic dispersion is also an attenuation mechanism because spreading of the contaminant reduces its concentration.

Waste Disposal

Disposal of wastes on and below land surface is the most significant cause of groundwater contamination. Waste categories include residential/municipal, human and animal, industrial, and natural resources development. A breakdown of waste-disposal methods is provided in Table 12-4.

TABLE 12-4

Sources of Waste and Waste-Disposal Methods

Domestic and municipal solid wastes
  Landfill
Domestic and municipal liquid waste
  Septic tank
  Cesspool
  Waste-stabilization lagoon
  Municipal treatment plant
  Land spreading
Industrial wastes (solid)
  Landfill
  Land spreading (sludges)
Industrial wastes (liquid)
  Lagoon or pit
  Disposal well
  Industrial treatment plant
Feedlot wastes
  Surface impoundment (lagoon)
Mining wastes
  Lagoon or tailings pond
  Tailings pile
Oil and gas field wastes
  Brine pond
  Drilling fluid pit
  Land spreading
  Injection well
Low-level radioactive wastes
  Landfill
High-level radioactive wastes
  Deep burial (proposed)

Ranked by amount of waste, the mining industry produces the largest quantity. This includes waste rock from mining and processing of ores, overburden materials removed in surface or strip-mining activities, and contaminated mine drainage. Most mining-waste problems are caused by bringing subsurface materials into contact with a different geochemical environment at the surface. The resulting chemical reactions release some of the chemical constituents contained in these wastes into shallow aquifers used for water supply.

Land disposal of solid waste includes residential, industrial, and some nuclear-waste sources. Residential waste alone amounts to 3.5 lb (1.6 kg) of waste produced per person per day in the United States. Location and design of safe sites for the burial of this waste is a staggering task. Various types of landfills are the most common method utilized for disposal of solid waste. These range from unregulated (and now illegal) dumps to carefully designed *secured landfills*. Secured landfills, as shown in Figure 12-29, prevent any contact between waste materials and natural materials at the site by the use of a liner composed of synthetic materials or compacted clay and a system of drains above the liner to collect and remove any leachate that may be generated. Groundwater samples from *monitoring wells* installed around the landfill are used to verify that the system is performing satisfactorily or to provide an early warning that the containment system has failed in some way. The most common type of landfill is a *sanitary landfill* (Figure 12-30). Sanitary landfills minimize groundwater contamination and other environmental problems by careful site selection and isolation of the waste in cells surrounded by compacted clay soils that will retard and attenuate waste migration. Favorable geologic settings for sanitary landfills are characterized by low-permeability materials and lack of aquifers beneath the site. Landfills are constructed in several ways. In the *trench method* (Figure 12-30a) trenches are excavated and progressively filled with compacted refuse. After each day's amount of refuse is added, suitable *cover soil* is placed over the refuse to form isolated cells of refuse. More cover soil is added to the surface after a trench is filled. Cover soil must have relatively low hydraulic conductivity to minimize infiltration into the waste. Ideally, soil excavated from the trenches can be used as cover soil. Finally, topsoil is added to promote revegetation of the completed landfill.

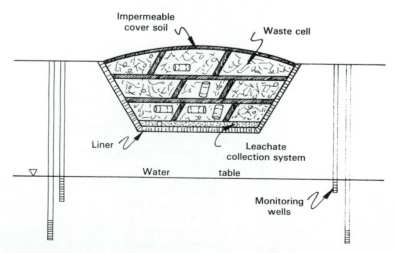

Figure 12-29 The design of a secured landfill utilized for hazardous wastes in which waste materials are totally isolated from the subsurface and surface environments.

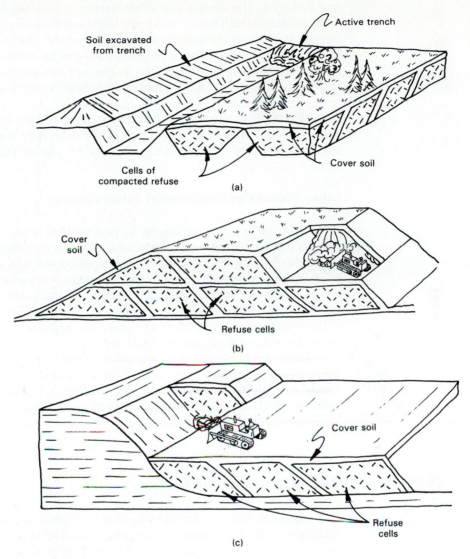

Figure 12-30 Types of sanitary landfills include (a) the trench method, (b) the area method, and (c) the slope method.

Landfills are also constructed above land surface. In the *area method* (Figure 12-30b) the completed landfill consists of a low mound of refuse cells. When a natural hillside is present at the site, the *slope method* (Figure 12-30c) can be used. Cover soil is often obtained from the hillside prior to the construction of the landfill.

Human and animal waste products that reach groundwater from sewage lagoons, septic tanks, animal feedlots, and other sources have contributed high levels of *nitrate* (NO_3^-) to aquifers in some areas. Nitrate levels above the recommended concentration limits can be harmful to health, particularly in infants.

Industrial wastewater impoundments contribute direct seepage of liquids to the subsurface when improperly designed. Although waste volumes are low compared to other waste types, the toxicity of these wastes, as in the case of the plating waste site on Long Island (Case Study 12-2) can be extremely high.

One of the most important scientific and engineering challenges facing our society is the design and construction of safe high-level nuclear-waste–disposal facilities. These wastes are highly radioactive and will remain so for hundreds or even thousands of years. Presently, the wastes are being stored at nuclear reactor sites until disposal sites can be constructed. The most likely solution appears to be deep burial in zones of low-permeability rock in tectonically inactive areas. Rock types under consideration include crystalline igneous and metamorphic rocks, volcanics, shales, and evaporite deposits. Thick beds of salt are particularly attractive because of their low strength and plastic deformational characteristics (Case Study 4-1). Any fractures or openings in salt deposits tend to be sealed naturally by plastic flow of the surrounding salt.

Other Sources of Groundwater Contamination

In recent years we have begun to realize that some seemingly innocuous human activities can lead to groundwater contamination. One such example is provided by agriculture. Fertilizers applied to crops contain nitrogen compounds that can be converted to nitrate in the subsurface. Even more serious, perhaps, are the synthetic organic compounds present in pesticides and herbicides. Little is known about the migration and possible attenuation of some of these toxic chemicals. More and more examples of their presence in groundwater have been documented, however.

Other human-related activities also lead to groundwater contamination. Application of salts to roads for deicing purposes is one example. In addition, accidental spills and leaks from trucks, railroad cars, storage tanks above and below ground, and pipelines have caused serious groundwater contamination in many areas.

household dumping

GROUNDWATER AND CONSTRUCTION

In contrast to hydrogeologists who devote their careers to finding and developing groundwater resources, engineers involved in construction of dams, tunnels, buildings, and highways spend much of their time trying to control or dispose of groundwater.

Dams illustrate the problem of groundwater movement, or seepage, quite well. Dams can be designed to prevent internal seepage, as in masonry or arch dams, or, as in the case of earth dams, to allow controlled seepage through the embankment at a safe rate. In the former case, excessive seepage beneath or around the dam is the major concern. Because the impoundment of the reservoir behind the dam increases the head within the groundwater flow system, extremely high fluid pressures can develop in the rocks or sediments under the dam. Excessive fluid pressures decrease the stability of the dam by generating uplift pressures on the base of the structure or by causing erosion of material near the downstream toe of the dam, thus tending to undermine it. Most dam failures have resulted from inadequate control of groundwater seepage through and beneath the structure.

Groundwater seepage beneath and through dams is studied by constructing flow nets of the type shown in Figure 12-31. From the flow net, seepage pressure and gradient can be calculated so that the dam design can be modified if necessary. Seepage control beneath dams is sometimes accomplished by injection of cement *grout* under high pressure into closely spaced boreholes before the dam is constructed. After the grout penetrates cracks and joints in the

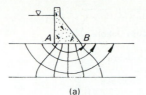

(a)

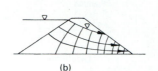

(b)

Figure 12-31 Flow nets showing groundwater flow systems beneath impermeable masonry dam (a) and through permeable earth dam (b). (From R. A. Freeze and J. A. Cherry, *Groundwater,* copyright © 1979 by Prentice-Hall, Inc., Englewood Cliffs, N.J.)

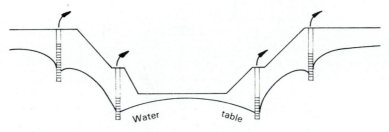

Figure 12-32 Dewatering by well points to lower the water table below the base of an excavation and prevent excessive inflow of water into the excavation.

rock, it sets up, providing a barrier to groundwater seepage. Various design measures are also incorporated in permeable earth dams for seepage control.

Groundwater inflow into tunnels and excavations often presents difficult engineering problems. In tunneling through rock, workers often suddenly encounter water in joints, fractures, or fault zones. Many lives are lost because of tunnel flooding when water under high pressure is suddenly encountered as the tunnel is advanced. Seepage into excavations often leads to slumping and sliding of the walls of the cut. In Chapter 13 we will consider the importance of groundwater in landslide phenomena of all types. If seepage into an excavation is determined to be undesirably rapid, *dewatering* techniques can be applied. One method, as shown in Figure 12-32, is to install a system of shallow pumping wells, or *well points*, around the excavation to lower the water table until the project can be completed.

CASE STUDY 12-1
GROUNDWATER MINING AND THE HIGH PLAINS AQUIFER

Problems arise during production of groundwater when an equilibrium condition in the cone of depression is not established. Under heavy pumpage, perhaps when multiple wells penetrate the same aquifer, steady declines in the water tables or potentiometric surface are noted. A good example of this unfortunate circumstance is the High Plains Aquifer, a shallow, mostly unconfined aquifer that occurs throughout much of the Great Plains region (Figure 12-33). The High Plains Aquifer has been extensively pumped to support irrigated farming in this region since the agricultural disaster caused by the 1930s drought. The development of irrigation has progressed to such an extent that annual groundwater withdraw-

Figure 12-33 Location of the High Plains Aquifer. (From U.S. Geological Survey Professional Paper 1400-B.)

als now exceed the annual recharge to the aquifer by 2 to 200 times. The result is a steady decline in water levels. These declines exceed 30 m in some areas. In effect, the water is being "mined" like any other nonrenewable resource. On a short-term basis the effect of water-table declines is to increase costs because of the greater pumping lift required. On a long-term basis, however, parts of the High Plains Aquifer will be in serious trouble within the next 20 years. The point may be reached where irrigated farming will no longer be possible in some areas. This will have severe economic consequences for the affected areas.

CASE STUDY 12-2
INDUSTRIAL WASTE DISPOSAL: A THREAT TO GROUNDWATER QUALITY

Liquid chemical wastes are sometimes discharged to surface impoundments, from which contaminants may leak downward to the water table. Perlmutter and Lieber (1970) presented an example of this type of contamination on Long Island, New York. During and after World War II an aircraft factory discharged aluminum-plating wastes into disposal ponds shown on the accompanying map (Figure 12-34). Because the area is underlain by a highly permeable unconfined aquifer with a high water table, the wastes rapidly infiltrated downward to the groundwater flow system.

From this point the waste formed a plume of contaminated groundwater about 1300 m long, 300 m wide, and 21 m thick. The gradual spreading resulting from hydrodynamic dispersion is evident on the map. Calculations indicate that the contaminated groundwater moves rapidly at an approximate rate of 165 m per year. Some of the contaminated water discharges into Massapequa Creek in the southern part of the area.

Figure 12-35 shows a subsurface cross section of the plume. It is apparent that contamination is confined to the upper glacial aquifer

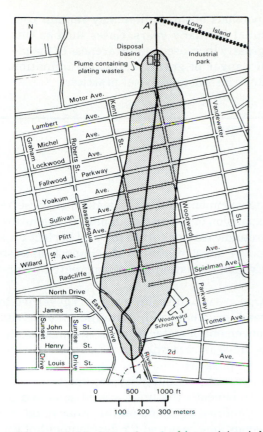

Figure 12-34 Map of part of Long Island, New York, showing size and extent of plume of contaminated groundwater originating from disposal pits at a metal-plating facility. (From N. M. Perlmutter and M. Lieber, 1970, U.S. Geological Survey Water Supply Paper 1879-G.)

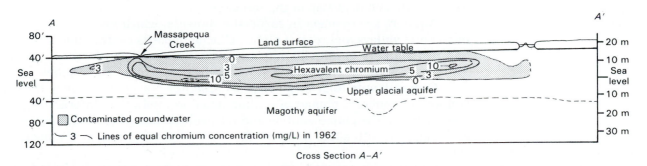

Figure 12-35 Cross section A–A (location shown on Figure 12-34) through the center of the contaminant plume. Contours show the concentrations of hexavalent chromium in milligrams per liter. (From N. M. Perlmutter and M. Lieber, U.S. Geological Survey Water Supply Paper 1879-G.)

and that movement is predominantly horizontal. The contours on the cross section show the distribution of hexavalent chromium in 1962. Hexavalent chromium is a toxic metal with a maximum permissible concentration in drinking water of only 0.05 mg/L (Table 12-2). At the levels present in the aquifer, use of the water for drinking would pose a very serious health hazard. In addition to chromium, the plume contains other toxic contaminants.

Fortunately, no public supply wells obtained water from the contaminated portion of the aquifer. Since the plume was detected, waste treatment has considerably reduced the amount of contaminants entering the aquifer. Even after a waste source is eliminated, however, portions of an aquifer may remain contaminated for many years. Cases such as this have greatly increased our awareness of the potential danger of groundwater contamination by industrial waste-disposal practices.

SUMMARY AND CONCLUSIONS

Groundwater is a vital resource; development and management of these resources are the keys to any major expansion of our water supplies in the future. In order to preserve undeveloped groundwater for future use and protect presently utilized groundwater sources, we must minimize the impact of waste disposal on groundwater and rectify contamination problems created by past waste-disposal practices.

Groundwater flow from recharge areas to discharge areas is governed by the parameters of Darcy's Law: head, gradient, and hydraulic conductivity. Recharge involves the downward movement of water through the unsaturated zone to the water table. The water table is recognized as the level at which fluid pressure is equal to atmospheric pressure.

Geologic materials with high hydraulic conductivity are known as aquifers. Unconfined aquifers lie directly beneath the unsaturated zone and are bounded above by the water table. Confined aquifers, on the other hand, can occur at considerable depths. The conditions required for development of confined aquifers include confining aquitards both above and below. Water levels in wells penetrating confined aquifers rise to the potentiometric surface and can rise above land surface in flowing artesian wells.

Pumping of wells creates a cone of depression in either the water table or the potentiometric surface. Excessive pumping can lead to steadily dropping water levels and depletion of the resource.

Aquifers are common in surficial sediments, which are especially thick in tectonic valleys and coastal plains. Glacial deposits contain important unconfined aquifers in alluvial outwash sediments and buried-valley deposits. Aquifers in rocks often are located within fractured and jointed zones.

Contaminants from waste-disposal practices travel very slowly through aquifers as coherent plumes of contaminated groundwater. Interactive processes between the plume solution and the geologic materials, including cation exchange, can attenuate many of the contaminant species in the degraded water.

Construction activities often require groundwater control. Uncontrolled groundwater seepage threatens the stability of dams and leads to hazardous conditions in tunnels and excavations. Dewatering methods may be necessary for completion of the project.

REFERENCES AND SUGGESTIONS FOR FURTHER READING

Freeze, R. A., and J. A. Cherry. 1979. *Groundwater*. Englewood Cliffs, N.J.: Prentice-Hall, Inc.

Gutentag, E. D., F. J. Heimes, N. C. Krothe, R. R. Luckey, and J. B. Weeks. 1984.

Geohydrology of the High Plains Aquifer in Parts of Colorado, Kansas, Nebraska, New Mexico, Oklahoma, South Dakota, Texas, and Wyoming. U.S. Geological Survey Professional Paper 1420-B.

HUBBERT, M. K. 1940. The Theory of Groundwater Motion. *Journal of Geology* 48:785–944.

PERLMUTTER, N. M., and MAXIM LIEBER. 1970. *Dispersal of Plating Wastes and Sewage Contamination in Ground Water and Surface Water, South Farmingdale–Massapequa Area, Nassua County, New York.* U.S. Geological Survey Water Supply Paper 1879-G.

PRINCETON UNIVERSITY WATER RESOURCES PROGRAM. 1984. *Groundwater Contamination from Hazardous Wastes.* Englewood Cliffs, N.J.: Prentice-Hall, Inc.

PROBLEMS

1. For what reasons would you need to use Darcy's Law in the field?
2. Why isn't the water table defined as the upper limit of the saturated zone?
3. Is it correct to say that groundwater flows parallel to the water table? Why or why not?
4. What factors control the amount of groundwater recharge?
5. How can you identify groundwater discharge areas?
6. What is an aquifer? Why aren't all saturated rocks and soils considered to be aquifers?
7. How do confined and unconfined aquifers differ in their response to pumping?
8. What rock units or sediments constitute the most productive aquifers in your area? How does groundwater influence the economy of the region?
9. What would be the chemical characteristics of groundwater from a limestone aquifer?
10. How are contaminants transported in groundwater?
11. How does groundwater contamination differ from contamination of rivers?
12. What is attenuation and how does it occur in an aquifer?
13. What geologic conditions are most favorable for waste disposal?
14. Why is groundwater sometimes a problem at construction sites? How can the problems be solved?

CHAPTER
13
MASS MOVEMENT AND SLOPE STABILITY

Mass movement is the collective name for a variety of processes involving the downslope motion of soil and rock materials under the influence of gravity. Damages resulting from such movements have been estimated to exceed $1 billion per year in the United States. An even greater cost is the loss of life that sometimes accompanies major slope movements. Prevention of damage from slope movements requires recognition and avoidance of potentially unstable slopes. Where construction does take place on slopes, a detailed geologic investigation coupled with a thorough engineering analysis of the stability of rocks and soils that underlie the slope must be the initial phase of the project.

TYPES OF SLOPE MOVEMENTS

Mass movements are classified according to the type of movement and the type of slope material involved (Figure 13-1). Slope materials are divided into bedrock, soil composed of predominantly coarse particles (debris), and soil composed of predominantly fine clasts (earth). Six types of movement are utilized in the classification. Each slope movement, therefore, is given a two-part name relating the type of movement to the type of material. Many slope movements cannot be assigned to a single process and thus must be included in the "complex" category, an indication that more than one type of movement has occurred.

Missing from the classification is any indication of the velocity of movement. Within each category, rates can range from imperceptibly slow to freight-train velocities (Figure 13-2). The prediction of the particular types of movement that can be expected in rocks and soils in different geological settings is a very important aspect of the analysis of slope stability that is carried out during site selection and design of construction projects.

Type of movement			Type of material		
			Bedrock	Engineering soils	
				Predominantly coarse	Predominantly fine
Falls			Rock fall	Debris fall	Earth fall
Topples			Rock topple	Debris topple	Earth topple
Slides	Rotational	Few units	Rock slump	Debris slump	Earth slump
	Translational		Rock block slide	Debris block slide	Earth block slide
		Many units	Rock slide	Debris slide	Earth slide
Lateral spreads			Rock spread	Debris spread	Earth spread
Flows			Rock flow (deep creep)	Debris flow	Earth flow (soil creep)
Complex			Combination of two or more principal types of movement		

Figure 13-1   Classification of slope movements. (From D. J. Varnes, "Slope movement types and processes." In *Landslides: Analysis and Control,* edited by R. L. Schuster and R. J. Krizek, TRB Special Report 176, Transportation Research Board, National Research Council, Washington, D.C., 1978.)

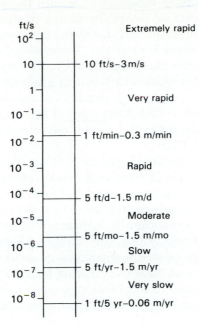

Figure 13-2 Velocity scale for slope movements. (From D. J. Varnes, "Slope movement types and processes." In *Landslides: Analysis and Control,* edited by R. L. Schuster and R. J. Krizek, TRB Special Report 176, Transportation Research Board, National Research Council, Washington, D.C., 1978.)

Falls and Topples

When a mass of rock, debris, or soil separates from a steeply sloping surface and rapidly moves downslope by free fall, bounding, or rolling, the movement is termed a *fall*. These phenomena range from massive bodies of rock on mountain peaks set in motion by earthquakes to small blocks of soil that fall down a river bank when lateral erosion by the stream undercuts the bank to the point where the overhanging section collapses. Once a fall is in progress, other types of movement can develop in the mass as it progresses down the slope.

A *topple* is a rotational movement that occurs as a block of material pivots forward about a fixed point near the base of the block (Figure 13-3). Topples develop in rock or soil slopes divided into blocks by vertical fractures or joints oriented parallel to the slope face. The thin slabs of material are pushed outward by lateral forces exerted by adjacent material or by water freezing and expanding in the cracks. If the base of the slab is fixed, the lateral forces cause an overturning moment on the slab, and a topple may occur.

Slides

One of the most common types of movement is sliding—that is, shearing displacement between two masses of material along a plane or within a thin zone of failure. The basic difference between slides and flow is that slides move as

Figure 13-3 Topples occurring in jointed bedrock. (Modified from D. J. Varnes, "Slope movement types and processes." In *Landslides: Analysis and Control,* edited by R. L. Schuster and R. J. Krizek, TRB Special Report 176, Transportation Research Board, National Research Council, Washington, D.C., 1978.)

a unit with little or no deformation within the sliding mass, whereas flows are thoroughly deformed internally during movement.

Slide types are further divided by the nature of the failure surface. Failure surfaces are often curved or circular. These slides, or *slumps*, are very common in soils or rocks of low shear strength (Figure 13-4). The rotational movement of the slump mass or block may result in a significant tilt of the upper surface of the moving block, or *head*, backward toward the exposed upper part of the failure plane, which is called the *scarp* (Figure 13-4). Trees, telephone poles, and other objects on top of the sliding blocks are also tilted by the rotational movement (Figure 13-5).

If failure surfaces are planar rather than curved, the resulting movement is a *translational slide*. In rock, *rock block slides* consisting of the motion of a single body of rock, can be distinguished from *rock slides* (Figure 13-6), which are composed of multiple rock fragments of various sizes. The actual configuration of the slide mass usually is determined by planes of weakness within the rock such as faults, joints, and bedding planes. The most hazardous situation occurs when predominant planes of weakness dip in the direction of the slope. A disastrous slide that at least began with translational movement localized along planes of weakness dipping toward the center of a valley occurred in the Italian Alps in 1963. During the slide 240 million cubic meters

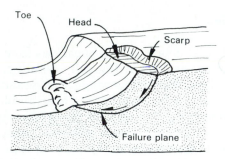

Figure 13-4 Slumps involve rotational sliding along a curved failure surface.

Figure 13-5 Telephone poles tilted by the rotational movement of a slump.

Figure 13-6 The scar of a rock slide that occurred in granitic igneous rocks.

of rock and soil traveled rapidly downslope into a reservoir impounded by Vaiont Dam, the world's highest thin-arch dam (267 m) (Figure 13-7). The slide, which filled the reservoir for a distance of 2 km, generated a huge wave of water that overtopped the dam and flooded the valley below. Nearly 3000 people were killed by a wall of water as much as 70 m high. The destruction would have been even greater had the dam not remained intact during the tremendous stress placed upon it. A major factor in the Vaiont slide was the presence of planes of weakness dipping toward the valley (Figure 13-7b). These planes included bedding planes in the sedimentary rock units at the site, as well as fractures produced by unloading of the rocks in the slopes, fractures similar to those caused by exfoliation. Other conditions contributing to the Vaiont disaster were increased pore pressures in the slopes resulting from seepage from the reservoir and heavy rains that preceded the final failure of the slope.

Debris slides are common on slopes where bedrock is overlain by colluvium (Figure 13-8). The failure plane in these situations is either the interface between less-weathered bedrock and the debris above or a plane within the debris. The common association of excess pore-water pressure and slope movement is recognized in this type of slope setting. The effect of increased pore-water pressure is to produce a buoyant force acting upward on the slide mass, thus decreasing the natural stability of the slope. In addition, heavy rainfall

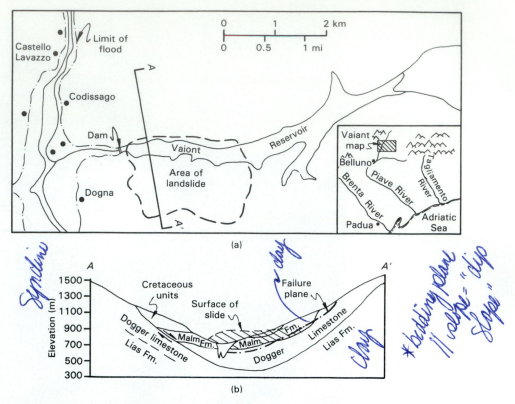

Figure 13-7 (a) Map of Vaiont Reservoir area showing extent of the landslide of October 1963. Black dots show the locations of villages in the flooded area. (b) Cross section of the valley showing the structure and stratigraphy of the slide area. (Modified from G. A. Kiersch, 1964, Vaiont Reservoir disaster. *Civil Engineering* 34:32–39.)

can increase the weight of the potential slide mass by saturating the upper part of the debris and creating a perched water table within the slope debris. The role of water will be examined in more detail later in this discussion.

A type of translational slide that caused major damage in Anchorage, Alaska, during the 1964 earthquake is illustrated in Figure 13-9. Translational movement was initiated above a plane in the weak Bootlegger Cove clay. Here the triggering mechanism was the violent ground shaking caused by a major earthquake, although the subsurface materials created a potentially unstable situation.

Lateral Spreads

The slow-to-rapid lateral extensional movements of rock or soil masses are known as *lateral spreads*. Liquefaction and flowage of a weak soil layer within a slope is the cause of most lateral spreads in the debris and earth categories of the slope-movement classification. The stronger material above the failure is rafted along without intense deformation, although it may be broken into blocks that can subside or rotate as the spread progresses (Figure 13-10). The most susceptible materials are sensitive clays (Chapter 6), like the Bootlegger Cove clay in Anchorage. These clays exhibit the tendency to instantaneously lose shear strength upon rearrangement of clay particles (remolding).

The danger of quick clays (clays with the highest sensitivity values) has

Figure 13-8   Shallow debris slide near Monterey, California, resulting from the severe winter storms of 1982/1983. (G. F. Wieczorek; photo courtesy of U.S. Geological Survey.)

been proven by many spreading failures in such places as the St. Lawrence River valley and the fjords of Scandinavia. A hypothesis for the mechanism of failure of the clays in these areas is based on their common geologic histories. When glaciers advanced to coastal regions, the crust was depressed under the great weight of the continental ice sheets. Offshore marine clays deposited during these times were later raised above sea level by the slow, but continuous, rebound of the crust after the glaciers retreated. Because of their marine origin, the clays were originally deposited with a flocculated structure. After the clay deposits become exposed to nonmarine conditions, the salty pore water was gradually displaced by fresh water derived from rainfall and snowmelt that began to leach the salt ions providing the bonds between particles. By this process a very unstable situation is created. In its flocculated state the clay has significant shear strength. However, when the bonding cations are removed, the clay fabric can be instantaneously changed from a flocculated state to a dispersed state. The dispersed clay has only a fraction of the shear strength it previously possessed and will behave as a liquid, with the ability to flow on extremely low slope angles. Thus lateral spreads are generated.

Figure 13-9 A complex translational slide in the Turnagain
Heights area of Anchorage, Alaska, resulting from the 1964
earthquake. Notice that some of the houses remained nearly
intact during the predominantly lateral movement. (W. R. Hansen;
photo courtesy of U.S. Geological Survey.)

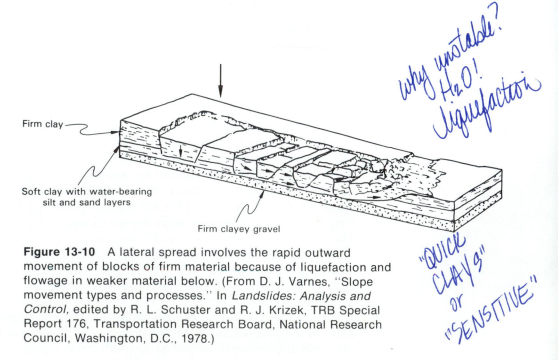

Firm clay

Soft clay with water-bearing
silt and sand layers

Firm clayey gravel

Figure 13-10 A lateral spread involves the rapid outward
movement of blocks of firm material because of liquefaction and
flowage in weaker material below. (From D. J. Varnes, "Slope
movement types and processes." In *Landslides: Analysis and
Control*, edited by R. L. Schuster and R. J. Krizek, TRB Special
Report 176, Transportation Research Board, National Research
Council, Washington, D.C., 1978.)

Earthquakes often provide the shock that induces the sudden liquefaction of quick clays. Other possible causes are blasting and vibrations caused by the movement of heavy equipment. Once a lateral spread is initiated, buildings, and everything else on the land surface, are rafted along above the flowing clay with the more coherent surficial material. A good example of this type of movement is the lateral spread that occurred in Nicolet, Québec, in 1955 (Figure 13-11). The costs of this event were three lives lost and several million dollars in property damage.

The lateral spreads just described involve the rapid flowage of earth materials. They are assigned to a separate category in the classification because of their predominantly lateral motion and because of the thick blocks of surficial material that are passively transported along above the flowing clay like boxes on a conveyor belt. There is really a complete transition from earth slides like the Anchorage slope failure, where remolding of sensitive clay occurred in very thin zones within more competent clay; to lateral spreads, where the zone of flowage is thicker; to certain types of rapid earth flows, in which the entire mass of sediment is flowing. It is the latter type of movement to which we now turn our attention.

Flows

The processes of flow are complex physical phenomena. Flow of various types of rock and soil may exhibit viscous or plastic behavior, as well as variations and combinations of both. The overriding criterion, however, is that flow must involve continuous internal deformation of the moving material.

Figure 13-11 Aerial view of the Nicolet, Quebec, lateral spread of 1955. Buildings and trees were rafted along in an upright position. (Henry Laliberté; photo courtesy of Governement du Québec, Ministère de l'Environnement.)

When the flow travels at velocities at the slow end of the velocity scale, the process is sometimes called *creep*. Creep may occur in rock or surficial debris. In rock, creeplike flow can be a very slow, steady process, persisting over long periods of time. Alternatively, slow creep may accelerate to the point where a dramatic failure results. Creep had been monitored in the slopes above the Vaiont reservoir for 6 months preceding the main slide. The rapid acceleration of the movement alerted officials to the impending disaster. Unfortunately, desperate attempts to draw down the reservoir were too late to avert the slide.

A common type of creep is often observed within weathered bedrock and soil on steep slopes (Figure 13-12). This process displays a pronounced seasonal variation characterized by expansion of soil materials perpendicular to the slope (Figure 13-13). The expansion, primarily due to freezing or swelling caused by increases in water content, is only the first phase of the movement. When either thawing or shrinkage takes place, the movement is vertically downward under the controlling influence of gravity. Thus during each cycle of expansion and contraction the soil undergoes a small downslope component of movement.

Figure 13-12 Creep in weathered bedrock, Maryland. The initially vertical beds at the top of the exposure have been tilted by slow downslope movement. (G. W. Stose; photo courtesy of U.S. Geological Survey.)

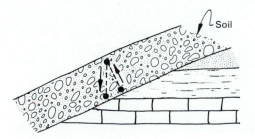

Figure 13-13 Creep is often a seasonal process in which freezing or swelling causes movement of particles perpendicular to the slope. Upon thawing or shrinkage, vertical movement results in a net downslope component of motion.

The more rapid examples of flow phenomena are variously termed *debris, earth, or mud* flows. These processes involve the rapid to very rapid flow of materials down steep slopes (Figure 13-14). The similarity in process among the flow of debris, lava, and glacial ice is reflected in the production of similar landforms by all three flow phenomena. Like glacial ice and lava, debris flows follow channels, spread out to form broad lobes when not confined to channels (Figures 13-14, 15), and develop conspicuous lateral deposits along the sides of the channels. In debris flows, the lateral deposits are called *lateral levees* (Figure 13-15).

Many observations of debris flows have confirmed that the flows have a high water content, and thus are highly fluid, yet can transport boulders and other large objects many times the size of anything that even the fastest rivers can move. While in transit, debris flows treat boulders and other surficial objects with remarkable care. For example, large boulders that apparently have been transported intact within a debris flow have broken apart along fractures by weathering processes within a relatively short time period after deposition by the flow. These observations require careful analysis to determine the mechanical nature of debris-flow processes.

The visual similarity of rivers and rapid debris flows seems to suggest a similar flow mechanism. Thinking in these terms, we could consider debris flows to be simply rivers of high-density, highly viscous fluid in which boulders are supported by turbulent eddies (random velocity fluctuations) in the flow. Laboratory tests have shown, however, that dense debris flows are laminar rather than turbulent. Laminar flow lacks high-velocity turbulent eddies

Figure 13-14 An earth flow that occurred in 1953 in Okanogan County, Washington. The flow lasted 7 min and destroyed a house. The source of the flow is the hill in the background. (F. O. Jones; photo courtesy of U.S. Geological Survey.)

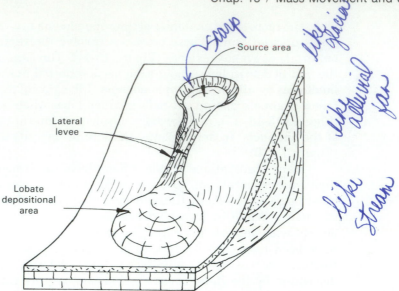

Figure 13-15 Rapid debris, or earth, flow. (Modified from D. J. Varnes, "Slope movement types and processes." In *Landslides: Analysis and Control,* edited by R. L. Schuster and R. J. Krizek, TRB Special Report 176, Transportation Research Board, National Research Council, Washington, D.C., 1978.)

(Chapter 14). Plastic behavior could be called upon to explain the concentration of boulders near the tops of the flow masses. Plastic flow could prevent the sinking of boulders into the center of the flow by the strength of the material. An ideal plastic, however, would accelerate infinitely once the yield strength had been exceeded. A composite model (Johnson, 1970) combines elements of plastic flow and laminar viscous flow. This type of behavior is the type exhibited by a Bingham substance (Chapter 6). It seems to best fit the observations of debris flows and debris-flow deposits with respect to their ability to transport boulders in a laminar flow regime.

A further implication of the flow of a Bingham material is the existence of a rigid "plug" near the center of the channel and at the top of the flow (Figure 13-16). This plug is rafted along as a nondeforming unit in the debris flow because the shear stress within the plug is less than the strength of the material.

Complex Slope Movements

Slope failures in nature involving a single type of movement are probably the exception rather than the rule. A very common occurrence is the initiation of a slope movement as a slide followed by conversion to a flow during the course of the event. A slump-earth flow, for example, is illustrated in Figure 13-17. Topographic features resulting from the slump include a *main scarp* separating the *crown* from the head, as well as one or more *minor scarps.* Beyond the

Figure 13-16 Cross section of channelized debris flow showing the development of a nondeforming plug at the center of the flow.

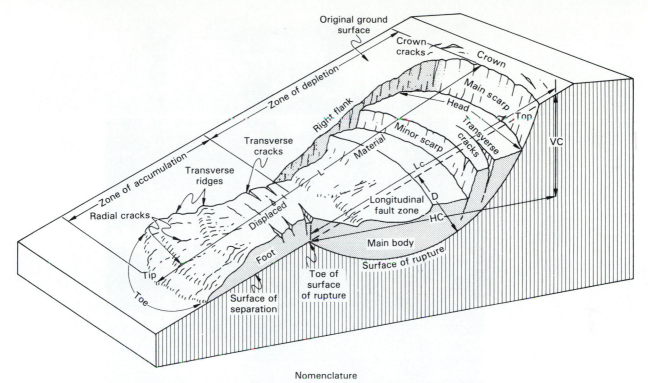

Nomenclature

Main scarp—A steep surface on the undisturbed ground around the periphery of the slide, caused by the movement of slide material away from undisturbed ground. The projection of the scarp surface under the displaced material becomes the surface of rupture.

Minor scarp—A steep surface on the displaced material produced by differential movements within the sliding mass.

Head—The upper parts of the slide material along the contact between the displaced material and the main scarp.

Top—The highest point of contact between the displaced material and the main scarp.

Toe of surface of rupture—The intersection (sometimes buried) between the lower part of the surface of rupture and the original ground surface.

Toe—The margin of displaced material most distant from the main scarp.

Tip—The point on the toe most distant from the top of the slide.

Foot—That portion of the displaced material that lies downslope from the toe of the surface of rupture.

Main body—That part of the displaced material that overlies the surface of rupture between the main scarp and toe of the surface rupture.

Flank—The side of the landslide.

Crown—The material that is still in place, practically undisplaced and adjacent to the to the highest parts of the main scarp.

Original ground surface—The slope that existed before the movement which is being considered took place. If this is the surface of an older landslide, that fact should be stated.

Left and right—Compass directions are preferable in describing a slide, but if right and left are used they refer to the slide as viewed from the crown.

Surface of separation—The surface separating displaced material from stable material but not known to have been a surface on which failure occurred.

Displaced material—The material that has moved away from its original position on the slope. It may be in a deformed or undeformed state.

Zone of depletion—The area within which the displaced material lies below the original ground surface.

Zone of accumulation—The area within which the displaced material lies above the original ground surface.

Figure 13-17   Diagram of a slump-earth flow. (From D. J. Varnes, "Slope movement types and processes." In *Landslides: Analysis and Control,* edited by R. L. Schuster and R. J. Krizek, TRB Special Report 176, Transportation Research Board, National Research Council, Washington, D.C., 1978.)

main body in the downslope direction the lobate *toe* indicates that the dominant type of movement in this section is flowage. *Transverse ridges* in bands parallel to the toe cross the earthflow.

Some of the greatest natural disasters in history have resulted from complex slope movements that began as rock falls or rock slides and then transformed into rapid debris flows, or avalanches, along their downslope paths. Particularly deadly examples of these events originated from the glacier-mantled Peruvian peak Nevados Huascaran in 1962 and 1970 (Figure 13-18). The

Figure 13-18 Aerial view of Nevados Huascaran area, Peru. Rock-fall-debris avalanches in 1962 and 1970 buried the villages of Yungay and Ranrahirca, killing thousands of people. (Photo courtesy of U.S. Geological Survey.)

first avalanche traveled 14 km in 5 minutes and buried the village of Ranrahirca, killing 3500 residents in the process. An even larger avalanche swept down the mountain in 1970, overtopping a ridge too high for the 1962 avalanche, and devastated the town of Yungay. More than 18,000 lives were lost on this tragic day. Unfortunately, rock-slide-debris avalanches are also common in the United States and Canada in the high mountain ranges along the western margin and western interior of the continent. The Madison Canyon slide of 1959 in Montana (Figure 13-19), which was triggered by an earthquake, created a lake as it flowed across a valley and blocked a river. A lake quickly developed in the valley behind the debris-flow dam, raising fears that the rapid rise of water level would soon cause a catastrophic failure of the natural dam. Quick work by the U.S. Army Corps of Engineers, which excavated an emergency spillway through the debris, prevented the lake from spilling over the debris-flow dam and causing massive floods downstream.

The mechanics of movement of rock-slide-debris avalanches have been

Figure 13-19 The Madison Canyon slide. Actually a debris avalanche, this devastating slope movement blocked the Madison River to form a lake and buried a campground at the base of the valley on the side opposite the avalanche.

debated by geologists in recent years. One theory, which was widely accepted in the 1960s and 1970s attributed debris flows to a sliding mechanism. According to this hypothesis the mass moved as a relatively intact block that slid above a thin layer of compressed air. The compressed air would sufficiently lower the coefficient of friction to allow the great runout distances observed in these movements (Figure 13-20b). A sliding mechanism was invoked to explain the observation that the distribution of rock types in the mass of material after it had come to rest was identical with the distribution of rock types in the source area of the avalanche high upon the mountain peak. In other words, the mass appeared to have moved without extensive internal disruption. This stratigraphic order within the moving mass could be explained if the mass moved as an intact block (Figure 13-20b). Stratigraphic order would not be preserved in a viscous flow because the material near the top of the flow moves faster than material near the base. Thorough mixing of rock types would therefore be expected in viscous flow.

An alternative flow hypothesis has recently been put forth (even though it was first suggested in the 1880s by Swiss geologist Albert Heim) to explain the problem of stratigraphic order. In this type of flow, called *grain flow*, solid grains dispersed throughout an avalanche achieve the ability to flow without being carried along by a flowing viscous fluid. Although both water and compressed air may be involved in the flow, their presence is not necessary for the flow to occur. This is best illustrated by the identification of giant debris flows on the moon, a body without an atmosphere or water. As grains collide in a grain flow, they transfer momentum to grains ahead of them (Figure 13-20c) and therefore do not pass their counterparts in the direction of flow. Thus stratigraphic order of rock types will be preserved in a grain flow. This model seems to fit evidence observed in rock-slide-debris avalanches better than the compressed air hypothesis because of the topographic form of the deposits. Debris-avalanche deposits have the physical appearance of flow, including lobate toes and transverse flow ridges. Perhaps a better understanding of these phenomena will lead to a better evaluation of hazards in mountainous areas.

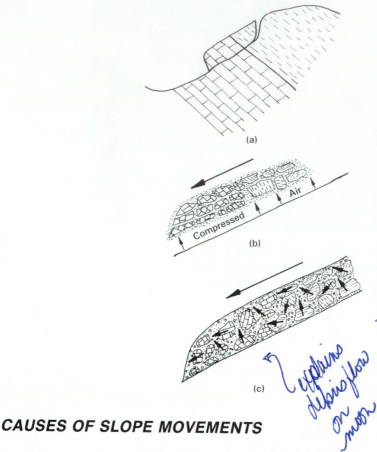

Figure 13-20 Possible flow mechanisms of rock-fall-debris avalanches. (a) A block of rock on a steep mountain slope is suddenly launched into motion by an earthquake or other triggering factor. (b) Travel of the debris by <u>sliding</u> on a layer of compressed air. <u>Initial stratigraphic order</u> within the moving mass is preserved. (c) Movement of the avalanche by grain <u>flow</u>. After suspension of blocks in the air, <u>motion and stratigraphic order are maintained</u> by inertial collisions between trailing blocks and those fragments ahead of them.

CAUSES OF SLOPE MOVEMENTS

Determining the cause of a slope movement may be more difficult than it appears. Rarely is there a single cause; in most cases, although there may be an obvious triggering mechanism, the interaction of many variables controls the initiation of a slope movement.

A basic approach to evaluating slope stability in terms of simple sliding mechanisms is to identify the forces tending to cause movement, or *driving forces*, and those that tend to resist failure, or *resisting forces*. The ratio of the resisting forces to the driving forces is a quantitative indication of stability known as the *safety factor*. Thus

$$\text{safety factor (S.F.)} = \frac{\sum \text{resisting forces}}{\sum \text{driving forces}} \qquad \text{Eq. 13-1}$$

The greater the safety factor, the less likely the slope is to fail. The minimum value of the safety factor is 1 because below this value the driving forces are greater than resisting forces, and the slide will be in progress. When the safety factor is exactly 1, the slope is in a state of incipient failure and any increase in driving forces or decrease in resisting forces will serve to trigger the movement.

Geologic Setting of Slopes

Before considering factors that effect changes in the safety factor of a slope, it is worthwhile to identify some of the conditions that render a slope susceptible to mass movement. The slope angle, for example, plays an important role

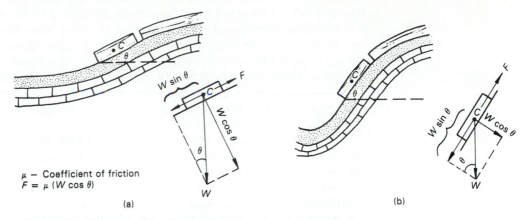

μ — Coefficient of friction
$F = \mu\,(W \cos \theta)$

(a)   (b)

Figure 13-21 Forces acting on a block of rock tending to slide downslope. In (a) the weight of the block is resolved into a downslope force ($W \sin \theta$) and a normal force ($W \cos \theta$). Downslope movement is opposed by the frictional force, F, which is equal to the coefficient of friction times the normal force. The same block is more likely to slide on a steeper slope (b) because the downslope force is greater and the normal force is less than in (a).

in establishing the initial ratio of resisting forces to driving forces. This can be illustrated by analyzing the forces acting on a block of rock resting upon a slope (Figure 13-21a). The weight of the rock mass, W, acting vertically downward can be resolved into a force acting parallel to the slope in the downslope direction, $W \sin \theta$, and a force acting normal to the slope, $W \cos \theta$. Opposing the block's tendency to slide down the slope is the frictional force F, which is dependent upon the normal force and the coefficient of friction (μ) between the rock and the slope, so that $F = \mu(W \cos \theta)$. As the slope angle increases (Figure 13-21b) the force acting downslope increases and the normal force decreases. When the downslope, or driving force, equals the frictional resisting force, motion of the block is incipient.

The lithology of the materials comprising the slope is probably the most important component of the slope-stability problem. Geologic materials of low strength or materials that tend to weather to materials of low strength present the greatest hazard. Included in this group are clays, shales, certain volcanic tuffs, and rocks containing soft platy minerals like mica. Clays, and other rocks and soils that contain clays, are particularly notorious for their association with slope failures. Clays have an initially low strength and tend to become even weaker with increases in water content. The ability of clays to absorb water and swell can lead to a significant loss in strength by the materials on a slope.

A special type of slope failure occurs in clays that are classified as *overconsolidated*. Overconsolidated clays have been subjected to large overburden loads during their geologic history. Under these loads clays become denser by compaction and expulsion of pore water. When overburden pressures are eased, either by erosion of overlying materials or by excavation, the compacted clays tend to expand and, in the process, develop a network of fractures, or *fissures*. The clay is then classified as overconsolidated because it currently supports a lower weight of overburden than it has at some time in its geologic past. The effect of these fissures is to lower the strength of the clay along the discontinuities. When small samples are taken for lab strength tests, they are frequently trimmed from intact blocks of clay between the fissures. The strength values obtained from these tests, described as the *peak strength*, give a misleading estimate of the strength of the clay mass because mass movements on the slope are initiated by shear failure along the fissures. The strength mobilized along these discontinuities, the *residual strength*, is significantly lower than the strength measured from intact samples (Figure 13-22). There are numerous examples of excavated slopes such as road and railway cuts that have failed because the slope angle was designed using the peak instead of the residual strength.

In rock slopes a situation similar to that with overconsolidated clays exists, because it is the discontinuities in the rock that control the stability of the rock mass. Discontinuities along which strength is reduced are common in almost every type of rock. In sedimentary rocks, bedding planes constitute significant planes of weakness within the rock mass. In terms of slope stability, the orientation of the planes is critical. Bedding that dips in the same direction as the slope is the most dangerous situation (Case Study 6-2). Road cuts and other excavations can decrease the stability of rock slopes by increasing the stress or decreasing strength along bedding planes in several different ways.

Planes of weakness other than bedding present in all rock types include joints and faults. Careful preliminary studies of slopes are necessary to determine the spacing, width, and orientation of fractures.

Influence of Water

Water ranks with material type as a major controlling factor in the occurrence of mass movements. Its importance can be seen in the fact that most rapid mass movements occur during and after periods of heavy rainfall.

Water can interact with slope materials in several different ways. An obvious result of the addition of water to a slope is an increase in the weight of a potential sliding block. With reference to Figure 13-21, the effect of addition of water is to increase both the downslope force $W \sin \theta$ and the normal force $W \cos \theta$. Although these changes will not drastically alter the safety factor of the slope, the water has a totally different effect on the normal force than is suggested above. To understand this relationship, we must recall the effec-

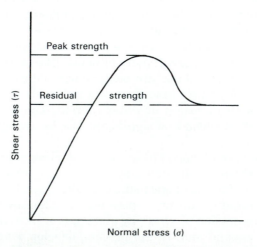

Figure 13-22 In overconsolidated clays, the residual strength— the resistance to failure along existing fissures in the clay—is less than the peak strength, the resistance to failure implied by tests on intact samples of clay.

tive stress equation presented in Chapter 12. In Figure 13-23 the stresses are shown on a potential failure plane within the slope material. Because the fluid pressure acts upward against the soil and rock load above the failure plane, the effective stress equation can be rearranged as

$$\sigma_e = \sigma_T - P \qquad\qquad \text{Eq. 13-2}$$

to show the amount of load borne by the soil particles. The name effective stress is appropriate because it determines the frictional resistance to sliding that could be mobilized by the soil particles along a potential failure plane.

Figure 13-23 illustrates the application of the effective stress equation to sliding-type slope movements. The fluid pressure is determined by the thickness of the saturated zone above the failure plane. So when the fluid pressure *increases*, as it would when the water table rises in response to heavy rainfall, the effective stress equation indicates that the effective stress will *decrease*. The rise in water table has therefore decreased the *effective* normal stress (σ_e) resisting sliding movement along the failure plane.

When effective stress is inserted into the Mohr-Coulomb equation for shear strength, the equation becomes

$$S = C + \sigma_e \tan \phi \qquad\qquad \text{Eq. 13-3}$$

It is obvious from Eqs. 13-2 and 13-3 that the shear strength of a material will decrease when the fluid pressure increases.

Another way to think of the effect of fluid pressure on sliding is to compare the process to the buoyant force exerted on an object submerged in water. Just as the submerged weight of any object is less than its weight in air, the effective weight of soil particles submerged by a water table rise will be less than their unsaturated weight. Thus, the resisting forces on the failure plane are decreased and the safety factor is reduced because the driving forces are unchanged or even increased by the weight of the water added to the soil.

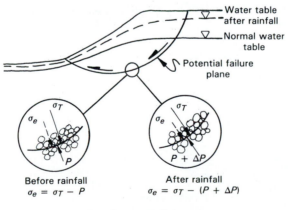

Figure 13-23 Effect of pore water on slope movements. Rainfall causes a rise in water table and increases pore pressure ($P + \Delta P$) throughout the subsurface. The stresses on a potential failure plane show the magnitude of effective stress (σ_e). Strength (S) is proportional to effective stress. After rainfall (right) effective stress decreases because of the increase in pore pressure. Strength of the soil along the potential failure plane is therefore lower.

A problem encountered with reservoirs provides an interesting application of the relationship of pore pressure and mass movements (Figure 13-24). The water table in the slopes adjacent to a reservoir rises and falls to adjust to the water level in the impoundment. When the water table rises, the decrease in resisting forces is offset by the lateral support exerted by water in the reservior on the reservoir banks (Figure 13-24a). If the water level in the reservoir is rapidly lowered, perhaps in anticipation of a flood that must be contained or partially contained in the reservoir, the lateral support provided by the water is removed. The water table in the adjacent slopes is much slower to respond, and the high pore pressures now tend to cause slumping of bank material into the reservoir (Figure 13-24b).

Processes that Reduce the Safety Factor

A natural or constructed slope exists under a certain ratio of resisting and driving forces (the safety factor) at any particular time. Many processes, however, acting over long or short periods of time, decrease the slope's safety factor. Recognition and identification of these processes are necessary for prediction of slope instability. In Table 13-1 processes are grouped into those that increase the shear stress on a failure plane and those that reduce shear strength within the materials on a slope.

One of the most common ways to increase the shear stress on a potential failure plane is to remove material from the base of the slope. As shown in Figure 13-25, construction of the road cut has removed material from an area where the failure plane is nearly horizontal or even curving upward. Material in this part of the potential slide mass acts against movement by resisting the tendency for rotational sliding. When the material is removed, the driving forces on the remainder of the slide mass are increased. In addition, the decrease in area of the failure plane after the excavation results in less shear strength developed along the plane. In this way, the resisting forces have been

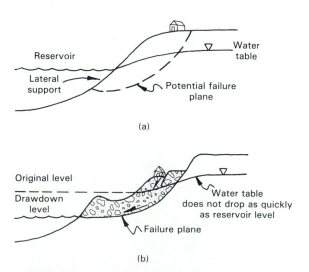

Figure 13-24 Reservoir bank failure due to rapid drawdown of reservoir level. When the reservoir is drawn down (b) lateral support provided by the water is lost. Pore pressure in the slope remains high because the water table adjusts to its new level more slowly than the reservoir. These conditions combine to cause slope failures.

TABLE 13-1

Processes Causing Changes in the Safety Factor

Processes that cause increased shear stress

1. Removal of lateral support
   a. Erosion by rivers
   b. Previous slope movements such as slumps that create new slopes
   c. Human modifications of slopes such as cuts, pits, canals, open-pit mines
2. Addition of weight to the slope (surcharge)
   a. Accumulation of rain and snow
   b. Increase in vegetation
   c. Construction of fill
   d. Stockpiling of ore, tailings (mine wastes), and other wastes
   e. Weight of buildings and other structures
   f. Weight of water from leaking pipelines, sewers, canals, and reservoirs
3. Earthquakes
4. Regional tilting
5. Removal of underlying support
   a. Undercutting by rivers and waves
   b. Construction of underground mines and tunnels
   c. Swelling of clays

Processes that reduce shear strength

1. Physical and chemical weathering processes
   a. Softening of fissured clays
   b. Physical disintegration of granular rocks such as granite or sandstone by frost action or thermal expansion
   c. Swelling of clays accompanied by loss of cohesion
   d. Drying of clays resulting in cracks that allow rapid infiltration of water
   e. Dissolution of cement
2. Increases in fluid pressure within soil
3. Miscellaneous
   a. Weakening due to progressive creep
   b. Actions of tree roots and burrowing animals

decreased. The combination of the increased shear stress and the decreased shear strength significantly lowers the safety factor of the slope.

The effect of surcharge applied upslope from the line of action of the center of gravity of the sliding mass is to increase the rotational tendency of the mass. Surcharge loads introduced by humans, including mine wastes and construction fill, frequently initiate slope failures (Figure 13-26).

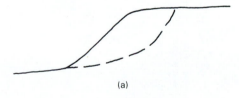

(a)

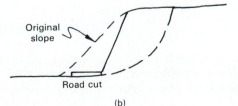

Original slope

Road cut

(b)

Figure 13-25 Stability of the slope shown in (a) is decreased by removal of material from the base of the slope and steepening the slope (b).

Figure 13-26 Slumping of surface mine spoils into a newly excavated mine pit was caused by piling too much spoil material at the edge of the pit.

Figure 13-27 A huge debris avalanche covering part of Sherman glacier, Alaska. The avalanche was triggered by the 1964 earthquake. (Austin Post; photo courtesy of U.S. Geological Survey.)

Processes that decrease the resistance to failure, with the exception of increases in fluid pressure, are usually long-term phenomena. These changes gradually modify the stability of the slope for many years prior to movement. The safety factor is then lowered to the point where a sudden triggering mechanism, an earthquake (Figure 13-27) or unusually wet period, for example, rapidly equalizes the driving forces and resisting forces, causing failure.

The large number of factors affecting both the driving and resisting forces supports the statement made earlier concerning the complexity of mass movements. Investigations of mass movements made after the fact, as well as those aimed at predicting stability relationships, must not overlook any of the long- or short-term geologic and human influences on slopes.

SLOPE STABILITY ANALYSIS AND DESIGN

The most important step preceding the construction of any engineering project that will alter or interact in any way with a slope is to determine the stability of the slope, both in its natural state as well as in its altered state after completion of the project. Embankments, earth dams, and other constructed slopes must likewise be designed with slope stability in mind. We must begin our discussion of stability analysis, therefore, with some comments about the existing state of stability of natural slopes in and near the area of investigation.

Recognition of Unstable Slopes

Areas of potential slope instability are most easily defined if evidence exists for previous slope movements. These areas can be delineated and evaluated by preliminary geologic studies. Several techniques can be used. First, all existing geologic reports of an area should be examined. Some geologic maps are specifically made for the purpose of identifying hazardous areas (Figure 13-28). Other maps show existing slope movements or the distribution of rock and surficial material types that are known to be associated with slope movements.

If more detailed studies of the project area are needed, these should be initiated by analysis of topographic maps and aerial photographs. Air photos are invaluable for the analysis and interpretation of landforms related to mass movements. Features associated with slides that can be recognized include scarps, slump shoulders, and disrupted drainage patterns. Hummocky topography, consisting of areas of hills and intervening depressions containing ponds or swamps, is a good indicator of recent slump or flow activity. Deposits of flows occur as lobate or tongue-shaped landforms in association with such other features as lateral levees and transverse flow ridges. Other relevant slope conditions, including fracture and joint patterns, the locations of springs and seeps, and the oversteepening of slopes by rivers and waves, can be noted on air photos.

Field studies complete the site investigation. Detailed geologic maps of the area can be made by examining rock and sediment exposures and plotting data on topographic base maps. Subsurface information is frequently needed for evaluation and design. Some types of data can be obtained using surface geophysical techniques like seismic refraction and earth resistivity. The most accurate method available for obtaining subsurface data (and also the most expensive) is test drilling. There are many methods and types of equipment that can be used for various subsurface conditions. Samples taken during drilling are useful not only for determining the distribution of soil and rock units

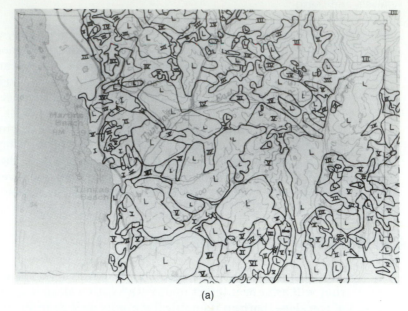

(a)

Explanation of Map Units

Least

I — Areas least susceptible to landsliding. Very few small landslides have formed in these areas. Formation of large landslides is possible but unlikely, except during earthquakes. Slopes generally less than 15%, but may include small areas of steep slopes that could have higher susceptibility. Includes some areas with 30% to more than 70% slopes that seem to be underlain by stable rock units. Additional slope stability problems; some of the areas may be more susceptible to landsliding if they are overlain by thick deposits of soil, slopewash, or ravine fill. Rockfalls may also occur on steep slopes. Also includes areas along creeks, rivers, sloughs, and lakes that may fail by landsliding during earthquakes. If area is adjacent to area with higher susceptibility, a landslide may encroach into the area, or the area may fail if a landslide undercuts it, such as the flat area adjacent to sea cliffs.

II — Low susceptibility to landsliding. Several small landslides have formed in these areas and some of these have caused extensive damage to homes and roads. A few large landslides may occur. Slopes vary from 5–15% for unstable rock units to more than 70% for rock units that seem to be stable. The statements about additional slope stability problems mentioned in I above also apply in this category.

III — Moderate susceptibility to landsliding. Many small landslides have formed in these areas and several of these have caused extensive damage to homes and roads. Some large landslides likely. Slopes generally greater than 30% but includes some slopes 15–30% in areas underlain by unstable rock units. See I for additional slope stability problems.

IV — Moderately high susceptibility to landsliding. Slopes all greater than 30%. These areas are mostly in undeveloped parts of the County. Several large landslides likely. See I for additional slope stability problems.

V — High susceptibility to landsliding. Slopes all greater than 30%. Many large and small landslides may form. These areas are mostly in undeveloped parts of the County. See I for additional slope stability problems.

VI — Very high susceptibility to landsliding. Slopes all greater than 30%. Development of many large and small landslides is likely. Slopes all greater than 30%. The areas are mainly in undeveloped parts of the County. See I for additional slope stability problems.

Most

L — Highest susceptibility to landsliding. Consists of landslide and possible landslide deposits. No small landslide deposits are shown. Some of these areas may be relatively stable and suitable for development, whereas others are active and causing damage to roads, houses, and other cultural features.

Definitions: Large landslide — more than 500 ft in maximum dimension
Small landslide — 50 to 500 ft in maximum dimension

(b)

Figure 13-28 A portion of a landslide-susceptibility map for San Mateo County, California. (From E. E. Brabb, E. H. Pampeyan, and M. G. Bonilla, 1972, *Landslide Susceptibility in San Mateo County, California*, U.S. Geological Survey Miscellaneous Field Studies Map MF-360.)

but also for conducting lab tests to characterize the properties of the materials. For slope-stability analysis, the shear strength of the soil, including its lateral and vertical variations, is a necessary type of data. After test holes have been drilled, piezometers can be installed for monitoring groundwater conditions.

Stability Analysis

The methods used for analyzing stability are based on finding the safety factor for a particular slope movement. For soil slopes composed of uniform, cohesive clays, rotational slump is considered to be the dominant failure mechanism. Successive circular failure surfaces are evaluated in order to determine the potential failure surface with the lowest safety factor. If this failure surface, the *critical circle*, has a safety factor near 1.0, the slope is considered to be unstable and must be redesigned or modified. Several trial surfaces and their computed safety factors are shown in Figure 13-29.

One of the most basic methods of calculating the safety factor for circular failure surfaces is shown in Figure 13-30. The safety factor for the indicated failure surface is expressed as the ratio of the resisting and driving moments about the center of rotation, point 0. Each cross section of the slope, such as the one illustrated, can be analyzed separately by treating the failure surface as a line rather than a plane. The resisting moments are the sum of the products of the soil shear strengths for each incremental length, L, of the failure circle acting against the downslope movement of the mass, and the moment arm, R, about the center of rotation. For the entire failure circle the resisting moments are therefore expressed as SLR. The driving moment is expressed as the weight, W, of the slump mass acting through its center of gravity times its moment arm, X, about the center of rotation. The safety factor for this

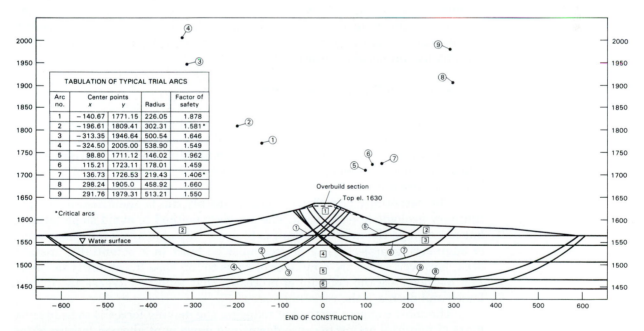

Figure 13-29 Slope-stability analysis of a proposed earth dam. The factor of safety is calculated for each trial arc. Centers of the arcs are plotted above the dam. Trial arcs 2 and 7 are the critical arcs. (From Flood Control Burlington Dam, Design Memo No. 2, U.S. Army Corps of Engineers, 1978.)

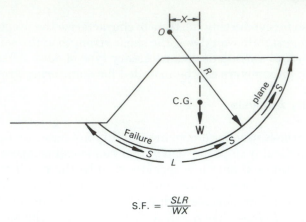

$$\text{S.F.} = \frac{SLR}{WX}$$

Figure 13-30   The Swedish circle method gives the safety factor of a slope for a particular failure plane as the ratio of the resisting moment (SLR) to the driving moment (WX).

particular slump would then be

$$\text{S.F.} = \frac{\sum \text{resisting moments}}{\sum \text{driving moments}} = \frac{SLR}{WX} \qquad\qquad \text{Eq. 13-4}$$

Higher or lower values of safety factors will be obtained by considering different slope angles and different trial failure surfaces. More sophisticated methods of analysis are used in most slope-stability studies. Slope-stability computer programs make the rapid evaluation of many trial surfaces possible.

The analysis of rock slopes first requires a detailed examination of discontinuities in rock because failures will be controlled by the spacing and geometry of these planes of weakness. The difficulty of determining an exact failure mechanism in these slopes, because of the complexity of the discontinuities, makes slope-stability analyses of rock slopes considerably less accurate than many soil-slope analyses.

Preventive and Remedial Measures

Once a slope fails, measures usually must be taken to correct the problems and stabilize the slope. These measures are very expensive and, in addition, often follow costly damages to highways, buildings, and other projects. It is much more desirable to prevent such losses by taking steps to stabilize slopes before failure occurs.

The potentially hazardous slopes commonly can be identified in the preliminary geologic investigation. The simplest method to avoid slope failures, but not always the most practical, is to avoid unstable areas for construction of engineering projects. If a project must be built in a potentially unstable area, however, there are a number of ways to increase the stability of slopes. The methods are the opposite of processes that tend to increase the driving forces and decrease the resisting forces.

The most common method of reducing the driving forces is to reduce the mass of material acting to cause downslope movement above a failure plane. This can be done by flattening the slope (decreasing the slope angle) or excavating material from the upper part of the slope (Figure 13-31). For rock slopes hazardous blocks of rock bounded by joints or bedding planes can be blasted and removed to decrease the possibility of rock slides at a later date.

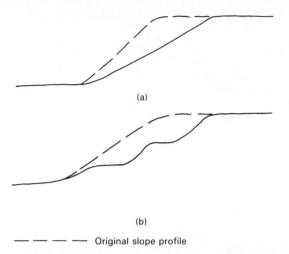

(a)

(b)

— — — — — Original slope profile

Figure 13-31 Techniques for increasing slope stability include (a) decreasing the slope angle and (b) benching the slope.

Highway cuts are often *benched* to decrease the potential damage from rock falls or rock slides (Figure 13-31b).

The resisting forces of a slope can be increased in several ways. Since one of the most important processes that reduces resisting forces is the increase in pore-water pressure, dewatering, or drainage, of a slope comprises one of the most effective mechanisms for increasing resisting forces. Drainage includes both controlling the surface-water movement across a slope as well as decreasing the internal water within a slope. Surface water and shallow subsurface water can be directed from the slope by drainage ditches and interceptor drains (Figure 13-32). Deeper drainage devices can also be installed. Horizontal drains (Figure 13-32) bring water to the slope face, where it can be safely removed from the slope. An alternative method of deep internal drainage is the use of wells that are continuously or intermittently pumped.

In addition to the beneficial effect of drainage, resisting forces can also be increased by the construction of various types of walls or fills at the base of the slope (Figure 13-33). Among several types of retaining walls used, walls composed of piles are particularly useful for resisting failure along planes located below the base of the slope (Figure 13-33b). Closely spaced piles made of timber, concrete, or other materials anchor the unstable material above the failure surface to more stable rock or soil units below.

Where more space is available, *buttress* or *counterweight* fills can be emplaced at the base of the slope (Figure 13-33a). The fills should be composed of rock or soils with good strength and drainage characteristics. The use of counterweight fills is sometimes called "loading the toe." Increases or decreases

Flatten Slope
Drain H₂O
Load Toe

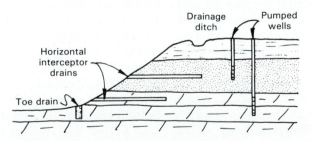

Figure 13-32 Various methods of drainage control and dewatering used to increase stability.

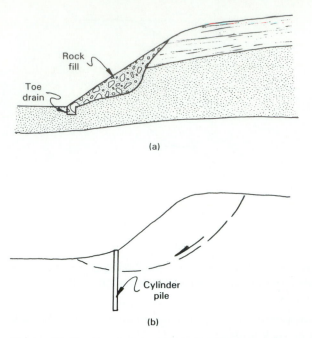

(a)

(b)

Figure 13-33 Retaining walls are often constructed at slope bases to prevent failure: (a) rock buttress, (b) cylinder-pile wall.

of mass can therefore be used to stabilize slopes. Removal of slope material from the upper part of the slope and addition of suitable fill to the lower part of the slope produce a more desirable safety factor.

CASE STUDY 13-1
LANDSLIDE REMEDIATION

The costs necessary to stabilize an unstable slope can be very high, particularly when the slope movement affects a highly developed area. Cincinnati, Ohio, is a city with serious slope-stability problems, which result from the presence of weak surficial materials along the steep sides of the Ohio River valley. The bedrock beneath the city includes shales that weather to form an unstable mantle of colluvium up to 15 m thick. Colluvium is a slope deposit created by the gradual weathering and slow downslope movement of debris derived from bedrock. The strength of the clay-rich colluvium is much lower than the shale from which it originates.

During preliminary construction of interchanges for an interstate highway, excavations in colluvium were made at the base of a hill known at Mt. Adams. Slowly moving slides soon developed in the colluvium upslope from the excavation, damaging numerous buildings,

streets, and utilities. In an attempt to stabilize the hillside, a 300-m-long retaining wall was constructed at the base of the slope (Figure 13-34).

The retaining wall utilizes a unique design involving cylinder piles and tiebacks to a tunnel constructed deep beneath the hill (Figure 13-35). Most of the cylinder piles in the wall are 2.1 m in diameter and are located on 2.4-m centers. The piles, which are constructed of reinforced concrete, extend to depths of more than 20 m, well into bedrock beneath the colluvium. Additional support is provided by high-strength wire tie backs that are anchored in a tunnel that runs parallel to the retaining wall. The $22 million total cost of the wall, which was completed in 1982, was shared between the federal government and the city. The retaining wall project was at that time the most expensive landslide-stabilization project in the United States.

Figure 13-34 Cylinder-pile retaining wall under construction at the base of Mount Adams in Cincinnati, Ohio.

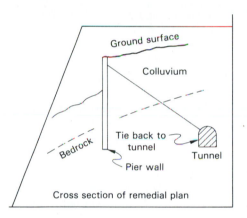

Figure 13-35 Cross section of the retaining wall showing the supporting tie backs anchored in the tunnel excavated in bedrock. (From Fleming et al., 1981, *Engineering Geology of the Cincinnati Area*, American Geological Institute.)

SUMMARY AND CONCLUSIONS

Mass movements include falls, topples, slides, lateral spreads, and flows. Combinations of more than one process are common. Rotational slides (slumps) have a circular failure surface and occur in slopes underlain by materials with low shear strength. Translational slides in rock or debris have a planar failure surface that often occurs at the contact of bedrock with overlying unconsolidated material. Processes involving liquefaction and flow of a zone within a slope composed of quick clay, accompanied by transport of more competent material above, like packages on a conveyor belt, are called lateral spreads. Flow—slope movements with internal deformation—are mechanically complex. Earth and debris flows approximate the behavior of a Bingham substance, a combination of viscous and plastic flow that is laminar and can transport large boulders within a plug of constant velocity at the center of the channel. Catastrophic rock-fall-debris avalanches may create a fluid composed of rock fragments, a phenomenon known as grain flow.

Determining the causes of specific slope movements involves the evaluation of a large number of interacting variables. Slopes that are intrinsically susceptible to mass movements contain materials with low shear strength. Clays are commonly involved; quick and overconsolidated clays present special problems. Failures of rock slopes are controlled by planes of weakness within the rock.

Slope stability can be described as a ratio between resisting and driving forces. Removal of lateral support, addition of weight, and earthquakes tend to increase the driving forces, while increases in pore pressure and various weathering processes decrease the resisting forces. If these processes can be identified by preliminary geologic investigations, various design and construction procedures can be used to counteract potential declines in the safety factor or actually increase the safety factor of the slope.

A better understanding of slope processes can help prevent loss of life and property in addition to minimizing the long-term costs of engineering projects.

REFERENCES AND SUGGESTIONS
FOR FURTHER READING

FLEMING, R. W., A. M. JOHNSON, and J. E. HOUGH. 1981. *Engineering Geology of the Cincinnati Area: GSA Cincinnati '81 Field Trip Guidebooks*, pt. 3, edited by T. G. Roberts. American Geological Institute.

HADLEY, J. B. 1964. *Landslides and Related Phenomena Accompanying the Hebgen Lake Earthquake of August 17, 1959*. U.S. Geological Survey Professional Paper 435, pp. 107–138.

HANSEN, W. R. 1966. *Effects of the Earthquake of March 27, 1964, at Anchorage, Alaska*. U.S. Geological Survey Professional Paper 542-A.

JOHNSON, A. M. 1970. *Physical Processes in Geology*. San Francisco: Freeman, Cooper, and Co.

KIERSCH, G. A. 1964. Vaiont Reservoir disaster. *Civil Engineering* 34:32–39.

PLAFKER, G., and G. E. ERIKSEN. 1978. "Nevados Huascaran avalanches, Peru." In *Rockslides and Avalanches: 1. Natural Phenomena*, edited by B. Voight. Amsterdam: Elsevier, pp. 277–314.

SCHUSTER, R. L., 1978. "Introduction." In *Landslides, Analysis, and Control*, edited by R. L. Schuster and R. J. Krizek. Transportation Research Board Special Report 176, National Academy of Sciences.

SCHUSTER, R. L., and R. J. KRIZEK, eds. 1978. *Landslides, Analysis, and Control*. Transportation Research Board Special Report 176. National Research Council.

U.S. Army Corps of Engineers, St. Paul District. 1978. Flood control Burlington Dam, Souris River, North Dakota. Design Memorandum no. 2, Phase 2: Project Design, Appendix B—Geology and Soils.

PROBLEMS

1. On what parameters is the classification of mass movements based?
2. Describe the surface features of a slump.
3. What conditions are conducive to the development of debris slides?
4. How can lateral spreads move long distances without excessive deformation of the blocks involved in the movement?
5. What is the basic mechanical difference between slides and flows?

6. Why can the movement of a debris flow be explained by the deformation of a Bingham substance?

7. Identify and explain the major explanations put forth for the movement of catastrophic rock-slide-debris avalanches.

8. What is meant by grain flow?

9. How does the safety factor measure the stability of a slope?

10. What are overconsolidated clays?

11. Explain why slope movements are often preceded by a decrease in effective stress.

12. What are the specific objectives of remedial actions taken after a slope failure occurs?

CHAPTER
14
RIVERS

A study of human history would be grossly inadequate if it failed to consider the impact of rivers on the development of civilization. Rivers have served as transportation routes, borders for political land divisions, and sources of water for drinking, industry, farming, and waste disposal since the earliest days of human existence. Geologists recognize the importance of river, or *fluvial*, processes throughout geologic time from the study of fluvial landscapes and from the abundance of sediments and rocks deposited by rivers of the past. One might suppose that our intimate association with rivers would have led to a thorough understanding of fluvial processes and the ability to harness the useful properties of rivers, while avoiding the losses of life and property that accompany river flooding. On the contrary, flood damages in the United States have increased during the past few decades despite the construction of flood-control projects on a massive scale in the present century.

RIVER BASIN HYDROLOGY AND MORPHOLOGY

The Drainage Basin

A river by itself is only one component of the hydrologic cycle. The potential energy possessed by water that falls as rain and snow on the continents has the capacity to accomplish a great deal of work during its return flow to the ocean. We have already considered the fraction of this water that moves through the subsurface; now we must examine the water that travels on the surface of the earth. The work done by surface water is truly immense—over millions of years, continents are gradually worn down by rivers along with other weathering and erosional processes and transported to the sea as small particles and dissolved constituents.

A convenient approach to the study of fluvial systems is based on the concept of a *drainage basin* (Figure 14-1). A drainage basin is an area containing an integrated network of stream segments that join together to form successively larger streams until one channel carries the entire surface flow out of the basin at its outlet. The outlet of the Mississippi Basin is, therefore, the point where the Mississippi River meets the Gulf of Mexico. Within the Mississippi Basin, however, many smaller basins can be defined. The Missouri Basin, for example, consists of the area drained by the Missouri River and all its tributaries upstream from its confluence with the Mississippi River. The boundaries of river basins are *drainage divides* (Figure 14-1), because they separate adjacent basins. Drainage divides are usually ridges or other topographically high areas. Thus a river basin can be thought of as a well-defined area bounded on all sides by divides and within which all streams and rivers contribute to the flow of the main, or *trunk*, stream that leaves the basin.

Stream Patterns

Numerous analyses of the spatial arrangement of streams in drainage basins have shown that stream networks conform to a small number of *stream patterns* (Figure 14-2). Specific stream patterns develop in response to the initial topography of an area and the distribution of rock types of varying erosional resistance. *Dendritic* patterns are characteristic of moderate slopes and rocks of fairly uniform resistance to stream erosion. These same rock types can develop a *parallel* pattern on steeper slopes. A different type of pattern results when the rock units are arranged in parallel bands of alternating weak and strong lithologies. Rocks with a low resistance to erosion are preferentially

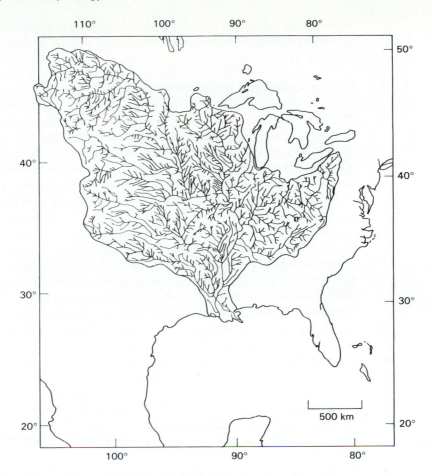

Figure 14-1 The Mississippi River drainage basin, which covers most of the central United States. (From R. K. Matthews, *Dynamic Stratigraphy,* 2d ed., copyright © 1984 by Prentice-Hall, Inc., Englewood Cliffs, N.J.)

exploited by streams, so that, in a *trellis* pattern, larger streams occupy zones of nonresistant rocks, while small, short stream segments drain the resistant, topographically higher beds. Folded sedimentary rock sequences commonly display trellis stream patterns. *Rectangular* patterns are characteristic of rocks of uniform resistance cut by perpendicular joint sets. Streams preferentially occupy joint traces because of the lowered resistance caused by fracturing and

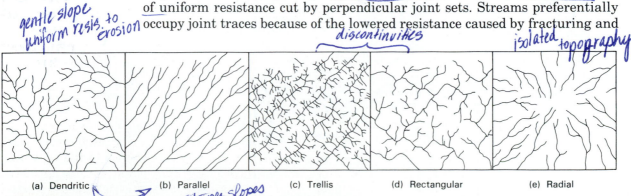

(a) Dendritic   (b) Parallel   (c) Trellis   (d) Rectangular   (e) Radial

Figure 14-2 Common drainage patterns. (From A. D. Howard, 1967, *American Association of Petroleum Geologists Bulletin* 51:2246–2259. Used by permission of the American Association of Petroleum Geologists.)

weathering along the joints. The final major stream pattern, *radial*, develops on isolated topographic uplands; volcanoes, domes, and buttes may display this pattern.

Stream Order and Drainage Density

The stream pattern is only one aspect of the stream networks that occupy drainage basins. Early researchers of drainage-basin relationships devised a system of ordering stream segments within a drainage basin (Figure 14-3). The system consists of progressively higher order numbers, starting from the small streams near the drainage divides to the larger streams at the center of the basin. *First-order* streams are defined as those stream segments with no tributaries. When two first-order streams segments join, a *second-order* stream is formed. The resulting stream ordering system is useful because of the consistent proportional relationships between stream order and other basin characteristics. For example, Figure 14-4 indicates that stream order is proportional to the logarithm of the number of stream segments, the length of stream segments, and the drainage basin area. These relationships, showing that the number of stream segments decreases and the length of stream segments increases with increasing stream order, are common enough to be considered empirical laws of drainage networks. In a similar fashion, the area of a particular drainage basin is directly proportional to stream order.

Another important relationship in drainage basins is the *drainage density*. This parameter is defined as the total length of all stream segments divided by the area of the basin; it is therefore a measure of the number of stream channels per unit area of a drainage basin. Drainage density is extremely important because it influences the conveyance of water through the stream network during floods and lesser runoff events. Drainage density is controlled by the interaction between climate and geology. The geologic factors include the permeability of the rocks and soils at the basin surface. High-permeability

#streams channels
area

hi density -
more channels
for runoff
(good for floods)

high stream
order =
- lot of streams
- bigger basin
 # longer stream
 # aren't as many
higher-order
streams

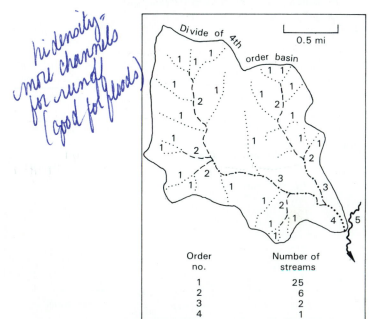

Figure 14-3 Strahler's method of determining stream order. (From A. N. Strahler, 1957, *Transactions,* American Geophysical Union. Used by permission of the American Geophysical Union.)

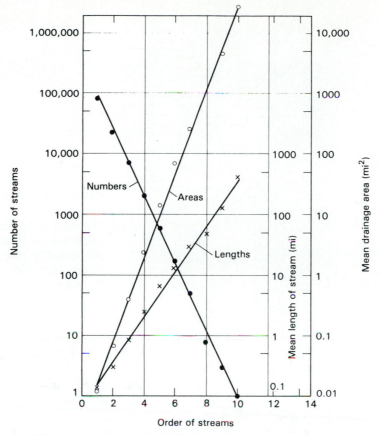

Figure 14-4 Relationship of stream order to other parameters in the Susquehanna River drainage basin. With increasing stream order, the number of streams decreases, whereas the drainage-basin area and stream length increase. (From L. M. Brush, Jr., 1961, U.S. Geological Survey Professional Paper 282-F.)

materials usually have high infiltration capacities. The more rain or snow melt that infiltrates, the less is available for surface runoff. A low-density, or coarse-texture, drainage basin is the result (Figure 14-5). Low infiltration capacity, on the other hand, causes more surface runoff. More drainage channels are eroded and a fine-texture drainage basin develops. Climate influences *H₂O is taken in* drainage density by controlling the amount and type of vegetation present. The heavy vegetation cover of humid climatic regions increases infiltration relative to the barren slopes of an arid or semiarid region, where overland flow is much more intense. The importance of drainage density lies in the percentage of rain that travels as overland flow versus that which infiltrates into the subsurface. A fine-texture drainage basin can transmit more water through the stream channels at a faster rate. Severe and frequent floods can be the result. *permeability → more H₂O in soil = less H₂O to erode & make channels*

The Hydrologic Budget

One important reason for establishing the concept of a drainage basin is to develop a quantitative accounting system for water in the basin. The *hydrologic budget* of a watershed relates the quantity of water supplied to the basin through precipitation to the quantity of water leaving the basin. Assuming

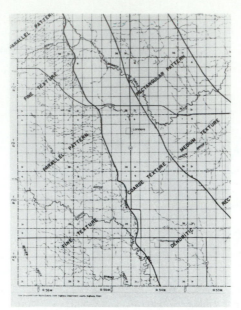

Figure 14-5 Stream drainage in a portion of Grand Forks County, North Dakota. The area of coarse texture is underlain by sand with high permeability. The area of fine texture is a slope underlain by material with low permeability. (From D. E. Hansen and J. Kume, 1970, North Dakota Geological Survey Bulletin 53, Part I.)

that there are no groundwater inflows or outflows from a basin and that the equality refers to a long-term average, the hydrologic budget equation can be expressed as

$$P = Q + E$$ Eq. 14-1

where P represents precipitation, Q is the runoff, and E is the evapotranspiration. The runoff term in the equation includes both the surface runoff from the basin and the groundwater component of runoff, which is the base flow of the streams in the basin (Chapter 12). Since precipitation and runoff can be measured fairly accurately, evapotranspiration is usually calculated as the difference between precipitation and runoff.

One of the major objectives of engineering hydrology is the prediction of the amount of flow in a river that will result from a particular precipitation event. This information, which is needed for design of bridges, dams, and flood-control structures, can be obtained only through an understanding of the processes operating in the drainage basin. The geology, topography, climate, and vegetation interact to produce a specific network of streams in a drainage basin. These factors also control the movement of water from a particular storm through the basin.

Development of Drainage Networks

Similar statistical relationships between variables in drainage basins throughout the world prove that rivers erode the valleys in which they flow over long periods of time. One of the major processes in the development of drainage networks is *headward erosion*. All streams exhibit the tendency to extend their channels in a headward, or upstream, direction. If streams are followed upstream to the point where first-order tributary channels just begin, the process of headward erosion can best be observed. The channel frequently ends on a hillslope with a concave profile (Figure 14-6). This part of the channel collects converging overland flow from three sides of the slopes above during heavy

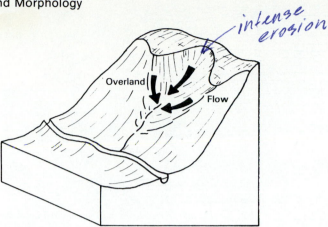

intense erosion

Figure 14-6 Focusing of overland flow at the heads of first-order tributaries. Headward erosion is the result.

rainfall events. Consequently, erosion is intense at this point and the channel is gradually extended upslope.

Headward erosion is related to the processes of *stream capture*. Streams erode headward at different rates because of the erosional resistance of the rocks or sediments they are flowing over or because one stream has a steeper slope than another and therefore deepens its channel at a greater rate. As a stream is actively eroding its channel headward, it may intersect another stream (Figure 14-7). Because of its greater slope and energy, the first stream may divert, or capture, the second stream and all its tributaries above the point of intersection. The river that has lost its upstream sections is said to have been *beheaded*. It is left with a much smaller discharge and drainage basin than it had prior to the capture. Drainage basins can by stream capture become larger or smaller through geologic time.

If a stream is actively downcutting and deepening its valley, it may encounter contacts between rock types of very different erosional resistance. These conditions can lead to the formation of a waterfall, particularly if a more

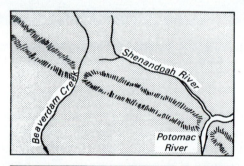

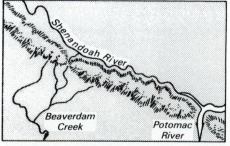

- rock on top is more resistant
- rock on bottom wears away

Figure 14-7 An example of stream capture in Virginia. Headward erosion by a tributary of the Shenandoah River captured the upper part of Beaverdam Creek. (From T. L. McKnight, *Physical Geography: A Landscape Appreciation*, copyright © 1984 by Prentice-Hall, Inc., Englewood Cliffs, N.J.)

Figure 14-8 A waterfall formed by erosion of a formation of low resistance underlying a resistant formation. (From S. Judson, M. E. Kauffman, and L. D. Leet, *Physical Geology*, 7th ed., copyright © 1987 by Prentice-Hall, Inc., Englewood Cliffs, N.J.)

Figure 14-9 Niagara Falls, a waterfall formed in the manner suggested by Figure 14-8. (J. R. Balsley; photo courtesy of U.S. Geological Survey.)

resistant rock is underlain by a less resistant rock. As shown in Figure 14-8, the underlying nonresistant rocks are rapidly eroded, thereby removing underlying support for the resistant rocks. Eventually, the resistant rocks collapse to restore a more stable slope profile. This process is continuous and, with time, the waterfall migrates upstream, while maintaining its form. Niagara Falls is a classic example of this process (Figure 14-9).

STREAM HYDRAULICS

Flow Types

We are all familiar with the variety of flow conditions that exist in river channels. Although the rapid rush of water down a boulder-filled mountain stream channel appears very different from the sluggish, almost imperceptible, move-

compute ⎡ Driving - gravity
⎣ Resisting - friction drag

340

ment of a winding coastal-plain river, the flow conditions in all rivers are the
result of a balance between the gravitational driving forces and the resistance
of the stream channel and the flowing water itself. Two basic flow conditions
can be defined by considering the flow paths of individual water particles. At
low velocities, the parallel flow paths of adjacent particles constitute *laminar*
flow. Figure 14-10 illustrates laminar flow using streamlines, which are imaginary lines parallel to the velocity vectors of representative water particles.
Resistance to flow under laminar conditions is generated by viscous interaction
between particles within the flow.

As velocity increases, the parallel laminar flow paths can no longer be
maintained. Instead, random velocity fluctuations in all directions arise within
the flow. These perturbations of flow are called *turbulence* and the resulting
flow is turbulent flow. Turbulence generates additional resistance by these
secondary motions, initiating a component of viscosity called *eddy viscosity.*

A dimensionless number called the *Reynolds number* (*Re*) is used to distinguish laminar from turbulent flow. Its defintion is

$$Re = \frac{\rho_w v R}{\mu}$$ Eq. 14-2

where ρ_w is the density of water, v is velocity, μ is dynamic viscosity, and R
is the hydraulic radius of the channel. Hydraulic radius, R, is a geometric
property of the river channel defined as the cross-sectional area of the channel,
A, divided by the wetted perimeter, P:

$$R = \frac{A}{P}$$ Eq. 14-3

In many natural channels, cross-sectional area can be approximated by the
product of width and depth (Figure 14-11), and the wetted perimeter as $P =$
$W + 2D$. The boundary between laminar and turbulent flow occurs within
the Reynolds-number range of 500–750. In nature laminar flow usually occurs
in groundwater flow systems, while even the slowest of rivers fall into the
range of turbulent flow.

A second dimensionless number used to characterize the flow of rivers is
the *Froude number* (*Fr*), which is defined as

$$Fr = \frac{v}{\sqrt{gD}}$$ Eq. 14-4

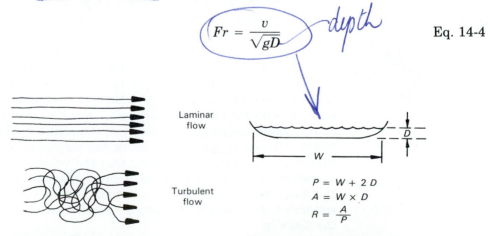

Figure 14-10 Laminar and
turbulent flow.

Figure 14-11 Hydraulic
geometry of a natural stream
channel.

where v is stream velocity, g is the acceleration of gravity, and D is depth of flow. Froude number values are used to classify flow into categories defined as subcritical ($Fr < 1$), critical ($Fr \sim 1$), and supercritical or rapid ($Fr > 1$). Since velocities of natural streams rarely exceed several meters per second, examination of Eq. 14-4 indicates that only shallow, high-velocity streams can achieve critical or supercritical flow. Gravelly mountain streams occasionally meet these criteria.

Froude numbers can also be used to predict the type of *bedform* that will develop on a stream bottom composed of loose, cohesionless sediment such as sand. Bedforms include wavelike accumulations of bed material that form in response to the frictional drag exerted by the flowing water on the particles lying on the stream bed. The continued flow of water over the bedforms erodes, deposits, and transports sand, so that the entire train of sand waves migrates across the bed, usually in a downstream direction, while maintaining its wave-like form. Figure 14-12 illustrates that, with gradually increasing flow velocities and Froude numbers, bedforms progress through a sequence of *ripples*, *dunes*, plane bed conditions, and *antidunes*. Ripples are small bedforms less than 60 cm in length from trough to trough, whereas dunes and antidunes are larger bedforms that usually range in size from 60 cm to several meters in length. Antidunes migrate in an upstream direction because particles are eroded from the downstream side of one antidune and deposited on the upstream side of the adjacent bedform.

Discharge and Velocity

The amount of water moving past a certain point in a channel in a given amount of time is a very important characteristic of flow. This quantity, known as *discharge*, is measured perpendicular to the direction of flow. The usual units of measurement for discharge are cubic meters per second. Flow is considered to be either *steady* or *unsteady*, depending on whether discharge is constant or variable with time at a certain point. *Uniform* flow exists when the water depth is constant along a stream course and the slope of the water surface is equal and parallel to the slope of the bed. Even though steady, uniform flow is never truly present in a natural river, this assumption is often made for short periods of time so that uniform flow equations can be used.

The velocity of a stream at any point in the channel is determined by the slope of the channel and the distance from the point in the flow to the sides or bed of the channel. The influence of the bed on velocity is shown in Figure 14-13. There is a very thin layer of water adjacent to the bed in which the water velocity is zero. The variation in velocity with vertical distance above the bed is shown in Figure 14-13a. The channel sides exert a similar influence on velocity (Figure 14-13b). These two effects are combined in Figure 14-13c, which shows contours of relative velocity (lines connecting points with an equal fraction of the maximum velocity). It is clear that the maximum velocity occurs at the water surface in the middle of the channel. The mean velocity of the stream (v) occurs somewhere below the water surface in the middle of the channel. It is sometimes assumed that the velocity at six-tenths of the depth as measured downward from the water surface in the middle of the stream is the mean velocity of the cross section. If this velocity can be measured or estimated, it is possible to determine the discharge of the river at a point using the formula

$$Q = Av$$ Eq. 14-5

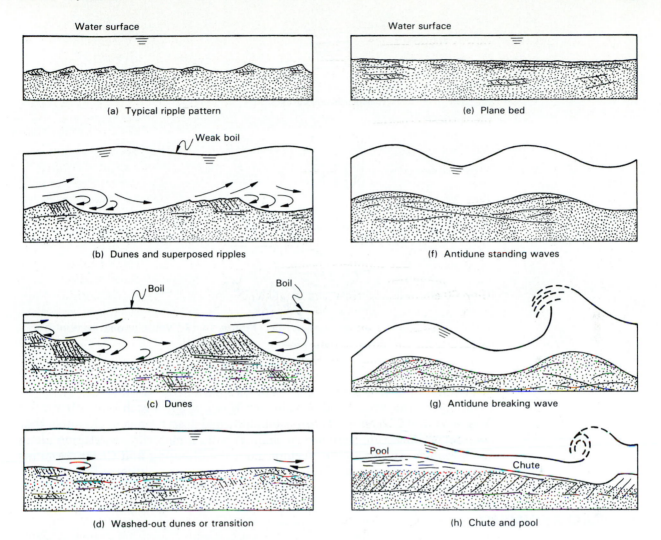

Figure 14-12   Sequence of bedforms associated with increasing flow velocities and Froude numbers. Froude number is less than 1 for diagrams A through C, approximately 1 for diagrams D and E, and greater than 1 for diagrams F through H. (From D. B. Simons and E. V. Richardson, 1966, U.S. Geological Survey Professional Paper 422-J.)

where Q is discharge in cubic meters per second, A is cross-sectional area in square meters, and v is mean velocity in meters per second.

Other commonly used flow formulas can be derived by equating the driving forces of stream flow (gravity) with the resisting forces (channel boundaries) and solving for the resulting velocity. For example, the Chezy formula,

$$v = C\sqrt{RS}$$ Eq. 14-6

expresses the velocity (v) in terms of the hydraulic radius (R), the slope (S), and a constant (C). The Manning equation (for SI units),

$$v = \frac{1}{n}R^{2/3}S^{1/2}$$ Eq. 14-7

Longitudinal cross section
(a)

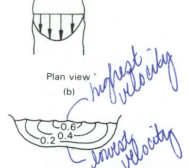

Plan view
(b)

Transverse cross section   **Figure 14-13** Velocity distribution in a
(c)   river channel.

is a similar approach, although it includes a parameter *n*, which describes the
roughness of the stream bed. A value for *Manning's n*, which typically ranges
from 0.01 to 0.07, can be chosen from published descriptions of channels that
consider variation in grain size of channel-bottom material, vegetation along
the stream banks, and other factors. Both the Manning and Chezy formulas
are used for natural streams even though they assume uniform flow.

STREAM SEDIMENT

Part of the work that streams accomplish is the transportation of sediment
produced by weathering and erosional processes. Over millions of years, entire
mountain ranges are leveled by these processes. In a particular reach of a
stream, sediment can simply be carried along by the stream. Alternatively,
the stream, by the forces it exerts on its channel boundaries, can *entrain* particles from the channel sides and bottom into the flow. In this way rivers can
widen and deepen their channels.

Entrainment

Erosion of a stream's sides or bottom can be accomplished in several ways. If
the boundary material is composed of soluble rock like limestone, *dissolution*
by the stream is possible. In addition, the sediment carried by the stream can
do erosional work. Particularly in the case of large particles that slide and roll
across the stream bottom, *abrasion* can occur. Laboratory experiments indicate
that streams are much more erosive if they contain a high sediment load relative to sediment-free water. When large particles are dropped to the stream
bottom as a result of velocity fluctuations, the *impact* can be great enough to
break off projecting fragments of the stream-bed material.

   Although streams may actively erode their channels by the preceding
mechanisms, the process of *entrainment* usually refers to the ability of the

river to pick up loose particles of material from the bed by the direct hydraulic action of flowing water. As the stream flows in its channel, the moving water exerts shearing forces upon the stationary bed. If these forces are larger than the gravitational forces tending to hold a particular particle on the stream bottom, the particle will be picked up off the bed and entrained into the flow.

Flume experiments, in which river processes are simulated by the flow of water in long tanks, demonstrate that as the velocity increases, progressively larger particles are entrained from the bed. The Hjulstrom diagram, shown in Figure 14-14, relates water velocity and particle size. The dashed lines divide the areas into fields of erosion, transportation, and sedimentation. One important point shown by the diagram is that higher velocity is required to erode or entrain a particle of a given size than is needed to transport the particle. Another aspect of the curves that initially may not seem logical is that the velocity needed for erosion increases as particle size decreases near the left side of the diagram. This effect can be explained by the properties of clay particles that fall in that size range. The cohesion between clay particles leads to the entrainment of clay-particle aggregates into the flow rather than individual clay particles.

The relationship between velocity and particle size is often difficult to work with because of the variation of velocity in the channel. For a particle of a certain size it is the flow velocity just above the bed, or the *critical bed velocity*, that must be large enough to move the particle. The relationship between the critical bed velocity and the mean velocity of the stream is not always clear. For these reasons the relationship between particle size and *critical shear stress* is often used instead of velocity. The critical shear stress, τ_c, or the shear stress just great enough to move a particle of a certain size, is expressed as

$$\tau_c = \rho_w g R S$$

Eq. 14-8

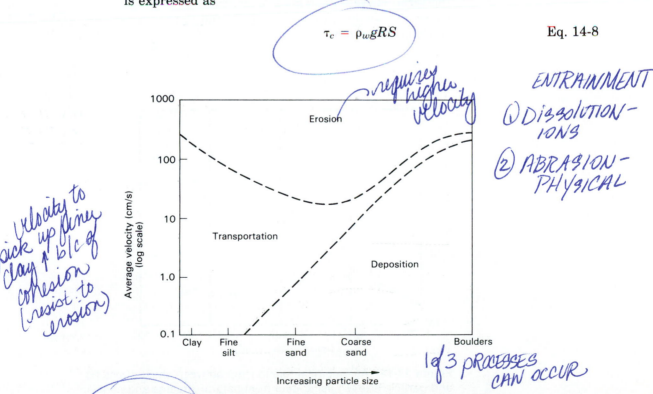

Figure 14-14 The Hjulstrom diagram, relating stream velocity to particle size for conditions of erosion, transportation, and deposition.

where ρ_w is the water density, g is gravity, R is hydraulic radius, and S is the energy slope of the stream. The energy slope can be approximated by the mean slope of the stream bed. This approach presents fewer problems in application because hydraulic radius and slope can be measured more accurately than critical bed velocity. An empirical relationship between critical shear stress and particle size is shown in Figure 14-15 for studies of natural rivers and canals.

Transportation

The total amount of sediment a stream carries is known as its *load*. Sediment load can be divided into three types based on the process by which it is transported.

Dissolved load is carried totally in solution by a stream. Much of the dissolved load of a stream is derived from the base flow—that is, the groundwater component of the total stream flow. During its relatively long residence time in the groundwater flow system, the base-flow component of stream flow chemically reacts with the solid materials it encounters through the various types of weathering processes. From these reactions it derives the dissolved load that it carries to the surface stream in a groundwater discharge area. The surface runoff component of a stream is generally lower in dissolved-solids content because of its short travel time to the stream channel.

Of the nondissolved load of a river, the smaller particles are carried as *suspended load*. This form of transport implies that particles are prevented from settling to the bottom of the stream by the movement of the water. The rate at which particles tend to fall through a static liquid column is known as the terminal velocity. This is a constant value that represents a balance between the opposing forces of gravity and viscous resistance by the fluid. The equation relating these forces is known as *Stokes' Law*, from which the ter-

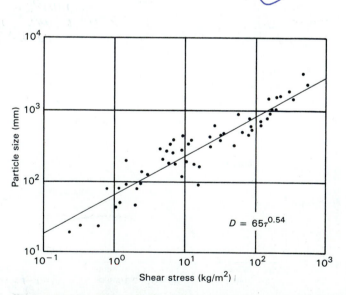

Figure 14-15 An empirical relationship between shear stress (τ) and particle size (D) using data from natural streams and canals. (Modified from V. R. Baker and D. F. Ritter, 1975. Geological Society of America Bulletin 86:975–978.)

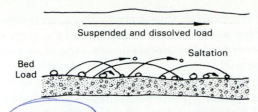

Figure 14-16 Transportation of stream sediment includes rolling and sliding of bed load, saltation of small bed-load particles, and suspended and dissolved load transport.

minal velocity can be stated as (SI units)

$$v = \frac{1}{18} d^2 \frac{(\rho_s - \rho_w)g}{\mu}$$ Eq. 14-9

where d is particle diameter, ρ_s and ρ_w are the densities of the particle and fluid, respectively, g is the acceleration of gravity, and μ is the dynamic viscosity of the fluid. Stokes' Law indicates that the terminal velocity of a particle is proportional to the square of the particle diameter.

In order for the particle to remain in suspension, an upward velocity greater than the terminal velocity must be imparted to the particle. Upward velocities are provided by turbulent eddies and other velocity fluctuations distributed throughout the stream. These turbulent effects are proportional to the mean velocity of the stream, so that the higher the stream velocity, the larger the particle size that can be carried in suspension.

The final type of load is the fraction of the total load that generally remains in contact with the bed during transportation. *Bed load* includes the particles that are too large to be transported by suspension. These particles move by rolling or sliding along the bed or by a process known as *saltation*, in which transient turbulent eddies can temporarily lift a particle off the bed. When the eddy dissipates, the particle can no longer be kept in suspension, and is dropped to the bed (Figure 14-16). The particle thus moves downstream in a series of small jumps.

Two general terms characterize the load transported by a stream. *Competence* refers to the largest particle a stream can carry. It is a function of the velocity and shear stress exerted by the stream. *Capacity* is the total amount of sediment that a stream can carry by all mechanisms.

DEPOSITIONAL PROCESSES

When the velocity of a stream decreases, particles begin to drop from suspension or cease movement along the stream bed. Deposition of this type is usually a very temporary situation. Sediment is merely stored in the channel for short periods of time until the velocity increases again. Periodic fluctuations in stream velocity are common.

More permanent deposition within and adjacent to a stream channel is also possible. When a river floods, for example, a river overflows its channel and inundates the flat *flood plain* adjacent to the channel. Velocity drops significantly once water leaves the channel. Flood plains are characterized by parallel layers of fine-grained sediment deposited by successive floods.

Climatic conditions or the existence of an excessive sediment supply can promote a gradual accumulation of sediment in a river valley. The nature of the deposits corresponds to the particular type of channel pattern developed by the river. Channel patterns, which can be classified as *straight, meandering,* and *braided,* will be discussed in detail in the next section.

A different type of depositional control is provided by decreases in channel gradient (slope). Gradient and velocity decrease where a stream valley extends from a mountain range out onto a plain and when a river enters a body of standing water like the ocean or a lake. Continuous sediment deposition occurs at these locations.

Meandering Streams

The factors that control the type of channel pattern include the gradient of the channel, the type of material the channel is incising, and the type and amount of sediment that the river carries. Straight channels are not very common, although meandering streams can have straight segments. Meandering streams, the most common type, tend to develop in gently sloping areas with low sediment supply and a high percentage of fine-grained sediment. The channel pattern also correlates with the cross-sectional shape of the channel (Figure 14-17), with meandering streams characterized by narrow, deep channels.

The intensity of meandering is quantified by the use of the parameter called *sinuosity,* which is measured by the ratio of channel length to valley length. Sinuosity is apparently related to the type of material that comprises the channel banks, the type of sediment load transported, and the gradient. The highly sinuous river shown in Figure 14-18 has a low gradient, is cut into sediment with a high percentage of silt and clay, and transports mostly fine-grained sediment as suspended load rather than bed load. Streams with the opposite characteristics have a much lower sinuosity.

The tendency of a stream to meander appears to be a basic law of fluid flow, although the factors that cause meandering are not totally understood. Even in a straight channel (Figure 14-19), the *thalweg,* the deepest part of the channel, establishes a meandering pattern. Straight excavated channels like canals or drainage ditches often erode their banks in order to develop a natural meandering pattern.

The flow of water in a meandering channel is shown in Figure 14-20. The position of the thalweg impinges on the outside banks of the meanders just downstream from the point of maximum curvature. Because the thalweg is

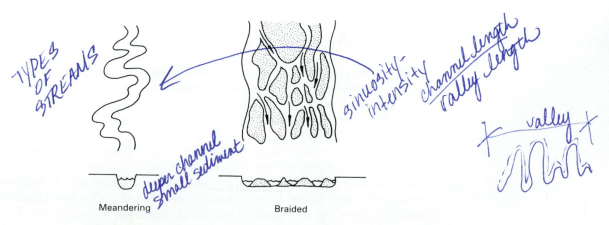

Figure 14-17 Map and cross-sectional views of meandering and braided streams.

Figure 14-18 Vertical air photo of a meandering stream with high sinuosity.

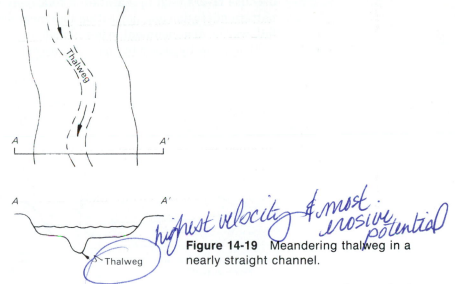

highest velocity & most erosive potential

Figure 14-19 Meandering thalweg in a nearly straight channel.

the position of greatest depth, it also represents the point of highest velocity in the channel. As a result of this higher velocity, the meandering channel tends to be eroded on the outside bank of the meander to form a *cut bank*. A

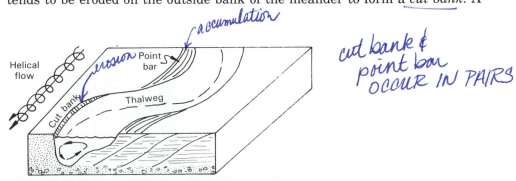

accumulation

erosion

cut bank & point bar OCCUR IN PAIRS

Figure 14-20 In a meandering channel, erosion occurs on the outer sides, or cut banks, of meander bends. Point-bar deposition is present on the inner bends. A secondary flow pattern superimposed on the downstream motion gives the overall flow a helical pattern.

secondary pattern of circulation is also imposed upon the flow in the vicinity of meanders. In the cross section of the lower part of Figure 14-20, the secondary flow is shown to move across the channel toward the cut bank at the water surface and then back along the stream bottom. Because the water is also moving downstream, the resulting overall motion resembles a corkscrew and is termed *helical flow.* On the inner bank of the meander, where the depth and velocity are less, deposition of the coarser fraction of stream sediment occurs. The resulting sediment accumulation is called a *point bar.*

This pattern of erosion and deposition of a meandering stream causes lateral migration of the channel meanders across the flood plain. The characteristic set of landforms that occur on the flood plains of meandering rivers (Figure 14-21) records the sequence of events of recent geologic time. Many of the topographic features are related to the lateral migration of meanders. The continuous process of cut-bank erosion and point-bar deposition produces a series of parallel concentric ridges and swales called *meander scrolls* (Figure 14-21). The ridges represent old point bars and therefore are composed of coarser sediment. The intervening swales are underlain by organic-rich, fine-grained sediment. Lateral migration of meanders cannot occur indefinitely because rivers tend to maintain a relatively constant value of sinuosity. The pattern of erosion and deposition in a meander leads to growth of the meander followed by abandonment. After abandonment, or meander *cutoff,* the old channel loop, now bypassed by the river, becomes an *oxbow lake* (Figures 14-22 and

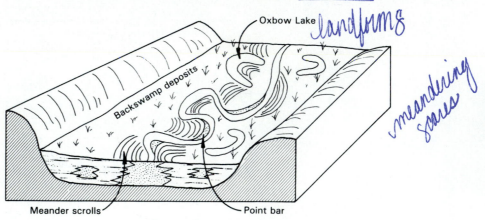

Figure 14-21 Landforms associated with a meandering river.

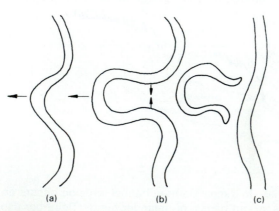

Figure 14-22 Meander migration leading to abandonment and formation of an oxbow lake. Arrows show direction of cut-bank migration.

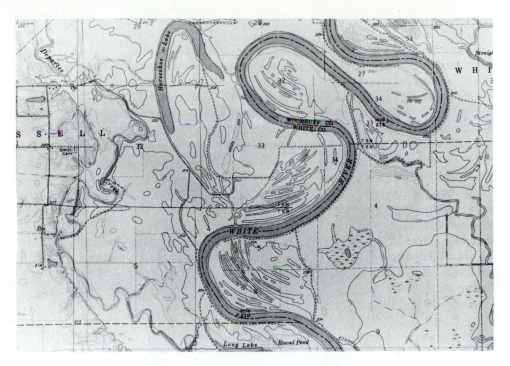

Figure 14-23 Topographic map of part of the White River flood plain in Arkansas, showing meander scrolls and an oxbow lake. The river will soon form another meander cutoff in the upper right part of the map. (Augusta SW Quadrangle, Arkansas, U.S. Geological Survey, 7.5-minute series.)

14-23). Eventually, the oxbow lake fills with highly organic silty and clayey sediments. These old channel fills, or *clay plugs,* provide foundation problems because of the high compressibility of the sediment.

The flood plain is named for good reason, for stream discharge periodically exceeds the channel capacity and flood water flows out of the channel and onto the flood plain. Typically, streams attain bankfull flow every 1 to 2 years on the average. The flood plain is the river system's natural storage area for the water and sediment available when the discharge exceeds bankfull flow.

Because flooding is intimately involved in the development of a flood plain, flood-plain sediment can be divided into two types. The dominantly coarser sediment in meander scrolls related to the lateral migration of meanders is classified as a *lateral accretion* deposit, and the sediment deposited by floods is known as a *vertical accretion* deposit. Vertical accretion deposits are fine grained because of the velocity decrease that accompanies flow out of the channel onto the broad flood plain. The coarsest available sediment is deposited adjacent to the channel margins to form *natural levees,* as shown in Figure 14-24. Grain size decreases with distance from the channel. The areas behind the natural levees are often called *backswamps* because of their organic-rich, fine-grained sediment, high water table, and swampy vegetation.

The complex distribution of sediments in the alluvial valley of a meandering stream presents many engineering problems because flood plains often are highly developed. Engineering problems arise both in the control of the periodic floods that inundate these areas and in construction on floodplains. High water tables frequently pose troublesome problems. In addition, the rapid lateral and vertical variation in sediment type requires a great deal of site investigation and soils testing for foundation design. The engineering prop-

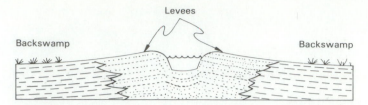

Figure 14-24   Natural levee deposits on a flood plain. (From S. Judson, M. E. Kauffman, and L. D. Leet, *Physical Geology*, 7th ed., copyright © 1987 by Prentice-Hall, Inc., Englewood Cliffs, N.J.)

erties associated with certain fluvial landforms and deposits are shown in Table 14-1.

Braided Streams

Streams that flow in broad, shallow channels of low sinuosity and consist of multiple subchannels separated by islands or bars are called *braided streams* (Figures 14-17, 14-25). The channel pattern within a braided stream—where subchannels continually divide and recombine around obstacles in the main channel—can be referred to as *anastomosing*. The anastomosing channel network of a braided stream forms in response to a different set of controlling variables from those associated with meandering streams. Conditions that favor the formation of braided streams include an abundant source of coarse-grained sediment, a high gradient, and cohesionless, nonresistant sediment in the channel banks. The cross-sectional profile that develops under these conditions is broad and shallow (Figure 14-26). In a broad, shallow stream higher velocities are closer to the stream bed, and transport of coarse particles is facilitated.

Figure 14-25   A braided stream. (T. L. Péwé; photo courtesy of U.S. Geological Survey.)

TABLE 14-1

Engineering Properties of Alluvial Deposits in the Lower Mississippi Valley

Environment	Soil texture (USC classification)	Natural water content (percent)	Liquid limit	Plasticity index	Cohesion (lb/ft²)	Angle of internal friction (degrees)
Natural levee	Clays (CL)	25–35	30–45	15–25	360–1200	0
	Silts (ML)	15–35	Nonplastic (NP)–35	NP–5	180–700	10–35
Point bar (ridges)	Silts (ML) and Silty Sands (SM)	25–45	30–45	10–25	0–850	25–35
Abandoned Channel	Clays (CL and CH)	30–95	30–100	10–65	300–1200	0
Backswamp	Clays (CH)	25–70	40–115	25–100	400–2500	0
Swamp	Organic Clay (OH)	110–265	135–200	100–165	Very low	
Marsh	Peat (Pt)	160–465	250–500	150–400	Very low	
Lacustrine	Clay (CH)	45–165	85–115	65–95	75–150	0
Beach	Sand (SP)	Saturated	NP	NP	0	30

SOURCE: From C. R. Kolb and W. G. Shockley, 1957. Used by permission of the American Society of Civil Engineers.

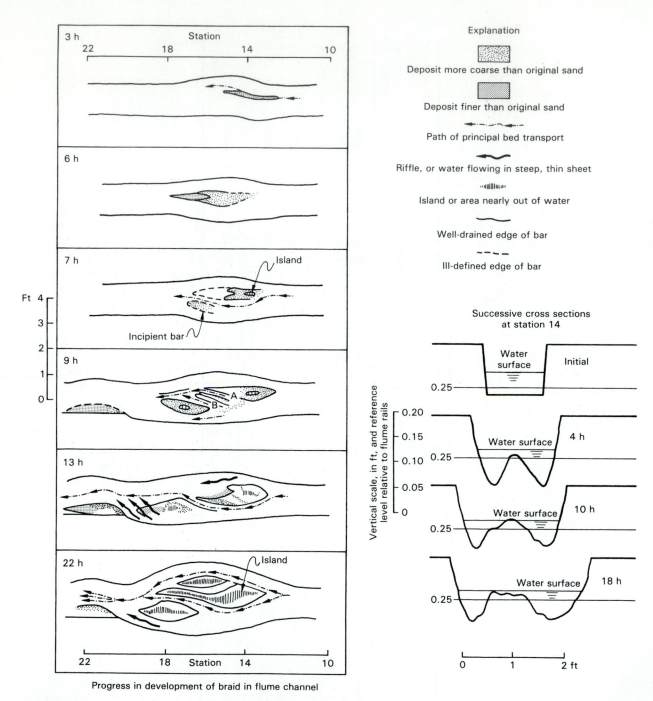

Progress in development of braid in flume channel

Figure 14-26 Development and modification of braid bars in a flume channel. Note changes in cross section during flow period. (From L. B. Leopold and M. G. Wolman, 1957, U.S. Geological Survey Professional Paper 282-B.)

The formation of braids involves the deposition of bars of sediment within the channel. A bar is a large-scale bedform. Once initiated, bars grow and migrate downstream because sediment is transported over the bar and then deposited on the downstream end, where the depth increases. As the bar grows, it occupies more of the channel cross-sectional area that was previously available for water. As a result, the stream must erode its banks laterally to enlarge the channel to maintain a sufficient capacity for the discharge (Figure 14-26). As the discharge drops from a period of very high flow when much of the bed load was in transportation, previously deposited bars are dissected by erosion and the sediment is redistributed through the system. The channel pattern in a braided stream constantly changes with fluctuations in discharge.

Alluvial Fans

If the gradient of a stream suddenly decreases, Eqs. 14-6 and 14-7 indicate that the velocity will decrease if other channel parameters remain the same. A river responds to a decrease in velocity by depositing the coarser fraction of its sediment load. A good example of a gradient decrease is the point where a stream emerges from a steep mountain front onto the floor of the adjacent basin. The stream gradient and velocity are suddenly decreased and sediment is deposited to form a conical land form called an *alluvial fan* (Figure 14-27).

The velocity decrease is also partly caused by the increase in width of the channel on the fan surface. The width can increase to the point where flow is no longer channelized but instead moves as a thin layer across a broad surface. This type of flow is known as *sheet flow*. If the fan surface is permeable, water rapidly infiltrates into the fan, and overland flow ceases. Other geomorphologic processes, such as debris flows, are also recognized as being important processes on many fans.

Alluvial fans are best developed where recent faulting has produced steep mountain fronts with adjacent basins (Figure 14-28). Arid climatic conditions, where flow in drainage networks occurs mainly during infrequent rainfall events of large magnitude, also facilitate the deposition of alluvial fans. The conical shape of some alluvial fans makes them easily recognizable on topographic maps.

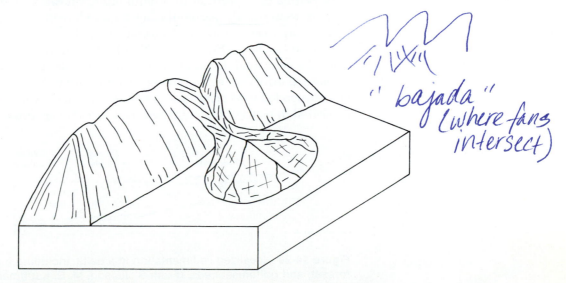

"bajada" (where fans intersect)

Figure 14-27 Alluvial fans form at places where stream gradients decrease and channel widths increase.

Figure 14-28 Alluvial fans in Death Valley National Monument, California. (H. E. Malde; photo courtesy of U.S. Geological Survey.)

Deltas

UNDERWATER ALLUVIAL FAN

SLOWING CURRENT

Where a river enters a lake or the sea, a *delta*—the subaqueous counterpart to an alluvial fan—is constructed. Deposition occurs in a delta because the river current slows and eventually dissipates as it moves progressively offshore. The gradual reduction in velocity is reflected in the distribution of sediment types in a delta, with deposition of coarse bed-load materials in the vicinity of the mouth of the river. As the river deposits its sediment load, the delta extends or *progrades* in a seaward direction.

Sediment deposition in a delta often follows a predictable pattern. The river flows across the accumulating delta sediments to a slope called the *foreset slope*. Coarse sediments are deposited on the foreset slope in long, dipping beds called *foreset beds* (Figures 14-29, 14-30). Progradation of the delta results from the continual accumulation of foreset beds. Beyond the foreset slope in a seaward direction, slow-moving currents carry the finer fractions of the river's suspended load across the ocean or lake floor, where they are eventually deposited in nearly horizontal beds called *bottomset beds*. As the river extends

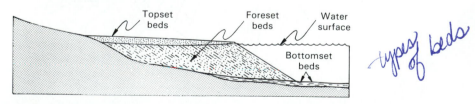

types of beds

Figure 14-29 Idealized sedimentation in a delta, including topset, foreset, and bottomset beds. (From S. Judson, M. E. Kauffman, and L. D. Leet, *Physical Geology,* 7th ed., copyright © 1987 by Prentice-Hall, Inc., Englewood Cliffs, N.J.)

Figure 14-30 Deltaic foreset beds. (C. S. Denny; photo courtesy of U.S. Geological Survey.)

its channel across the prograding delta to the foreset slope, it deposits horizontal beds called *topset beds*. This idealized sequence can be modified by many processes, including waves and tides.

As a river approaches a delta, the single main channel divides repeatedly into a network of *distributary* channels, which transport the sediments across the delta (Figure 14-31). Distributary channels are maintained by natural levees that project slightly above the delta surface. Sediment is transported through the distributary channel unless a break in the natural levee is present. These breaks, called *crevasses*, allow the diversion of water and sediment through the natural levee to a lateral position between distributary channels. The minidelta deposited as water flows through the crevasse is called a *crevasse splay*.

Sediment is deposited within a particular distributary network for a certain period of time. Eventually, that network becomes raised in elevation because of sediment deposition. The river will then abandon the active part of the delta in preference to an adjacent part of the delta with a steeper gradient.

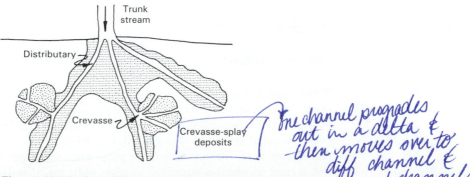

Figure 14-31 In a delta, the trunk stream branches into distributaries. When distributaries are breached, crevasse-splay sediments are deposited.

One channel prograder out in a delta & then moves over to diff channel & subchannels develop

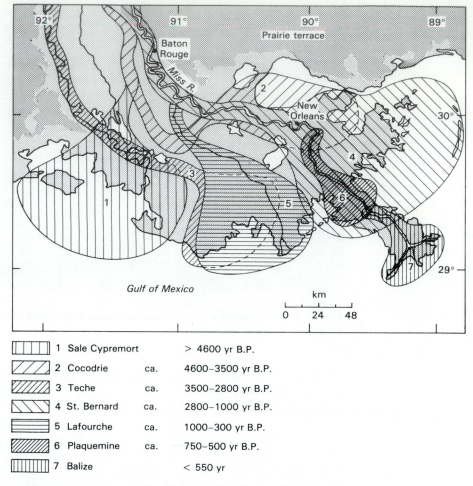

1	Sale Cypremort		> 4600 yr B.P.
2	Cocodrie	ca.	4600–3500 yr B.P.
3	Teche	ca.	3500–2800 yr B.P.
4	St. Bernard	ca.	2800–1000 yr B.P.
5	Lafourche	ca.	1000–300 yr B.P.
6	Plaquemine	ca.	750–500 yr B.P.
7	Balize		< 550 yr

Figure 14-32 Evolution of the Mississippi delta by the growth and abandonment of delta lobes. (From J. P. Morgan, 1970, Deltas— a résumé, *Journal of Geologic Education* 18, National Association of Geology Teachers.

The Mississippi River delta is a good example of this process. Over the past 5000 years, active sediment deposition has occupied seven major lobes (Figure 14-32).

EQUILIBRIUM IN RIVER SYSTEMS

The Graded Stream

River systems provide the primary means by which continents elevated to great heights by tectonic processes are gradually worn down to low-relief plains projecting only slightly above sea level. Throughout this process, rivers transport sediment from the topographically high areas of continents to the ocean. Along the way, stream channels progressively grow in size to accommodate the larger discharge supplied by the addition of tributary streams carrying the flow from drainage basins of lower stream order. Many studies of river systems have indicated that sediment grain size decreases in a downstream direction. The sediment transported by a stream ranges from large boulders in mountainous areas to silt and clay at the river's mouth. The slope, or gradient, of a stream also decreases progressively in the downstream direction. The change in gradient of a stream from headwaters to mouth is known as the *longitudinal profile*. As shown in Figure 14-33, longitudinal profiles have a characteristic concave shape. *Base level* is the elevation that the downstream segment of the curve gradually approaches. Sea level is the ultimate base level for all streams, although temporary base levels may be established by lakes or reservoirs at intermediate positions along a stream's course.

It has been generally believed that the concave longitudinal profiles of most streams represent a state of equilibrium in the river system. Under this assumption the longitudinal profile of a river adjusts to provide the velocity necessary to transport the sediment supplied to the channel. An equilibrium, or *graded*, stream tends to maintain its profile for a period of time during which there are no drastic changes in the prevailing climatic conditions affecting the basin. Although the idea of equilibrium in stream systems has merit, it is now known that the interactions between the numerous variables controlling the discharge, sediment load, and profile of rivers is complex. For example, the velocity of a river actually increases in the downstream direction despite the decrease in slope evident in the longitudinal profile.

Causes and Effects of Disequilibrium

The concept of equilibrium must be considered in terms of the period of time over which it is proposed. The graded stream represents equilibrium during some intermediate-length period of time (tens to hundreds of years), during which the slope and channel characteristics are adjusted to the discharge and sediment load supplied to the channel segment. There are many factors that tend to disturb the fragile balance that may exist.

Tectonic activity can be a major cause of disequilibrium in stream systems. An example of tectonic initiation of disequilibrium is shown in Figure 14-34. A fault has down-dropped part of a stream valley in relation to the upstream branch. The response of the stream to this disturbance is to reestablish an equilibrium condition. By erosion and downcutting upstream from the fault and deposition downstream from the fault, the stream can eventually develop a new equilibrium profile. Gradient is not the only variable that can

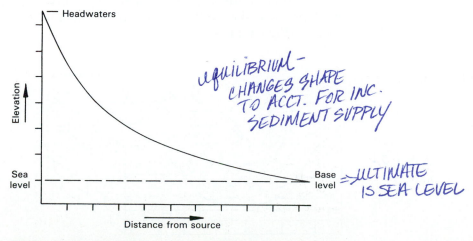

Figure 14-33 Idealized longitudinal profile of a stream. The gradient decreases as the stream approaches base level.

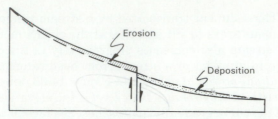

Figure 14-34 Alteration of a stream profile after fault displacement. The stream reestablishes a smooth, concave profile by erosion above the fault and deposition downstream.

be adjusted, however. Channel characteristics such as width and depth may also be altered. Other tectonic influences on river equilibrium involve changes in base level. If a coastal region is uplifted so that the elevation of sea level relative to that particular coastline drops, streams must steepen their gradients in the vicinity of the coast in order to adjust to the changing conditions.

Climatic change is of vital importance to stream equilibrium. The climate in a drainage basin, along with the geology, controls the discharge and sediment load supplied to a river system. Increases or decreases in precipitation within a drainage basin, aside from short-term variations, will cause changes in vegetation and will also cause the streams to adjust their gradient and channel characteristics. The extreme manifestation of climatic change—an advance of a glacier into the drainage basin—can lead to drainage-system disruptions that persist for thousands of years after the glacier has retreated. The Missouri River and many of its tributaries, for example, flowed northward to Hudson Bay before the Pleistocene Epoch. The southward advance of glaciers blocked and diverted the course of the Missouri into the Mississippi River drainage basin.

Stream Terraces.   One of the best types of evidence available for inferring past changes in river system equilibrium are *stream terraces*, which are remnants of former flood plains that stand above modern flood plains in stream valleys. Terraces attest to a former state of equilibrium that was abandoned during a period of erosional instability. Figure 14-35 illustrates the general morphology of terraces. Terraces may be *paired* or *unpaired*, depending upon whether corresponding surfaces at the same elevation exist on opposite sides of the valley. Morphologically, the terrace consists of a flat surface, the *tread*, bordered by a slope called the *scarp*. Multiple terraces are common in many valleys (Figure 14-36).

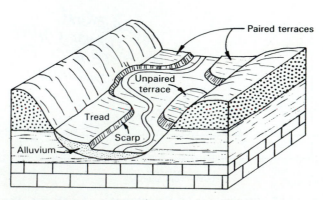

Figure 14-35 Morphology of river terraces.

Figure 14-36 Multiple river terraces in the Madison River valley, Montana.

In terms of genetic processes, terraces consist of two main types. During the formation of an *erosional terrace* (Figure 14-37), stream downcutting is dominant. Possible explanations for downcutting could include tectonic or climatic factors. The origin of *depositional terraces* includes a time interval in which the stream fills its valley with sediments. This period of *aggradation* can be caused by decreases in discharge, increases in sediment load, or changes in base level. A frequent cause of aggradation in northern latitudes was the abundant sediment load supplied to the stream systems by Pleistocene glaciers.

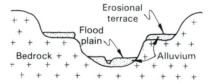

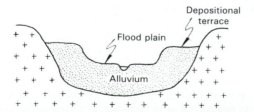

Figure 14-37 Differences between erosional and depositional terraces. Depositional terraces are underlain by thick deposits of alluvium.

Following the period of aggradation, a stream must then erode its channel to form a depositional terrace. The sudden decrease in sediment load accompanying the retreat of a glacier or other climatic factors can lead to the incision of a previously deposited valley fill.

Human Interaction with Equilibrium.   The changes in equilibrium we have considered thus far have not included those that are primarily induced by human activity. Among these the most important examples are changes in river systems brought about by land-use changes in the drainage basin. The increased sediment yield caused by conversion of natural areas to agricultural production and urban zones was discussed in Chapter 10 and illustrated in Figure 10-9. The consequences of increased sediment yield in a drainage basin can be severe. Sediment is temporarily stored in river channels, causing a decrease in channel capacity, which in turn results in more frequent and severe flooding. The effect of the artificial drainage network in a city (storm sewers) is to supply more water to river channels in a shorter time as compared to nonurbanized areas. The magnitude of flooding from a given rainstorm is also increased in an urbanized area. The river's response to greater sediment load, as well as greater discharge, is to enlarge its channel to accommodate the increased flow. Serious bank erosion may result, with flood-plain landowners sustaining property losses.

Another class of human adjustments to equilibrium includes engineering

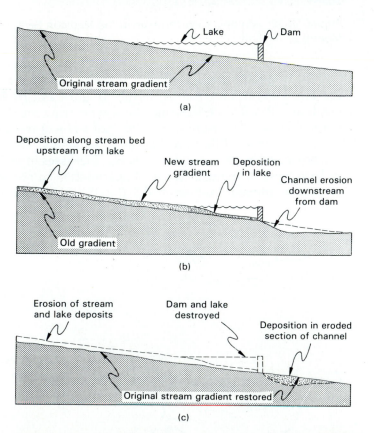

Figure 14-38   Construction or removal of reservoirs causes streams to adjust their gradients in order to approach a new condition of equilibrium. (From S. Judson, M. E. Kauffman, and L. D. Leet, *Physical Geology,* 7th ed., copyright © 1987 by Prentice-Hall, Inc., Englewood Cliffs, N.J.)

works that seek to control the river for such purposes as flood protection, water supply, and recreation. Reservoirs interrupt the longitudinal profile of streams by establishing temporary base levels. Figure 14-38 shows some of the effects. The sediment load of the stream is dropped as it flows into the reservoir. The decrease in gradient causes aggradation upstream from the impoundment as well, as the stream attempts to reconstruct an equilibrium profile utilizing the reservoir as base level. The outflow from the reservoir, having deposited its sediment load, is likely to erode as it flows down the remainder of the river channel and thus dissipate energy available for sediment transport. Dams along the Missouri and other rivers have required expensive downstream bank-stabilization projects to limit loss of flood-plain area by bank erosion.

Flood-control projects often involve alterations to river channels. These may include levees and other structures to contain the flow within the channel so that flooding will be minimized. Alternatively, the channel may be enlarged, or its gradient steepened, in order to convey flood water at a greater discharge or higher velocity through a particular reach. Unfortunately, these changes often are answered by undesirable and costly adjustments by the river system. Levees may decrease flooding along one reach of the channel, but because natural flood-plain storage of water is decreased, they may aggravate the downstream flood problem. The increase in channel size or gradient for attempted control of the river is called *channelization*. This disturbance of the river's equilibrium is propagated both upstream and downstream from the channelized reach. An interesting example documented by Daniels (1960) is the Willow River in Iowa. After artificial straightening of the channel for flood-control purposes, the gradient was increased to the point where the river began to incise its channel. Eventually, the destabilized channel was twice as wide and deep as its original size. Changes of this type are transmitted upstream in a river system. Tributary down cutting following channelization can lead to the destruction of farm land throughout the basin.

FLOODING

Flooding is the most serious hazard posed by river processes (Figure 14-39). In 1972, 238 people were killed by flooding in Rapid City, South Dakota. During the same year, flooding in the eastern United States caused by precipitation from Hurricane Agnes claimed 113 lives and resulted in more than $3 billion in property damage.

Figure 14-39   Flooding periodically damages structures built on unprotected flood plains.

Flooding typifies many hazardous geologic processes in that the obstacles to mitigating the threat to human life and property are largely psychological rather than of a hydrologic or engineering nature. The public perception of floods is that they are "acts of God" that are impossible to predict or explain. People are reluctant to relocate their homes and businesses out of flood-prone areas because of events that they believe may never happen in their lifetime. In reality, the magnitude and frequency of floods can usually be predicted with some degree of certainty, at least on a statistical basis. In addition, flood-prone areas can be accurately identified and mapped. In fact, flood-plain maps exist for most developed areas in the United States.

In terms of the river system, floods are part of the equilibrium condition we have been discussing, rather than exceptions to it. The size of a river channel is adjusted to discharges that occur, on the average, every 1 to 2 years. The flood plain is the river system's storage area for water and sediment from higher discharges that occur less frequently than every 1 to 2 years. The presence of the flood plain is undeniable evidence that river flooding is a periodic, expectable phenomena.

Flood Magnitude

Flood magnitude can be measured as the elevation to which a river rises during a flood but more commonly is reported as the maximum discharge of the stream during the event. Changes in river height, or *stage*, are measured by continuous-recording gauges installed along streams throughout most developed countries (Figure 14-40). At each gauging station, measurements have been made of cross-sectional area and stream velocity through the range of possible stages. Through the use of Eq. 14-5, a *rating curve* can be constructed for each station relating river stage to discharge. With the rating curve and the stage measurement made during a flood, a stream hydrograph can be made to show the relationship between discharge and time (Figure 14-41).

Variables that control the magnitude of a flood can be divided into two types: transient factors that depend on the meteorological conditions and per-

Figure 14-40   Streamflow gauging station. (B. W. Thomsen; photo courtesy of U.S. Geological Survey.)

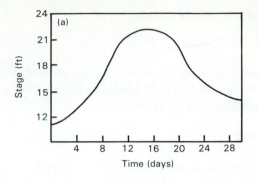

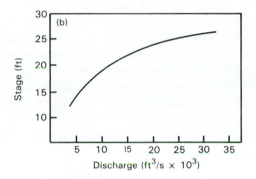

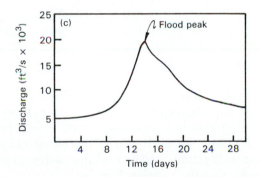

Figure 14-41 (a) Hydrograph of river stage vs. time during a flood. (b) A rating curve for a particular gauging station. (c) A hydrograph showing discharge vs. time constructed from (a) and (b).

manent factors that deal with the hydrology and geology of the drainage basin. There are a large number of meteorological conditions that must be considered. If the flood is caused by a rainfall event, the factors include the size, direction, and rate of movement of the storm; the intensity, duration, and location of the rain; and the moisture conditions prevailing in the soil before the storm. If the flood is caused by a snow-melt event, transient factors include amount of snow, rate of melting, and the soil-moisture conditions. Permanent controls within a basin include type and distribution of vegetation, topography, type of soil and rock at the surface, and type of drainage network present in the basin. From the number of factors involved, it is easy to understand why it is so difficult to predict the magnitude of a particular flood so that safety precautions can be taken. The most basic method of predicting the peak discharge from a particular rainfall event is by use of equations such as the *Rational*

Equation, which can be stated as

$$Q = CIA \qquad\qquad \text{Eq. 14-10}$$

where Q is peak discharge in cubic feet per second, C is a runoff coefficient whose values are given in Table 14-2, I is the rainfall intensity in inches per hour, and A is basin area in acres.

Flood Frequency

Perhaps as important as predicting the magnitude of a flood generated by a particular rainfall event is the problem of predicting the frequency of large floods. Because it is impossible to predict when a flood of a certain magnitude will occur, the only alternative is to attempt to determine the probability of a particular flood within a given year. A general relationship that seems to hold true is that the magnitude of natural processes is inversely proportional to the probability of their occurrence. Thus the larger the flood under consideration, the less likely it is to happen in a particular amount of time.

Another way of defining frequency is in terms of *recurrence interval* (R.I.), which is the average period of time between floods of a certain magnitude. It is related to probability through the following relationship:

$$\frac{1}{p} = \text{R.I.} \qquad\qquad \text{Eq. 14-11}$$

Thus if a flood has a probability (p) of 0.01 of occurring in a single year, its recurrence interval is 100 years. It cannot be overemphasized that the recurrence interval refers to a statistical average over a long period of time. A town that experiences a 100-year flood has the same risk of a similar flood in every following year.

Flood-frequency curves are established from gauging-station records to show the relationship between magnitude and frequency. A basic way of constructing these diagrams is to plot magnitude versus recurrence interval for the maximum discharge for each year of the historical record (Figure 14-42). Recurrence interval is established by ranking each yearly discharge from highest to lowest and using the relationship

$$\text{R.I.} = \frac{N + 1}{M} \qquad\qquad \text{Eq. 14-12}$$

TABLE 14-2

Values of *C* for the Rational Equation

C	Land cover or land use
0.1–0.3	Forest
0.1–0.5	Grasslands
0.3–0.6	Row crops, plowed ground
0.5–0.8	Bare ground, smooth
0.5–0.9	Suburban lands
0.7–0.95	Urban lands
0.9–0.99	Water body, full
0.0	Water body, empty

Source: From C. C. Mathewson, *Engineering Geology*, copyright © 1981 by Charles E. Merrill Publishing Co., Columbus, Ohio.

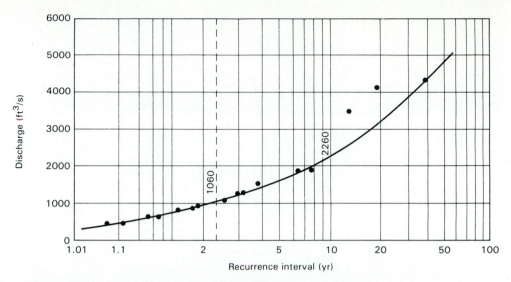

Figure 14-42 A flood-frequency diagram for Big Piney Run near Salisbury, Pennsylvania. (From T. Dalrymple, 1960, U.S. Geological Survey Water Supply Paper 1543-A.)

where N is the number of years of record and M is the magnitude rank. For example, in a city with 49 years of record, the second largest flood, whatever its magnitude, would be assigned a recurrence interval of 25 years. This method can be successful only after many years of records are available. Mathematical functions are also used to construct flood-frequency diagrams.

Effect of Land-Use Changes

The flood-frequency analysis presented involves the assumption that flood magnitudes are random samples from a population of possible events and that the underlying factors that cause floods do not change. In fact, however, we know that yearly mean temperature and precipitation amounts can change through time, making floods more or less frequent and severe for a certain period of time. The land-use changes that were discussed previously also influence flood magnitude and frequency. Most important among possible land-use changes is urbanization. The cumulative effects of urbanization, including less infiltration and more rapid runoff, can alter a flood hydrograph in the manner shown in Figure 14-43. The result is flooding of greater magnitude and shorter lag time. Such floods are termed *flashy*. The flood-frequency analysis made from natural conditions in the stream basin before urbanization may not accurately predict the magnitude-frequency relationships after urbanization.

Flood Control

The unpredictability of floods and the great damages that they can cause present a major dilemma to society. How can we best protect our population against such a nebulous and poorly understood threat? In answer to this question there are basically two alternatives that can be considered. First, river systems can be controlled and regulated by means of channel improvements, levees, and dams. This is known as the *structural approach*. Second, land use on flood plains can be regulated by zoning ordinances so that flood-prone areas

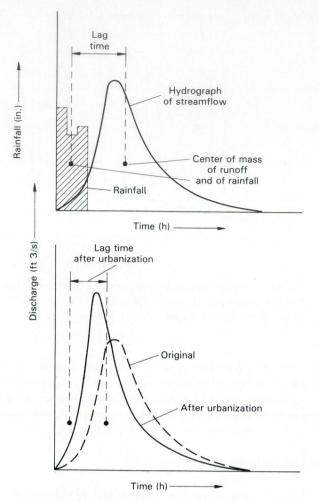

Figure 14-43 Changes in a flood hydrograph after urbanization of a drainage basin. (After L. B. Leopold, 1968, U.S. Geological Survey Circular 554.)

are not occupied by buildings susceptible to great damage during flooding. This alternative is known as the *nonstructural* approach.

Until recently, the United States has chosen to combat floods through the structural approach. Costa and Baker (1981) report that the federal government has spent more than $14 billion on flood control since 1936. The advantage of the structural approach is that flood-control projects can offer protection against floods of a certain magnitude (the design flow) if they are properly designed and constructed. In addition, they are politically attractive to legislators because they create jobs in an area and are perceived by the public as a positive step toward solving the problem. Unfortunately, there are a number of disadvantages to flood-control structures. Most important, a flood greater than the magnitude for which the structure is designed can cause greater damage than would have occurred without the structure. If a dam fails, for example, the resulting flood can assume catastrophic proportions. Most dam failures have been caused by inadequate site investigations or design relative to the geologic conditions at the site. The Teton Dam in Idaho, for example, failed during the initial filling of the reservoir it impounded.

We have already discussed the potential effect of altering river channels by levees, channelization, or other projects. These projects change the natural balance of the stream and cause it to adjust other variables in the system.

A final comment about structural measures is that they may actually encourage development of flood-prone areas. If people believe that there is no danger from floods, they will readily develop flood plains. This only increases the potential for destruction in the event of a rare flood of large magnitude.

The nonstructural approach is an attempt to integrate geologic processes into land use and development. The method of flood protection in this approach is to delineate flood-prone areas and restrict development in those areas to uses that are compatible with periodic flooding. In urban areas, these uses include parks, golf courses, and other natural or open areas. In addition to flood protection, these zones enhance the aesthetic value of a river in a crowded urban setting. The costs for land acquisition may even be less in the long run than the costs of repeated flood damage. The limitations to this approach are that it requires the relocation of homes and businesses. This is considered by some to be an infringement of their right to live wherever they please. In highly developed areas, it may be possible to apply this approach only after a major flood has occurred.

CASE STUDY 14-1
FUTURE SHIFT OF THE MISSISSIPPI RIVER?

The history of the Mississippi River delta indicates that the position of the river at any particular time is a transient situation. As a river and its distributary system build up a delta lobe by sediment deposition on natural levees and crevasse-splay deposits, it gradually becomes higher in elevation than adjacent, inactive parts of the delta. The river finally reacts by changing its course to begin deposition of a new lobe of the delta.

For some years, it has been evident that the Mississippi River delta is tending to shift its course to one of its distributaries, the Atchafalaya River (Chen 1983). If this event occurred,

the Atchafalaya (Figure 14-44) would become the main channel of the Mississippi and the current channel would become a distributary with a greatly reduced flow. The consequences of such a shift would be disastrous to the Louisiana cities of New Orleans and Baton Rouge because of the loss of shipping trade and also fresh water. The greatly reduced flow of the main channel would not provide the cities with a sufficient supply of fresh water and, in fact, salt water from the Gulf of Mexico would move up the river channel. Salt-water encroachment would occur because the Mississippi channel bottom is below sea level for more than 100 km

Figure 14-44 Map of Louisiana showing the Mississippi River and the Atchafalaya River, the valley to which the Mississippi may alter its course.

upriver from Baton Rouge. Dramatic changes would also occur in the Atchafalaya Basin. The town of Morgan City, already a flood-prone area, would be flooded out of existence.

In anticipation of a channel shift, the U.S. Army Corps of Engineers constructed the Old River Control Structure (ORCS) in the 1950s, a series of locks, dams, and levees at the head of the Atchafalaya to maintain flow down the main channel. Damage to the ORCS by severe floods in 1973 necessitated construction of the $219 million Auxiliary Control Structure to preserve the ability of the ORCS to block the channel shift into the Atchafalaya.

Thus the stage is set for a continuing struggle between the Mississippi River and the U.S. Army Corps of Engineers. As the present course of the Mississippi becomes more and more unstable with respect to a course to a different part of the delta, more and more extensive (and expensive) control measures will be required. Only time will reveal the outcome of this situation.

CASE STUDY 14-2
FLOODING AND LAND USE

Two examples that illustrate differing approaches to flood control are provided by the cities of Winnipeg, Manitoba, and Rapid City, South Dakota.

The city of Winnipeg lies at the confluence of the Assiniboine River and the Red River of the North (Figure 14-45), both of which flood regularly. Topographically, the area is extremely flat because it is the floor of a glacial lake basin. Large areas of the city are inundated by floods because of the shallow valleys of the rivers and the low relief of the areas adjacent to the rivers. Floods in Winnipeg generally occur only during the spring snowmelt. Snow depths and the rate of melting are critical factors in determining whether or not a flood will be forthcoming. After the devastating flood of 1950, the citizens of Winnipeg decided to construct a floodway to divert flood waters from the Red River around the city. An inlet control structure in the channel of the Red River consists of a gate that can be raised to block the flow of the river. Water then rises behind the structure and flows into the floodway. The artificial channel, constructed in 1966, has an average depth of 9 m, a maximum width of 18 m, and can carry a maximum discharge of 1700 m³/s (Figure 14-46). The floodway, along with other structural and nonstructural flood-protection measures has been successful in preventing flood damage in the city.

Rapid City, South Dakota, lies along Rapid Creek at the base of the Black Hills. On the night of June 9, 1972, heavy thunderstorms

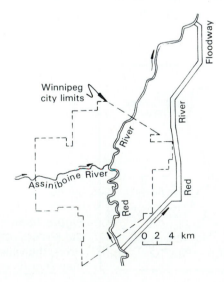

Figure 14-45 Map of the Winnipeg, Manitoba, area showing the location of the Red River floodway.

Figure 14-46 The Red River floodway.

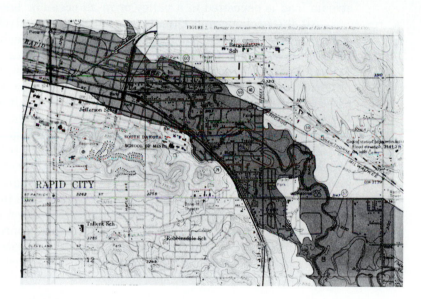

Figure 14-47 Map of part of Rapid City, South Dakota, showing extent of flooding (dark area) June 9–10, 1972. (From O. J. Larimer, 1973, U.S. Geological Survey Hydrologic Investigations Atlas 511.)

over the Black Hills dumped as much as 38 cm of rain in less than 6 h. The result was a 100-year flood that killed 238 people and cost $128 million in property damage (Figure 14-47). In response to this event, the city adopted a nonstructural plan for flood control (Rahn 1984). The flood plain today consists of parks, natural areas, golf courses, and other recreational development. A few large commercial buildings were allowed to remain. The main objective was to remove private dwellings from the flood-prone area. As the mayor of Rapid City remarked, "Anyone who sleeps on a flood plain is crazy." Unfortunately, outside of the city limits, areas along Rapid Creek that were devastated by the 1972 flood were rebuilt with the same type of structures in the same hazardous locations. New construction also has occurred in the flood plain. The "act of God" mentality is difficult to combat.

SUMMARY AND CONCLUSIONS

Individual rivers are components of an integrated stream network in a drainage basin. The stream network develops a particular spatial stream pattern that is controlled by the geology, soils, topography, climate, and vegetation in a drainage basin. The concept of stream order has led to quantitative analysis of the variables characterizing stream networks. A hydrologic budget can be established for the drainage basin as a whole.

The flow of water in rivers can be studied using hydraulic parameters such as Reynolds number and Froude number. The Chezy and Manning equations relate the velocity of the stream to the gradient, channel shape, and boundary roughness. The velocity and shear stress, in turn, control the entrainment and transportation of stream sediment. Sediment is transported as bed load, suspended load, and dissolved load until the velocity drops and deposition occurs. Deposition and other stream characterictics depend on whether the river is meandering or braided. These stream types result in different sediment distributions and landforms. Alluvial fans and deltas develop at points of major change in gradient in stream systems, where much of the stream load is deposited in a valley or in an ocean or lake.

All the geologic and climatic variables that control the discharge and sediment load can be considered as a system that can achieve a state of equilibrium when considered over a period of years. If natural or artificial changes in equilibrium are induced, the system attempts to reestablish an equilibrium condition. In this process many of the variables may be affected. Stream terraces constitute evidence that natural changes in equilibrium have occurred. Urbanization, dams, and other human alterations to the system may produce far-reaching changes to the channel and to the drainage basin.

Flooding is a natural, periodic process in river systems. Flood magnitude depends on a large number of geologic and climatic factors. Flood-frequency graphs show the inverse relationship between flood magnitude and recurrence interval. The problem of flood control can be approached in two ways. A certain degree of protection can be achieved by the construction of dams, levees, and other projects. This method is called the structural approach. A better alternative in most cases is the nonstructural approach, whereby land use on flood plains is restricted to those uses that are most compatible with periodic inundation.

REFERENCES AND SUGGESTIONS FOR FURTHER READING

BAKER, V. R., and D. F. RITTER. 1975. *Competence of Rivers to Transport Coarse Bedload Material.* Geological Society of America Bulletin 86:975–978.

BLOOM, A. L. 1978. *Geomorphology.* Englewood Cliffs, N.J.: Prentice-Hall, Inc.

BRUSH, L. M., Jr. 1961. *Drainage Basins, Channels, and Flow Characteristics of Selected Streams in Central Pennsylvania.* U.S. Geological Survey Professional Paper 282-F.

CHEN, A. 1983. Dammed if they do and dammed if they don't. *Science News* 123:204–206.

COSTA, J. E., and V. R. BAKER. 1981. *Surficial Geology: Building with the Earth.* New York: John Wiley.

DALRYMPLE, T. 1960. *Flood-frequency Analyses*. U.S. Geological Survey Water Supply Paper 1543-A.

DANIELS, R. B. 1960. Entrenchment of the Willow drainage ditch, Harrison County, Iowa. *American Journal of Science* 225:161–176.

HANSEN, D. E., and J. KUME. 1970. *Geology and Ground Water Resources of Grand Forks County, North Dakota*. Part 1: Geology. North Dakota Geological Survey Bulletin 53, and North Dakota State Water Commission County Groundwater Studies 13.

HOWARD, A. D. 1967. Drainage analysis in geologic interpretation: A summation. *American Association of Petroleum Geologists Bulletin* 51:2246–2259.

KOLB, C. R., and W. G. SHOCKLEY. 1957. Mississippi Valley geology: Its engineering significance. *Journal of Soil Mechanics and Foundations*, American Society of Civil Engineers 83(no. SM3)1–14.

LARIMER, O. J. 1973. *Flood of June 9–10, 1972, at Rapid City, South Dakota*. U.S. Geological Survey Hydrologic Investigations Atlas 511.

JUDSON, S., M. E. KAUFFMAN, and L. D. LEET. 1987. *Physical Geology*, 7th ed. Englewood Cliffs, N.J.: Prentice-Hall, Inc.

LEOPOLD, L. B. 1968. *Hydrology for Urban Land Planning: A Guidebook on the Hydrologic Effects of Urban Land Use*. U.S. Geological Survey Circular 554.

LEOPOLD, L. B., and M. G. WOLMAN. 1957. *River Channel Patterns: Braided, Meandering, and Straight*. U.S. Geological Survey Professional Paper 282-B.

MATTHEWS, R. K. 1984. *Dynamic Stratigraphy*, 2d ed. Englewood Cliffs, N.J.: Prentice-Hall, Inc.

MATHEWSON, C. C. 1981. *Engineering Geology*. Columbus, Ohio: Charles E. Merrill Publishing Co.

McKNIGHT, T. L. 1984. *Physical Geography: A Landscape Appreciation*. Englewood Cliffs, N.J.: Prentice-Hall, Inc.

MORGAN, J. P. 1970. Deltas: a résumé. *Journal of Geologic Education* 18:107–117.

RAHN, P. H. 1984. Flood-plain management program in Rapid City, South Dakota. *Geological Society of America Bulletin* 95:838–843.

RITTER, D. F. 1986. *Process Geomorphology*, 2d ed. Dubuque, Ia.: Wm. C. Brown.

RENDER, F. W. 1970. Geohydrology of the metropolitan Winnipeg area as related to groundwater supply and construction. *Canadian Geotechnical Journal* 7:244–274.

SIMONS, D. B., and E. V. RICHARDSON. 1966. *Resistance to Flow in Alluvial Channels*. U.S. Geological Survey Professional Paper 422-J.

STRAHLER, A. N. 1957. Quantitative analysis of watershed geomorphology. *Transactions, American Geophysical Union* 38:913–920.

PROBLEMS

1. Why do hydrologists study river systems in terms of drainage basins?
2. How is drainage density related to the geology of a drainage basin?
3. How does stream capture modify river systems?
4. How do the Reynolds and Froude numbers characterize flow in rivers?
5. What factors determine the velocity of flow in a stream?
6. Describe the two parameters that are used to predict entrainment of particles as flow conditions change.
7. What are the ways in which sediment load is transported in rivers?
8. Compare and contrast braided and meandering stream systems.
9. What are the similarities and differences between deltas and alluvial fans?
10. How can a stream be considered to be in a state of equilibrium when discharge, velocity, and other parameters are constantly changing?

11. How do human activities affect equilibrium?

12. What parameters are used to measure flood magnitude and frequency?

13. Obtain flood-frequency data for a stream in your area and evaluate the land use and development on its flood plain using maps or field visits. Is land use compatible with flood hazards or will there be extensive damage during the next major flood?

CHAPTER
15

OCEANS
AND COASTS

The importance of the oceans can be appreciated when we realize that 80% of the earth's surface is covered by water. The significance of the oceans, however, goes well beyond their areal extent. Life on the earth began in the oceans. Today, microscopic organisms in the sea form the base of the oceanic food chain, which supports the population of fish and other higher organisms in the oceans. This huge reservoir of food is desperately needed by the exponentially growing human population.

Circulation of the oceans exerts an important influence on the earth's climate. Our ability to sustain agriculture on land can be directly related to interactions between oceanic and atmospheric circulation.

In recent decades we have begun to tap the vast storehouse of mineral and energy resources contained in the oceans. Throughout geologic time the remains of tiny marine organisms have accumulated to form the earth's supplies of petroleum. Exploration for petroleum is now focusing on offshore targets. The problems involved in the construction of offshore drilling platforms and other structures make an understanding of the oceans a matter of practical importance.

Coastlines, where the oceans and continents meet, are among the most dynamic areas of the earth. The combination of active geologic processes and dense development of coastal areas leads to a wide spectrum of engineering and environmental problems.

THE OCEAN BASINS

Oceanic Circulation

Large-scale motions of seawater are partly driven by prevailing winds over the sea surface. Surface currents, the Gulf Stream for example, travel long distances and transport huge quantities of water. Individual currents are actually parts of huge loops called *gyres* (Figure 15-1). These currents are important mechanisms for transferring heat, in the form of warm water, from the equatorial regions toward the poles.

The oceans also undergo vertical circulation. When surface currents diverge, or flow away from a continent, deep water moves upward to replace the surface water. This movement, known as *upwelling* (Figure 15-2), is of great economic importance because deep water is rich in nutrients. Organisms, including fish, proliferate in areas of upwelling. The opposite of upwelling is sinking. Vertical movement is also driven by density differences. Cold polar water sinks and moves very slowly toward the equator. North Atlantic deep water is forced to flow over the even denser Antarctic bottom water (Figure 15-2).

Topography

The modern study of the oceans began with the voyage of the research vessel H.M.S. *Challenger* in 1872. Prior to this voyage, the ocean basins were assumed to be broad, featureless plains. Instead, the *Challenger* expedition brought back evidence of huge submarine mountain ranges, deep trenches, and other topographic features. Detailed study of the ocean basins accelerated during and after World War II. This work provided much of the evidence for plate tectonics.

The topography of ocean basins is dominated by midoceanic ridges (Figure 15-3). These ridges rival the size and extent of continental mountain ranges.

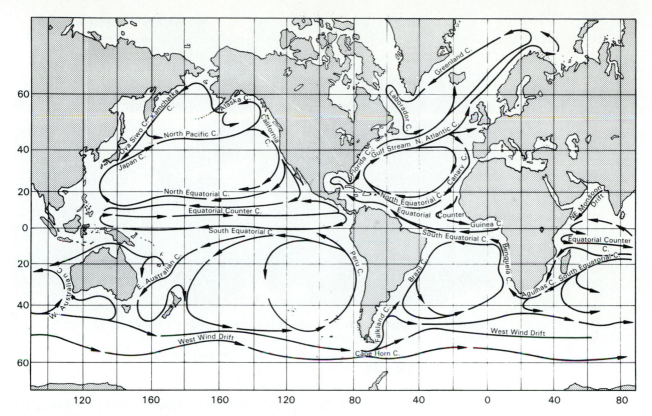

Figure 15-1 Current movement in the oceans. (From S. Judson, M. E. Kauffman, and L. D. Leet, *Physical Geology*, 7th ed., copyright © 1987 by Prentice-Hall, Inc., Englewood Cliffs, N.J.)

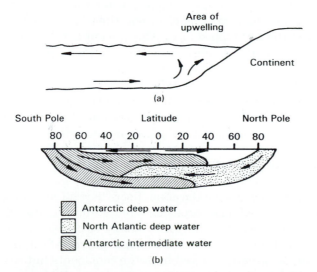

Figure 15-2 (a) Upwelling caused by surface flow away from a continent. (b) Vertical distribution of selected water masses.

Figure 15-3 Topographic features of the ocean basins. Notice position and size of midoceanic ridges. (Reproduced from the "World Ocean Floor" panorama by Bruce C. Heezen and Marie Tharp, copyright © 1977 by Marie Tharp.)

At the crest of the midoceanic ridges lies the faulted rift valley, where new oceanic crust is formed in the sea-floor spreading process.

The ocean bottom slopes downward away from the rugged midoceanic ridges into a region of more subdued relief containing *abyssal hills*. The decrease in elevation is the result of lithospheric cooling as newly formed crustal material moves away from the crest of the midoceanic ridge. Isolated volcanic mountains, or *seamounts*, rise above the floor of the oceans in many places. Some seamounts have flat tops rather than the typical conical shape of a volcano. Drilling of these flat-topped seamounts, termed *guyots*, has produced evidence for their emergence at one time in the geologic past. Wave erosion produced the flat-topped form. The present depth of guyots is an indication of the great amount of subsidence that has taken place.

The greatest depths of the ocean basins are located at the trenches, long narrow troughs associated with subduction zones.

The surface of the ocean bottom changes considerably in the vicinity of the continents. Rapid rates of deposition of continental sediment bury the landscape, producing *abyssal plains* of low relief.

Oceanic Sedimentation

Sediments deposited on the ocean floors beyond the influence of the continents, called *pelagic sediments*, consist of inorganic and organic types. The inorganic pelagic sediments are composed mostly of clay-size (<0.002 mm) particles of quartz, feldspar, clay, and other common rock-forming minerals. These particles are originally derived from the continents but can be transported great distances by ocean currents or winds because of their extremely small size. Large areas of the ocean floors are covered by red or brown muds composed of the inorganic pelagic sediments. Sedimentation rates are very low, often less than 1 cm per 1000 years.

Organic sediments are also deposited on the ocean floor. These include

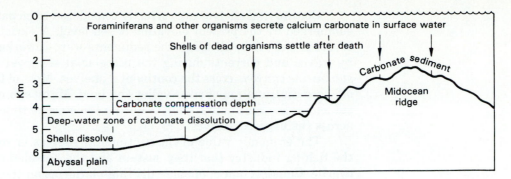

Figure 15-4 Carbonate compensation depth in the oceans. (From F. Press and R. Siever, *Earth,* 3d ed., copyright © 1982 by W. H. Freeman and Co., San Francisco.)

shells of microscopic organisms living near the surface of the ocean, as well as fecal pellets and other organic remains. The shells secreted by organisms are composed of calcium carbonate or silica. Shells of organisms that secrete calcium carbonate, like the one-celled foraminifera, are most abundant in the oceans, although these shells are never found in sediments deposited below a water depth of approximately 4 km. The reason for this is that deeper waters tend to be undersaturated with respect to calcium carbonate and therefore steadily dissolve shells that fall to the bottom. The depth at which seawater becomes undersaturated is called the *carbonate compensation depth* (Figure 15-4). Carbonate-rich *foraminiferal oozes* are common where the ocean bottom lies above this critical level. *Radiolarian* or *diatom oozes* are deposited where organisms that secrete siliceous shells are abundant. These sediments are not affected by the carbonate compensation depth.

CONTINENTAL MARGINS

Topography

The margins of the continents have several topographic divisions (Figure 15-5). Extending seaward from the coastline are gently sloping platforms called *continental shelves*. Continental shelves vary in width from less than 2 km to more than 1000 km. Although arbitrary depth limits have been used to define the extent of the shelves, the best method is to define the limit of the continental shelf as the point where the slope suddenly increases. A water-depth range of 200 to 650 m will include a large percentage of the shelves.

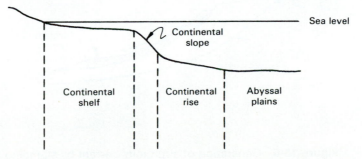

Figure 15-5 Topographic divisions of the continental margins.

Continental shelves are correctly considered to be parts of the continents submerged by the present elevation of sea level. A variety of sediment types covers the shelves, although the sediments were reworked and redistributed by waves and currents during the postglacial sea-level rise. Valleys called *submarine canyons* cross the continental shelves. Most of these valleys are the offshore continuation of river valleys on land. Many submarine canyons were cut across the continental shelves during Pleistocene time when rivers flowed across the exposed shelves.

The economic value of the continental shelves is very great because of the fishing industry that they sustain and the petroleum reserves that they overlie. Construction of offshore drilling platforms on the continental shelves is an engineering problem that requires detailed knowledge of the distribution and engineering properties of shelf sediments. The platforms must be designed to withstand the huge waves that are generated as storms pass over the continental shelves.

The continental shelves terminate at the upper margins of the *continental slopes*. The most steeply sloping segments of the continental slopes average 70 m per kilometer in comparison to the mean value of about 2 m per kilometer for the shelves. Submarine canyons dissect the continental slopes, although these valleys could not have been eroded by rivers because sea level has never dropped low enough to expose them.

At the base of the continental slopes, the slope angle again decreases to a more gentle decline along the surface of the *continental rises*. The continental rises gradually merge seaward with the flat abyssal plains of the ocean floors.

Turbidity Currents

The continental slopes and rises are blanketed by thick sedimentary sequences composed of continental sediment that was transported to the edge of the continental shelves by rivers and wave action during the recent geologic past. On the continental slopes these unconsolidated sediments are somewhat unstable at the greater slope angle on that part of the continental margin. Slumps are common, sometimes triggered by earthquakes.

When slope movements occur on the continental slope, a very important type of current can be generated (Figure 15-6). The current consists of a dense cloud of suspended sediment that accelerates and flows rapidly downslope because it is more dense than the surrounding seawater. The name given to this type of flow is *turbidity current*. When turbidity currents flow out onto the continental rises, the velocity decreases because of the decrease in slope. Par-

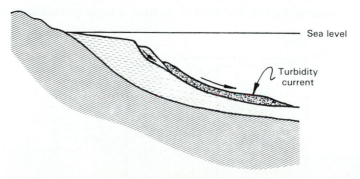

Figure 15-6 Generation of a turbidity current by slumping in the sediments of the continental slope.

ticles begin to drop out of suspension in the order of decreasing size as the current slows, so that graded bedding develops in the layer of sediment deposited by the turbidity current.

Dramatic evidence for the existence of turbidity currents was provided when data were analyzed from the 1929 earthquake of the Grand Banks of Newfoundland. The data included times of breakage of submarine transatlantic telephone cables following the earthquake. Some of the cables were broken hours after the event, so the earthquake was ruled out as a direct cause. Cables were broken sequentially with increasing distance from a point on the continental slope. The only reasonable explanation for this pattern of cable breakage is that the earthquake generated a powerful turbidity current that flowed down the continental slope at a velocity of 40 to 55 km/h and systematically broke each cable it crossed along its path.

Turbidity currents are important because they transport huge quantities of continental sediment onto the adjacent abyssal plains. The submarine canyons incised into the continental slopes most likely were eroded by these dense, sediment-laden flows. Ancient sequences of marine sedimentary rocks exposed on land have now been interpreted to be the result of turbidity-current deposition (Chapter 4). These sequences of turbidities consist of alternating layers of sandstone or siltstone and shale (Figure 4-17). The coarse-grained beds display graded bedding and are interpreted to be deposits of turbidity currents. The interbedded shales represent normal pelagic sedimentation between turbidity currents. The wide distribution of these sedimentary sequences attests to the importance of turbidity currents in marine sedimentation along continental margins.

SHORELINES

Waves

Among the processes that shape our coastlines, those involving waves are probably the most important. The impingement of waves upon a shoreline governs coastal erosion, the formation of beaches, and the distribution of sediment in the shoreline environment. Coastal engineering and management therefore must be based upon an understanding of wave processes.

Generation and Movement. The basic parameters used to describe wave motion are illustrated in Figure 15-7. *Wavelength* is the distance between adjacent crests or troughs. The amount of time between passage of successive crests relative to a fixed point is the *wave period*. The *wave velocity, C,* then, is simply the wavelength, L, divided by the period, T:

$$C = \frac{L}{T}$$

Eq. 15-1

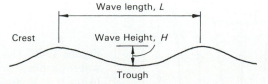

Figure 15-7 Definition of wavelength and wave height.

Although waves move through the water at a particular velocity, the water through which the waves propagate does not travel forward in the direction of wave movement. Instead, particles of water beneath a passing wave describe circular orbits with decreasing diameters as depth increases below the surface (Figure 15-8). This pattern of motion can be observed in the behavior of a cork bobbing in the water as waves pass beneath. When we speak of wave velocity, then, we are referring to the velocity of the moving waveform rather than the water. Additional parameters needed to characterize wave movement are *wave height*—the vertical distance from trough to crest—and water depth. The relationship between water depth and wavelength is an important aspect of wave motion.

Waves are produced by the transfer of energy from wind to the water surface. The amount of energy transferred, which is reflected in the size and period of the waves generated, is dependent on the three parameters illustrated in Figure 15-9. The storm duration and wind speed are functions of the size and intensity of the storm. The *fetch,* the distance over which the wind blows, is a function of the storm and the body of water. Fetch is a limiting factor for the generation of waves in small lakes and ponds.

The generation of waves by storms at sea produces a complex group, or spectrum, of waves, with large variation in period. Under the influence of the generating storm, these *sea* waves are highly irregular. Individual wave crests are random summations of smaller waves of different period.

The amount of energy transferred to the water from the wind is a function of the wave period and the square of the wave height. Long-period waves, those with periods of 20 seconds or more, are generated only when wind speed, storm duration, and fetch are all at their maximum values. Wave heights of 30 m or more have been observed in the open ocean. As waves move out of the area of generation, wave groups of similar periods sort themselves out and the wave motion becomes more regular. These waves, which travel great distances from their area of generation, are termed *swell*. Although the equations that describe wave motion are fairly complex, simplifications can be made under certain conditions. For example, in deep water—in which the depth is greater than one-half the wavelength—both wavelength and velocity can be expressed as functions of period. Thus

$$L = \frac{gT^2}{2\pi}$$

Eq. 15-2

where g is the acceleration of gravity. By substituting Eq. 15-2 into Eq. 15-1,

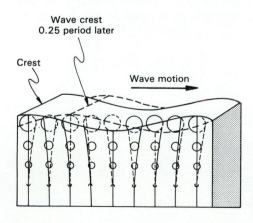

Figure 15-8 The circular motion of water particles in waves in deep water. (After U.S. Hydrographic Office Publication 604, 1951.)

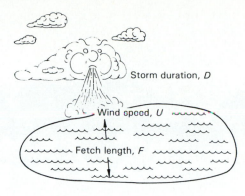

Storm duration, *D*

Wind speed, *U*

Fetch length, *F*

Figure 15-9 Factors that determine the size and period of wind waves. (From P. D. Komar, *Beach Processes and Sedimentation*, copyright © 1976 by Prentice-Hall, Inc., Englewood Cliffs, N.J.)

wave velocity can be expressed as a function of period:

$$C = \frac{gT}{2\pi}$$

Eq. 15-3

After conversion to swell, wave groups of similar period travel forward until they encounter a coastline. Along this path the waves lose energy by several processes, including lateral spreading of the waves from the area affected by the storm to a larger area of the open ocean. The potential coastal erosion or damage to structures depends upon the characteristics of the waves originally generated by the storm and the changes to the wave groups that occur along their path to the coast. Charts are available for prediction of wave heights and periods from known storm characteristics.

Breaking Waves. As waves approach shore, they undergo a gradual transition from sinusoidal profiles with low, rounded crests to waves with sharp peaks and higher wave heights (Figure 15-10). This transitional process begins when the water depth decreases to about one-half the wavelength. As the waveform begins to interact with the bottom, the circular orbits of water particles (Figure 15-8) become compressed to elliptical shapes. The wavelength and the velocity both decrease, while the period remains the same. As the wavelength decreases, the wave crest must build in order to maintain a constant mass of water within a single wavelength. The ratio of the wave height to wavelength, called *wave steepness,* gradually increases until a critical value of stability is exceeded. This occurs when the steepness reaches a value of about $\frac{1}{7}$. At this point the wave breaks (Figure 15-11) because it has become too steep to maintain its shape. In addition, at the point of breakage the hor-

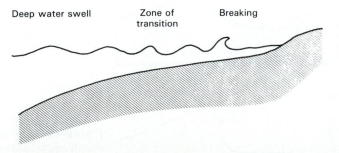

Deep water swell   Zone of transition   Breaking

Figure 15-10 Changes in waves as they approach shore. (From P. D. Komar, *Beach Processes and Sedimentation*, copyright © 1976 by Prentice-Hall, Inc., Englewood Cliffs, N.J.)

Figure 15-11 Waves breaking on shore. (R. Dolan; photo courtesy of U.S. Geological Survey.)

izontal orbital velocity at the crest exceeds the velocity of the waveform, and the crest collapses ahead of the wave.

Most breakers can be classified as either *spilling* breakers or *plunging* breakers (Figure 15-12). Spilling breakers, in which the crest collapses into bubbles and foam that rush down the shoreward face of the wave, occur over gently sloping bottoms. The plunging breaker creates a spectacular visual effect as the crest curls outward and downward in front of the wave, forming a tunnel for a few seconds prior to total collapse of the wave.

Wave Refraction.   The decrease in velocity experienced by waves as they enter shallow water results in a change in direction of the wave called *wave refraction.* Along a straight shoreline in which the offshore bottom contours parallel the shore, wave refraction is evident when waves approach the coast at an angle (Figure 15-13). The portion of a wave crest that reaches shallow water first begins to decrease in velocity compared with the portion of the same wave that has not yet reached shallow water. Consequently, the seaward

Spilling breakers

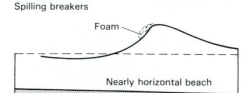

Plunging breakers

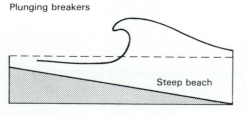

Figure 15-12 Form of spilling and plunging breakers. (From P. D. Komar, *Beach Processes and Sedimentation,* copyright © 1976 by Prentice-Hall, Inc., Englewood Cliffs, N.J.)

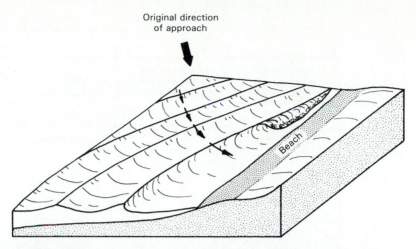

Original direction
of approach

Beach

Figure 15-13 Refraction of wave crests approaching shore at an
angle to become parallel with beach.

part of the wave travels faster and the wave crest appears to bend, so that it
becomes nearly parallel to the beach by the time it breaks.

Wave refraction has an important influence on erosional and depositional
processes affecting irregular coastlines (Figure 15-14). In a coastline composed
of headlands, or points jutting out into the ocean between adjacent bays, the
bottom contours often parallel the shoreline. In other words, waves approaching the shore first begin to interact with the bottom directly offshore from the
headland because water is shallower there relative to the bay on either side.
As shown in Figure 15-14, wave refraction tends to focus the wave rays (lines
drawn perpendicular to wave crests) toward the headland. The spacing of the
wave rays, which are equally spaced and parallel in open water, is an indication
of the amount of wave energy expended against a particular part of the coastline. Close spacing in the vicinity of headlands indicates that wave energy is
concentrated in those areas. The rate of shore erosion, therefore, is much
greater on headlands than in adjacent bays. The low-energy environment of
bays, in fact, often results in net deposition of sediment. This explains why
headlands are often rocky and lack good beaches, while nearby bays contain
broad expanses of sandy beaches. If the shoreline materials are nonresistant
to erosion, these processes tend to straighten out an initially irregular coast
by erosion of headlands and deposition in bays.

Tides

In coastal areas, tides are an important factor in many processes. This is particularly true where the conditions and forces that influence tides combine to
produce a large difference in elevation, or *tidal range,* between high and low
tides. The Bay of Fundy in Newfoundland, for example, has a maximum tidal
range of 16 m, the highest in the world.

Although many factors influence tides, the dominant control is the gravitational attraction between the earth and the moon. In terms of Newton's law
of gravitation, the net attractive force can be expressed as

$$F = \frac{M_m M_e}{R^2} G$$ Eq. 15-4

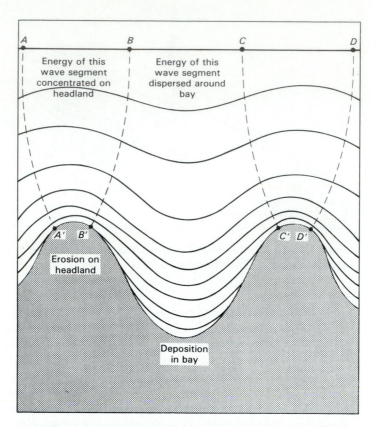

Figure 15-14 Concentration of wave energy on headlands of an irregular shoreline. The energy of wave segments *AB* and *CD* is distributed over a short distance at the headland, whereas the energy of segment *BC* is dispersed over a long shoreline distance in the bay. (From S. Judson, M. E. Kauffman, and L. D. Leet, *Physical Geology,* 7th ed., copyright © 1987 by Prentice-Hall, Inc., Englewood Cliffs, N.J.)

The terms M_m and M_e represent the masses of the moon and the earth, respectively, and R is the distance between the center of the earth and the moon. G is the universal gravitation constant.

Although Eq. 15-4 represents the net attractive force between the earth and the moon, the attractive force between the moon and a unit of mass on the earth's surface is slightly different from the net force because the distance R between the earth's surface and the moon is not the same as the distance between the centers of the two bodies. It is the difference between the total force of attraction and the local attractive force that is responsible for tides.

If the earth were covered with a uniform layer of water over its surface, tidal forces would produce the effect shown in Figure 15-15. Two tidal bulges would be created. The bulge facing the moon would be generated because the water on the moon's side of the earth is attracted more than the earth as a whole. The bulge on the opposite side of the earth is produced in a similar way—the moon's attraction for water on the opposite side of the earth is less than its attraction for the main body of the earth.

As the earth rotates, the tidal bulges travel over the surface of the earth. A fixed point on the earth would therefore experience two high and two low tides in a little more than 24 h because the moon revolves around the earth

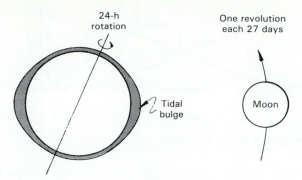

Figure 15-15 Tidal bulge produced on opposite sides of the earth by the gravitational attraction of the moon.

in the same direction as the earth's rotation. It therefore takes an additional 50.47 min for the fixed point on the earth to "catch up" to its original position relative to the moon because the moon also has moved along its revolutionary path during the day.

A number of factors complicate this simplified explanation of tide-generating forces. One of these factors is the influence of the sun. Despite its huge mass, the sun's great distance from the earth lessens its effect upon the tides. The influence it does have, however, must be considered in terms of the positions of the three bodies. When the earth, moon, and sun are aligned (Figure 15-16), the size of the tidal bulge on the earth is maximized. These periodic high tides are called *spring tides,* although they have nothing to do with the seasons. When the sun and the moon are aligned at right angles with respect to the earth (Figure 15-16) the lower tides produced are called *neap tides.*

The factors discussed above still do not explain the huge tidal ranges of the Bay of Fundy and other areas. For the reasons that cause these tides we must consider the distribution of continents and ocean basins and the topography of bays and estuaries along the coastline.

Because the earth is not covered by a uniform layer of water, tidal movements must occur in the confined basins that contain the world's oceans. An ocean basin can be likened to a rectangular container of water that is suddenly disturbed by being lifted at one end (Figure 15-17), forming a wave of water that moves toward the opposite side of the tank. When it reaches the side, it is reflected back toward its original side. The process continues and a standing wave is generated as the wave moves back and forth across the tank. The period of the wave is determined by the length and depth of the tank. In other

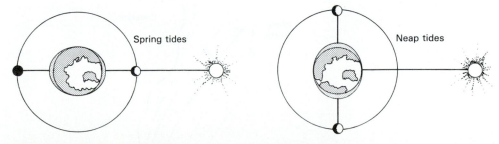

Figure 15-16 Origin of spring and neap tides. (From E. A. Hay and A. L. McAlester, *Physical Geology: Principles and Perspectives,* 2d ed., copyright © 1984 by Prentice-Hall, Inc., Englewood Cliffs, N.J.)

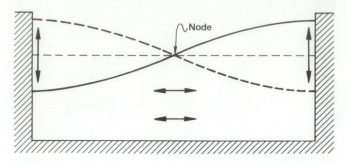

Figure 15-17 Standing wave in an enclosed basin. Certain bays and estuaries experience very high tides due to resonant amplification when the natural period of the basin and the period of the tidal wave are similar. (From P. D. Komar, *Beach Processes and Sedimentation,* copyright © 1976 by Prentice-Hall, Inc., Englewood Cliffs, N.J.)

words, every container, including ocean basins, bays, and estuaries, has a characteristic natural period. As the earth rotates, tides move across ocean basins as long-period waves. If the period of a tidal wave (not to be confused with the misnamed "tidal waves" that cause great destruction along coasts) is similar to the natural period of the basin in which it moves, resonance occurs and the wave is amplified. A process of this type takes place in bays like the Bay of Fundy; the natural period of the bay is close to the tidal period, so a resonant standing wave moves back and forth across the bay. This accounts for the very large tides observed there.

Longshore Currents

Although wave refraction tends to change the direction of wave movement so that waves approach the shore more directly, an oblique approach angle is often observed (Figure 15-18). This movement results in a net flow of water, or *longshore current,* parallel to the shore. Longshore currents are apparent to swimmers in the surf zone who may suddenly realize that they have drifted along the shore for some distance from the point on the beach at which they swam out into the water. Rather than continuing indefinitely along the shore, longshore currents are often broken into cells (Figure 15-19) that are bounded by rapidly moving currents flowing directly away from the beach in a seaward direction. These *rip currents* are dangerous because of their ability to carry

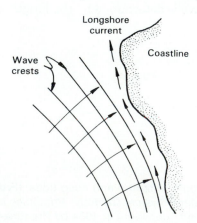

Figure 15-18 Longshore current associated with waves approaching coast at an angle.

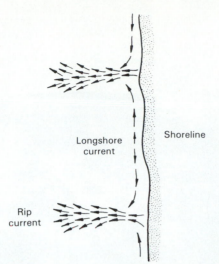

Figure 15-19 A longshore current cell bounded by rip currents.

swimmers out to sea. Many swimmers make the mistake of swimming directly against the current toward shore. Instead, a person caught in a rip current should swim laterally, because rip currents are narrow zones that can be evaded by swimming along the shore for a short distance. Bottom topography can control the position of rip currents, as, for example, near the mouth of a submarine canyon. The direction and characteristics of incoming waves, however, can also influence the development of longshore cells. Rip currents have been observed on shores with a very regular bottom topography.

One of the most important effects of longshore currents is the transportation of sand along the shore. This *longshore drift* is the movement of sand in the shore zone by longshore currents. The interaction of coastal structures with longshore drift is one of the most important problems in coastal engineering.

Coastal Morphology

Coastlines are characterized by continuous change. The changes that occur present a great challenge to coastal engineers because coastal processes can be rapid and powerful. In addition to the drastic changes that can be caused by a single damaging storm, long-term processes are also at work. These include worldwide changes in sea level or changes in the elevation of the land relative to sea level caused by tectonic forces. A study of the geomorphology of a coast can frequently indicate the types of processes that are currently influencing the coast.

Erosional Features. Coastlines rapidly eroding under the relentless attack of wave erosion provide examples of steep *wave-cut cliffs* (Figure 15-20). The cliffs maintain steep slope angles as they retreat because they are undermined at their bases by wave erosion that removes support for the overlying material. A rockfall, slump, or other type of mass movement can occur when the safety factor of the slope decreases to a value of 1.0 (Figure 15-21). The debris that accumulates at the base of the slope is further eroded and transported away from shore by wave action. The flat surface at the base of wave-cut cliffs beveled by wave erosion is called a *wave-cut platform* (Figure 15-22).

Figure 15-20 Wave-cut cliffs along the Oregon coast. A stack remains as an erosional remnant of the cliff. (J. S. Diller; photo courtesy of U.S. Geological Survey.)

Figure 15-21 A mass movement caused by coastal erosion. (J. T. McGill; photo courtesy of U.S. Geological Survey.)

As the coastline retreats, resistant rock units may project out of the water as *stacks* or *arches* (Figure 15-20). Eventually, these remnants are also worn down and removed by wave erosion.

Depositional Features. In addition to erosional processes, shorelines undergo deposition from waves, wind, and longshore currents. The landforms that result from these processes are dominant along some coastlines. Some of these features are illustrated in Figure 15-23. A detailed discussion of the beach, probably the most important depositional landform, will be reserved for the following section.

Figure 15-22 Wave-cut platform in Pleistocene carbonate reef rock, Puerto Rico. (C. A. Kaye; photo courtesy of U.S. Geological Survey.)

The transportation of sand by longshore currents is responsible for the deposition of a *spit*, a thin strip of beach that extends from a point or headland across the mouth of a bay or other coastal indentation. Frequently, the ends of spits curve inward toward the bay, forming a *hooked spit*.

The presence of an offshore island close to the shore creates a low-energy zone shoreward of the island that is somewhat protected from wave processes. A spit often forms in such a protected location, and when the spit connects the beach and the offshore island, it is given the name *tombolo* (Figure 15-23).

Most depositional coastlines are characterized by long, narrow islands that lie just offshore and trend parallel to the coast. Such *barrier islands* are common off the East and Gulf coasts of the United States (Figures 15-23 and 15-24). Periodic breaks in the barrier islands, or *tidal inlets,* allow tidal flow between the ocean and the narrow *lagoons* that lie between the barrier islands and the mainland.

The origin of barrier islands has been a matter of considerable geologic debate. One of the most likely recent hypotheses relates the formation of barrier islands to the recent rise in sea level accompanying the melting of Pleistocene glaciers. Linear ridges may have been built along coastal plains by wave or wind processes and isolated as barrier islands when the low-lying

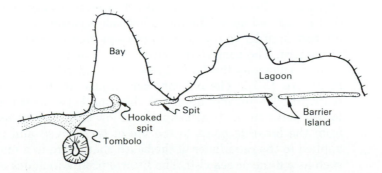

Figure 15-23 Depositional landforms along coasts.

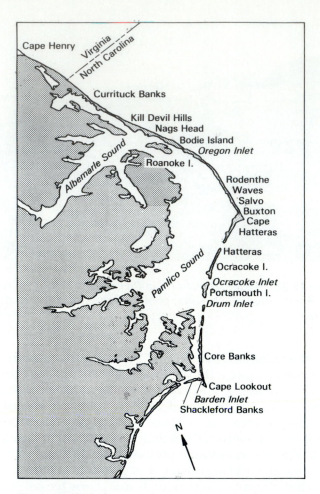

Figure 15-24   Barrier islands off the North Carolina coast.

areas shoreward of the ridges were flooded to form the lagoons. Once formed, barrier islands can migrate in response to changes in sea level and other factors.

Beaches

The *beach* is the actual point of interaction between the sea and the land. As such, its form adjusts constantly to the energy expended upon it by tides, wind, and waves. Because the term beach describes only the part of the shoreline that is exposed above the mean low-tide position, we must expand our zone of interest in order to consider all the significant processes in the nearshore environment. The zone that includes the beach as well as the shallow offshore area is known as the *littoral zone*. Its subdivisions and landforms are shown in Figure 15-25. The *offshore* begins just at the breaker zone and extends seaward. The *inshore* describes the zone between the breaker zone and the lowest point on the beach exposed at low tide. The *foreshore* includes the area from the low-tide point to the top of *berm crest,* and the term *backshore* is applied to the remainder of the beach shoreward to a major topographic break such as a dune or sea cliff. The littoral profile includes one or more flat *berms* on the shoreward extent of the beach, the sloping *beach face* that is under

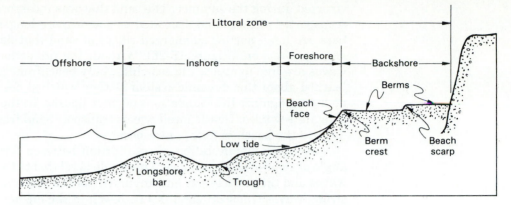

Figure 15-25 Topographic divisions of the beach profile. (From P. D. Komar, *Beach Processes and Sedimentation,* copyright © 1976 by Prentice-Hall, Inc., Englewood Cliffs, N.J.)

constant wave action, and one or more *longshore bars* and *troughs* in the vicinity of the breaker zone.

The profile illustrated in Figure 15-25 is a very generalized representation. The beach, or littoral, profile is constantly changing in response to changing wave and tide conditions. The beach must therefore be thought of as a dynamic system in which changes in wave inputs initiate responses that tend to establish a new condition of equilibrium under those particular wave conditions. The adjustments that take place in order to reach equilibrium are changes in the beach profile. The size of the particles making up the beach is also an important factor in this process.

An example of this system at work is the change in profile that occurs from the predominantly swell-wave condition that exists in summer to the influence of storm waves that are more common in winter. Figure 15-26 shows the profiles that develop under each of these conditions. The broad berm that characterizes the summer profile is a depositional landform that is gradually built up by the onshore sand transport by swell waves. A berm is a very desirable attribute for a beach, not only because it provides wide expanses for sun bathers, but because it protects the sea cliff or dunes from wave erosion. The powerful storm waves of winter attack and erode the berm that was con-

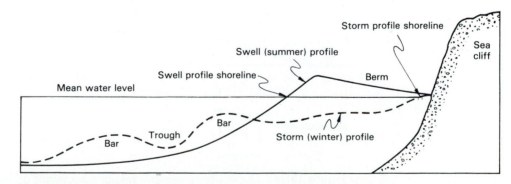

Figure 15-26 The summer and winter beach profiles—the response of a beach to different energy conditions. (From P. D. Komar, *Beach Processes and Sedimentation,* copyright © 1976 by Prentice-Hall, Inc., Englewood Cliffs, N.J.)

structed during the summer. The sand that was incorporated into the berm is transported offshore and deposited as one or more longshore bars. Longshore bars are long, narrow submerged ridges of sand that develop in the vicinity of large breakers (Figure 15-26). Without the protection of the berm, storm waves are free to attack the sea cliff. Many beachfront property owners have learned about this dynamic system as they watched destructive storm waves pounding against the foundations of their houses. In the summer the process is again reversed: Gentle swell waves transport sand from the longshore bars up onto the beach to form a new berm.

Another manifestation of equilibrium between form and process is the angle of slope of the beach face. This angle is related to the size of the incoming waves and the particle size and angularity of the beach material. Even though strong waves transport sand from a beach in the offshore direction, they also have the ability to transport large particles toward the beach. Thus an exposed beach that is frequently subjected to strong waves is commonly composed of coarse particles.

The movement of a wave onto the beach face is illustrated in Figure 15-27. The forward movement of the wave up the beach face, or *swash*, transports particles until they are deposited as the velocity decreases. Before the return flow down the beach face, or *backwash*, can occur, some of the water infiltrates into the beach face. More water infiltrates into the beach as the particle size increases. Because of this loss of water, the backwash tends to be weaker than the swash. In order for an equilibrium profile to develop, the same amount of sand must be transported down the beach slope as up. Backwash must be assisted by an additional force, namely, the downslope component of gravity. The larger the beach particles are, the steeper the slope must be, because of infiltration losses. Beaches composed of large particles, like the beach illustrated in Figure 15-28, must therefore develop steeper slopes in order to establish an equilibrium state.

Types of Coasts

Classification of coasts is useful because the particular type of landforms present on a coast gives an indication of the geologic processes that are currently active or have been significant during the development of the coast.

Depositional coasts (Figure 15-23) are characterized by deltas, barrier islands, and other depositional landforms. Gently sloping coastal plains un-

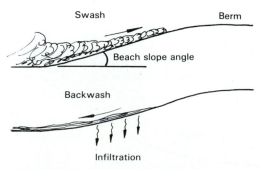

Figure 15-27 The stronger swash transports sediment up the beach face. Backwash is weakened by infiltration of water into the beach. In order for an equilibrium profile to develop, equal amounts of sediment must be transported up and down the beach face. The coarser the particles, the steeper the beach slope angle must be.

derlain by nonresistant materials favor the development of depositional coasts. The East and Gulf coasts of the United States are good examples of depositional coasts.

Glacial coasts are produced by glacial erosion of valleys that meet the coastline. Glacial valleys tend to have very steep walls and a U-shaped cross-sectional profile. Scouring of these valleys took place at a time when sea level was much lower than at present, so the valleys extend well below present sea level. The postglacial rise in sea level drowned the glacial valleys, producing long, narrow bays with steep valley sides called *fjords*. The coast of Norway is famous for the picturesque scenery of its fjords (Figure 15-29).

Some coasts have been raised or uplifted relative to sea level. These *emergent coasts* can be produced in several ways. Tectonic uplift associated with faulting is an important mechanism. Parts of the California coast have a series of raised wave-cut platforms (Figure 15-30), indicating that the coast has been subjected to a series of tectonic episodes. Each platform represents a temporary period of stability. Glaciated regions may also have emergent coastlines. In this instance, the mechanism producing emergence is *isostatic rebound*. The huge weight of a glacier several kilometers thick causes the rigid crust to sink slightly deeper into the plastic mantle of the earth. Isostatic rebound is the readjustment of the crust after the glaciers retreat. The fact that glaciated regions are still rising relative to sea level 10,000 years after the last ice retreat demonstrates that isostatic rebound is a very slow process. A series of beaches or other shoreline features above sea level is an indication of an emergent coastline produced by postglacial rebound.

A final type of coastline is the *biogenic coast,* characterized by the presence of carbonate reef structures in the shallow offshore zone along the coast. Carbonate reefs are composed of the shells of living organisms such as corals that grow in colonies where individual corals are connected to form a three-dimensional framework.

Coral reefs are classified according to their position relative to the mainland. Reefs that actually connect with the shoreline are called *fringing reefs*. If the reef is separated from the coast by a lagoon, the name *barrier reef* is used. A circular reef without the presence of land is known as an *atoll*. More than 100 years ago, Charles Darwin provided a perceptive hypothesis for the origin of atolls that has been confirmed by deep-sea drilling only in the past several decades. Noting that most islands in the tropics of the Pacific Ocean

Figure 15-28 A beach composed of large boulders because of the high wave energy.

Figure 15-29 The steep valley walls of the Norwegian fjords were produced by glacial erosion. (Photo courtesy of Norwegian Tourist Board.)

were of volcanic origin, Darwin proposed that fringing reefs originally developed around these volcanoes (Figure 15-31). Gradual subsidence of the ocean crust caused the volcanoes to sink until only a small island remained surrounded by barrier reefs. Upward growth of the reef structure is necessary to counteract the subsidence of the crust, because corals can live only in warm, shallow water. With continued subsidence, the eroded remnants of the volcano disappeared altogether, leaving only a circular atoll to mark its original position.

Sea-Level Changes

Far from being a static surface, the elevation of sea level has undergone major fluctuations in geologic time. One of the major causes for *eustatic* sea level changes, those that are synchronous throughout the world, is a change in the

Figure 15-30 Wave-cut terraces on the west side of San Clemente Island, California. (John S. Shelton.)

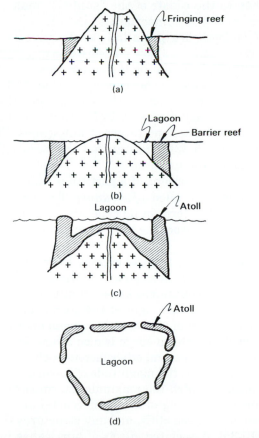

Figure 15-31 Evolution of atolls. (a) Fringing reefs form around an active volcano. (b) Reefs grow vertically upward to form barrier reefs as volcano subsides and is eroded. (c) Further subsidence leaves only reefs above sea level. (d) Map view of atoll.

mass of the polar ice caps. When polar ice caps expand and advance into the midlatitudes as they did repeatedly during the Pleistocene Epoch, sea level drops accordingly. During the last glaciation, sea level dropped an estimated 100 m. While sea level was lowered, a greater area of the continental shelves was exposed than at present. Rivers eroded valleys across these broad coastal plains underlain by unconsolidated sediment. Later, as glaciers retreated and released huge volumes of water to the oceans, the rising sea level inundated these newly formed valleys, creating estuaries. An irregular coastline in which estuaries extend up previous river valleys is sometimes referred to as a *drowned* coastline. Chesapeake Bay along the eastern coast of the United States is an example of a estuary that was formed by the drowning of a river valley during the postglacial sea-level rise.

COASTAL MANAGEMENT

Coastal Hazards

Throughout human history, a large percentage of the population of coastal nations has chosen to reside along the coasts. Unfortunately, various coastal processes have produced events of extreme magnitude that have inflicted great disasters upon coastal population centers. Today, the steady migration of the U.S. population to the coasts of the "sunbelt" makes an understanding of coastal hazards even more important.

Tropical cyclones are potentially destructive storm systems in which winds move around centers of low pressure in a circular fashion. In the Northern Hemisphere, wind circulation is counterclockwise. Regional names for intense tropical cyclones include hurricane, typhoon, and monsoon. These storms are generated over the ocean in the tropics and then migrate thousands of kilometers before they dissipate. In North and Central America, the Gulf Coast region is the prime land target of hurricanes. The map of part of this coastal area (Figure 15-32) indicates that the twentieth-century distribution of storms has been quite uniform.

The arrival of a hurricane at a particular coastline is preceded by a 1- to 1.5-m rise in water level called a *forerunner*. The forerunner is caused by swell waves produced by the storm in its generating area as much as several thousand kilometers distant from the coast at the time of arrival of the abnormally high swell.

Most damage from a tropical cyclone is produced by the *storm surge,* a rapid rise in water level generated by the winds of the storm as it approaches shore. Because of the counterclockwise circulation of winds, the storm surge reaches its maximum to the right of the storm center, where the winds are blowing in the direction of the storm's movement (Figure 15-33). The actual rise in water level in a storm surge is also a function of the tides at the time the storm strikes the shore and other factors such as the long-term rise in sea level (Figure 15-34). Thus the magnitude of storm damage depends on the time of day and month at which the maximum storm surge arrives. Effects of the storm surge include flooding of low-lying coastal areas and intense wave erosion of beaches, dunes, sea cliffs, and any structures that happen to be in the way (Figure 15-35). Associated effects of hurricanes include wind damage to structures and coastal flooding from the intense rainfall that occurs as the storm moves onshore.

Barrier islands are particularly susceptible to damage from large storms. In the absence of human modification, barrier islands are often narrow, low-

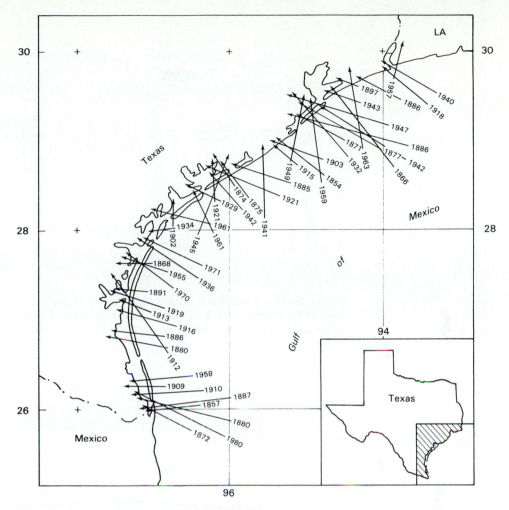

Figure 15-32   Map of hurricane landfalls along the Gulf Coast of Texas between 1850 and 1980. (From J. R. Giardino, P. E. Isett, and E. B. Fish, 1984, *Environmental Geology and Water Sciences 6*. Used by permission of Springer-Verlag, Inc., New York.)

lying ridges. During large storms, water can flow over the island to the lagoon, thus dissipating the energy of the storm waves. The erosion that accompanies these *washovers* is often intense enough to cut a new inlet through the island. Severe damage is inflicted to roads, bridges, buildings, and utilities in washover zones. An example of storm effects on barrier islands is shown in Figure 15-36.

Perhaps the most dreaded coastal hazard is the arrival of a wave or group of waves that can rise above the normal water level by 10 m or more. These waves are sometimes incorrectly referred to as tidal waves; however, they have nothing to do with astronomical tides. The actual cause of the waves is the rapid displacement of water by submarine earthquakes, volcanic eruptions, or landslides. Because of their common association with earthquakes they are called seismic sea waves or, from the Japanese term, *tsunamis*. Tsunamis move through the deep water of the ocean basins at velocities of several hundred meters per second. At these rates they can cross major oceans in a matter of hours. In deep water, tsunamis are barely detectable because their amplitudes reach only several meters. They are also characterized by wavelengths ranging

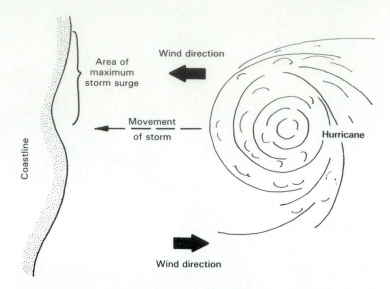

Figure 15-33 A hurricane approaching shore in the Northern Hemisphere will produce maximum storm surge to the right side of the storm center as viewed in the direction of movement of the storm.

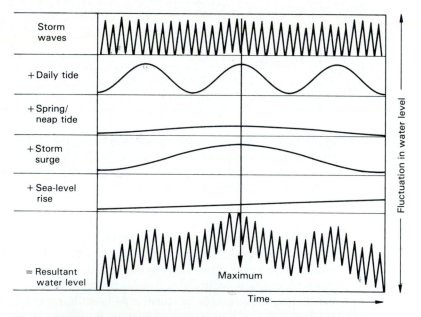

Figure 15-34 Dependence of water-level rise in storm surge on tides and other factors. (From R. Dolan and H. Lins, U.S. Geological Survey Professional Paper 1177-B.)

from 150 to 250 km and periods of 10 to 60 minutes. When tsunamis encounter shallow water, however, they undergo the same transformations as wind waves. The wave height increases to the point where the run-up is able to inundate nearshore areas and cause great damage (Figure 15-37). A tsunami generated during the eruption of the volcano Krakatoa in the Pacific in 1883 killed 36,000 people in Java and Sumatra. Catastrophies with similar losses of life at other times occurred in Japan and Portugal. In the Pacific Ocean a

Figure 15-35 Severe beach erosion by a storm. (Scott Hardaway; photo courtesy of Virginia Sea Grant at VIMS.)

Figure 15-36 Washover damage near Cape Hatteras, North Carolina. (R. Dolan; photo courtesy of U.S. Geological Survey.)

Figure 15-37 Damage at Seward, Alaska, from the tsunami generated by the 1964 Alaska earthquake. (NOAA/EDIS.)

large destructive tsunami has an average return period of about 10 years. Smaller tsunamis are generated several times per year in the Pacific.

Coastal Engineering Structures

Coastal engineering works date back to biblical times. Recently, attempts have been made to control and manage coastal processes on a larger scale. Some of these projects have failed because of a lack of understanding of coastal processes. Coastal systems tend at all times to establish a condition of equilibrium between form and process. Whenever natural processes are interfered with, adjustments throughout the system are a result. These adjustments often take the form of undesirable erosion or deposition of sediment.

Coastal engineering structures are designed to enhance or maintain navigation, to prevent erosion, or to promote deposition. *Jetties* are walls built perpendicular to the shoreline in pairs at the mouths of inlets (Figure 15-38). Their purpose is to prevent blockage of the inlet by littoral drift. In doing so, they may prevent the movement of sand by longshore currents past the jetty. This will cause deposition of sand along the up-drift side of the jetty and erosion of sand from the down-drift side.

Inlets serve as the location for the inflow and outflow of tidal currents. Therefore, the size and channel geometry imposed by jetties must be carefully designed to provide the proper velocity of tidal currents. If the channel opening is too large, deposition of sand and shoaling of the inlet channel occurs and dredging is required to maintain the necessary channel depth. If the opening

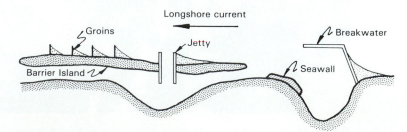

Figure 15-38 Types of coastal engineering structures.

is too small, the high current velocity causes scouring of the channel and potential undermining and failure of the jetties.

When adequate natural harbors are lacking, *breakwaters* are sometimes constructed (Figure 15-38). Breakwaters provide protection from waves, but they also interact with the longshore drift in the same manner as jetties. A classic example of this interaction is the Santa Barbara, California, breakwater. As illustrated in Figure 15-39, enlargement of the beach on the up-drift side of the breakwater occurred until longshore currents were able to transport sand along the breakwater and deposit it directly into the harbor. Dredging was then required to maintain the harbor. At the same time deposition was occurring, erosion of the beach was in progress on the down-drift side of the harbor. Eventually, shoreline dwellings were destroyed by wave erosion because the sand source for the beaches had been eliminated.

Detached breakwaters have also been constructed (Figure 15-40) in an attempt to provide protection from waves while at the same time allowing the passage of longshore drift. These structures usually fail to achieve their purpose because the protected area behind the breakwater becomes a low-energy zone for the longshore drift, and sand deposition ensues.

Seawalls are structures built parallel to the shore to prevent erosion and slumping of sea cliffs. Construction materials range from concrete to timber to riprap (Figure 15-41). Seawalls provide short-term protection against moderate events, but erosion continues on both sides of the protected zone.

Groins (Figure 15-38) are walls built perpendicular to the beach for the purpose of trapping sand from longshore drift. Although a beach can be con-

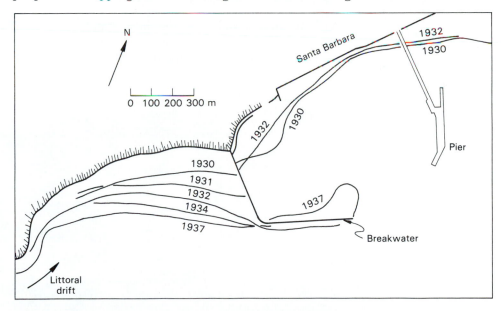

Figure 15-39 Patterns of erosion and deposition produced by the Santa Barbara breakwater. (After J. W. Johnson, 1957, *Journal of Waterways and Harbors* Div. 83, American Society of Civil Engineers.)

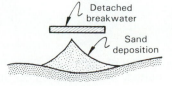

Figure 15-40 Construction of detached breakwaters usually results in sand deposition between the structure and shoreline.

Figure 15-41   Tiered seawall designed to prevent erosion along
the shore of Lake Michigan.

siderably widened by the construction of a groin, down-dift erosion is also
induced. This often necessitates a series of groins along a particular segment
of the coastline (Figure 15-42).

Implications for Coastal Development

With an understanding of the natural processes occurring along a particular
coastline, development that minimizes the potential for damage from hazardous processes can be implemented. Individual processes and the area that they

Figure 15-42   Aerial view of groin field along the Maryland coast.
(R. Dolan; photo courtesy of U.S. Geological Survey.)

may affect must be identified. When this is accomplished, zoning regulations can be utilized to prevent development of particularly hazardous areas. In some cases, development of marginal areas may be allowed if specific protective design criteria are met. In assessment of risks, however, the possibility of extreme events must not be neglected. The magnitude and frequency of these occurrences is a major uncertainty in any risk analysis.

In evaluation of coastline processes, historical records are of invaluable assistance. Maps and charts show changes in shoreline configuration during the period of historical record. Air photos are particularly useful in this respect.

The lessons of past coastal management projects must not be lost when considering future plans. In the following case study some of the problems of large-scale coastal management schemes are illustrated.

CASE STUDY 15-1
HUMAN INTERFERENCE WITH COASTAL PROCESSES

The coast of North Carolina is a depositional coast with a nearly continuous system of barrier islands (Figure 15-24). An illustrative comparison of altered and unaltered sections of these islands has been provided by Dolan, Godfrey, and Odum (1973).

A typical profile of a barrier island in its natural state is shown in Figure 15-43. This profile, consisting of a broad berm and low irregular dunes, is adjusted to the impact of large storms. Wave energy is efficiently dissipated by frequent overwash over the low natural dunes. Salt-tolerant vegetation species well adapted to periodic overwash inhabit the dunes and overwash terraces behind the dunes.

Although the barrier islands are well adjusted to extreme events that characterize their environment, frequent overwash is not com-

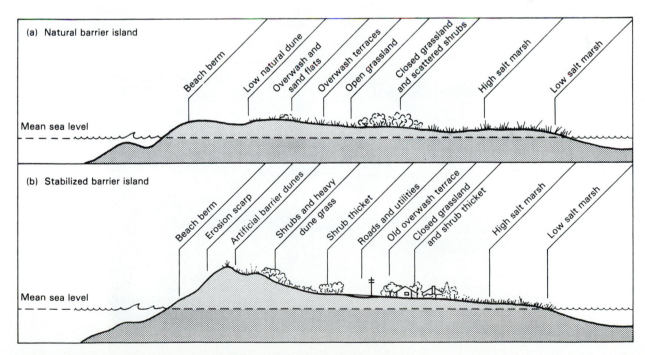

Figure 15-43 Cross-sectional profiles of barrier islands in their natural state (a) and after construction of barrier dunes (b). (From R. Dolan and H. Lins, U.S. Geological Survey Professional Paper 1177-B.)

patible with human occupation of the islands. To rectify this situation, nearly 1000 km of sand fence was constructed behind the beach berm in the 1930s. The purpose of the fences was to trap sand, thereby creating a system of high, artificial barrier dunes to protect the islands from overwash. Non-salt-resistant vegetation was introduced behind the dunes and fertilized in order to stabilize the dunes.

These measures accomplished their objective of making the islands safe for development. Unfortunately, the disturbance of the natural equilibrium state initiated serious long-term changes in the beach profile. Figure 15-43 shows the altered profile 30 years after construction of the artificial dunes. The major change has been the destruction of the original beach berm. The artificial barrier to wave overwash concentrated wave energy upon the beach. Beach widths decreased from an average of 150 m to less than 30 m. In some areas wave attack has removed the beaches altogether and has begun to destroy the artificial dune system. This process will continue in the future. When the artificial dune barrier is severed, overwash will again occur, only now roads and buildings located behind the dunes will be affected. The vegetation will die when flooded by saltwater.

The attempted stabilization of the North Carolina barrier islands is an excellent example of the effect of interference with coastal processes. The structures that were built behind the dunes intended to provide protection are now increasingly threatened by large storms. The cost of maintaining the developed areas will include the cost of building even more extensive and expensive protective structures.

CASE STUDY 15-2
WHERE NOT TO BUILD A SWIMMING POOL

Location of buildings in areas of active geologic processes can lead to destruction. The elaborate pool in Figure 15-44a extends too far out onto a beach along the Virginia coastline. A storm of moderate intensity in 1979 demonstrated the foolishness of such a site (Figure 15-44b).

(a)

Figure 15-44   (a) A structure that extends out onto a beach is vulnerable to storm damage. (b) The same house after a storm of moderate intensity. (C. Alston, photos courtesy of Virginia Sea Grant at VIMS.)

(b)

Figure 15-44 (*continued*)

SUMMARY AND CONCLUSIONS

The oceans support a great quantity and diversity of living organisms. Life on land is affected by the ocean's influence on climate. The circulation pattern of the ocean is composed of large gyres driven by atmospheric circulation that assist in heat transfer between the poles and equator.

Ocean-basin floors have a varied topographic configuration. Midoceanic ridges slope downward through regions of abyssal hills to sediment-covered abyssal plains. Pelagic sediments include inorganic muds as well as organic oozes composed of either calcium carbonate or silica. Calcium carbonate oozes are absent below the carbonate compensation depth.

The continental margins are covered with thick sedimentary sequences. The continental shelves slope gently seaward to a break in slope at the tops of the continental slopes. The continental slopes, along with the less steep continental rises, are characterized by turbidity-current transport and deposition of sediment.

The shoreline is a dynamic system in which the coastal landforms are adjusted to the waves, tides, and currents that interact with the nearshore sediments. The shoreline profile responds rapidly to changes affecting any component of the system. Waves begin as sea waves induced by storms over the ocean and travel toward the shoreline as more regular swell waves. As the circular orbits of wave motion begin to interact with the bottom, wave steepness increases until the waves become unstable and break. Although waves are refracted by differences in water depth, the oblique approach of waves initiates longshore currents that flow along the shoreline. The longshore drift of sand accompanying longshore currents is an important variable in the shoreline system.

The action of waves, tides, and longshore currents controls the depositional and erosional landforms that occur along a coastline. The beach responds quickly to changes in wave energy, developing different profiles under summer and winter conditions. Coastlines can be classified according to the dominant processes that have shaped their morphology.

The dynamic nature of coastal geologic processes makes coastal management difficult. The possibility of such extreme events as hurricanes and tsunamis is not always taken into consideration in coastal development. Coastal engineering structures are utilized in an attempt to control certain coastal processes, but the long-term effects of these structures often cause damage to the beach or coastline because of a lack of understanding of the way in which natural processes interact within the coastal system.

REFERENCES AND SUGGESTIONS FOR FURTHER READING

DOLAN, R., P. J. GODFREY, and W. E. ODUM. 1973. Man's impact on the barrier islands of North Carolina. *American Scientist* 61:152–162.

DOLAN, R., and H. LINS. 1985. *The Outer Banks of North Carolina.* U.S. Geological Survey Professional Paper 1177-B.

GIARDINAO, J. R., P. E. ISETT, and E. B. FISH. 1984. Impact of Hurricane Allen on the morphology of Padre Island, Texas. *Environmental Geology and Water Sciences* 6:39–43.

HAY, E. A., and A. L. MCALESTER. 1984. *Physical Geology: Principles and Perspectives,* 2d ed. Englewood Cliffs, N.J.: Prentice-Hall, Inc.

JOHNSON, J. W. 1957. The littoral drift problem at shoreline harbors. *Journal of Waterways and Harbors Division* 83:1–37, American Society of Civil Engineers.

JUDSON, S., M. E. KAUFFMAN, and L. D. LEET. 1987. *Physical Geology,* 7th ed. Englewood Cliffs, N.J.: Prentice-Hall, Inc.

KOMAR, P. D. 1976. *Beach Processes and Sedimentation.* Englewood Cliffs, N.J.: Prentice-Hall, Inc.

PRESS, F., and R. SIEVER. 1982. *Earth,* 3d ed. San Francisco: W. H. Freeman and Co.

PROBLEMS

1. Give a general overview of the topography of the ocean basins.
2. What is the cause and mode of travel of turbidity currents?
3. Why do waves break?
4. What effects could wave refraction have upon shoreline development?
5. Why is there such a large tidal range from place to place and over time at a particular place?
6. What are the engineering implications of longshore currents?
7. Describe the typical beach profile and explain how it changes throughout the year.
8. What are the hazards associated with tropical cyclones?
9. If you were asked to design an artificial harbor, what types of geologic, hydrologic, and oceanographic data would you need to gather before design and construction began?

CHAPTER
16
GLACIAL PROCESSES AND PERMAFROST

Those of us who live in the northern states of the United States or in high-latitude regions would be hard pressed to select a process more important in shaping our present landscape than glaciation. Although glaciers currently cover only about 10% of the earth's land surface, more than 30% of the surface was buried by ice during the repeated advances of continental glaciers during the Pleistocene Epoch (Figure 16-1). The importance of glaciers, however, goes well beyond the areas that actually were glaciated. The modern drainage system in many nonglaciated parts of the United States owes its origin to glaciation. For example, the evolution of the Mississippi Valley is intimately related to the advances and retreats of continental glaciers. During retreat, cataclysmic floods of meltwater were funneled through the Mississippi Valley into the Gulf of Mexico. In addition, the modern Mississippi would be a much smaller stream without the contribution of water from the Missouri River, which flowed north to the Hudson Bay before glaciation. The present valley of the Missouri was established by the diversion of meltwater streams along the margin of the ice sheet.

The great coastal cities of the world are dependent upon the current position of sea level for their economic existence. Any sustained climatic change involving advances or retreats of existing glaciers will either leave these ports high and dry or submerge parts of them. Currently, a rising trend in sea level is causing concern in many parts of the world.

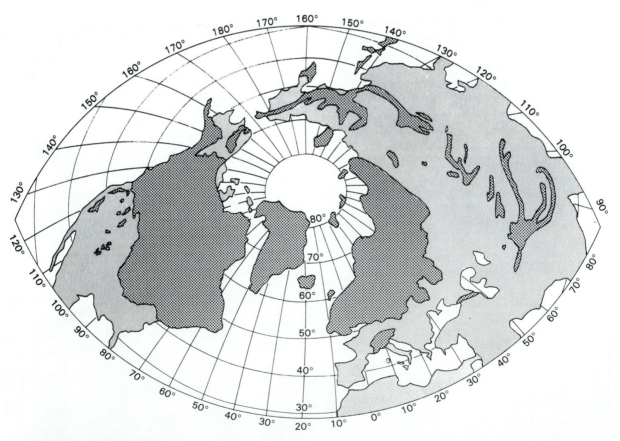

Figure 16-1 Extent of Pleistocene glaciations (darker areas) in the Northern Hemisphere. (After E. Antevs, 1929, Geological Society of America Bulletin 40.)

GLACIERS

Snowflakes fall to the surface of the earth in a variety of delicate, complex crystalline forms. On the ground, several processes begin to transform the new-fallen snow. Melting and sublimation (the direct conversion of ice to water vapor) occur. Compaction by new layers of snow reduces the volume and increases the density of the underlying material. Gradually, snow is converted to *firn*, a denser granular substance. Above an elevation called the *snow line*, some of each winter's snow remains after the summer melting season. If the net gain in snowfall is significant over a period of years, a growing body of firn and snow may become *a glacier*. A glacier is composed of an interlocking network of ice crystals formed by additional compression and recrystallization of firn. In addition, a glacier is sufficiently massive to flow downslope under its own weight or to have previously moved in this manner.

FIRN —

GLACIER = FIRN &
SNOW

Mass Balance

As a glacier flows downslope from the area in which there is a net accumulation of mass, it experiences progressively warmer temperatures, which result in a net loss of mass. Thus a glacier can be divided into two zones: a *zone of accumulation* and a *zone of ablation* (Figure 16-2). Ablation is a term that includes both melting and sublimation in the zone of net loss of mass. The boundary between the zones of accumulation and ablation is called the *equilibrium line*.

The mass balance of a glacier is the relationship between accumulation and ablation. This relationship determines the position of the glacier's *terminus* (Figure 16-2). When total accumulation exceeds total ablation, the terminus advances; conversely, the terminus retreats when ablation is greater than accumulation. A condition of equilibrium, or exact balance between accumulation and ablation, is indicated by a stable terminus. It must be emphasized, however, that a glacier in equilibrium may be in continual motion even though the position of the terminus remains stable. In this situation, the loss in mass from the ablation area is just compensated by the downslope flowage from the zone of accumulation.

SUMMER =
HIGH TEMP. &
MORE ABLATION

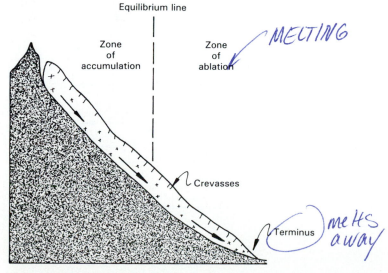

Figure 16-2   Profile of a valley glacier showing the components of its mass balance.

Movement

The flow of glaciers is a complex phenomenon involving a combination of several physical processes. One component of movement is derived from the visco-plastic internal deformation of ice, or the tendency of ice to flow downslope under its own weight when a sufficient thickness is achieved. Glacial flow involves elements of both viscous and plastic behavior (Chapter 6). In order for a plastic material to deform, a critical (yield) stress must be exceeded. The yield stress for ice is about 100 kPa (1 bar). When glacier ice is subjected to this magnitude of stress, which can be imposed by the downslope component of the weight of the glacier, the ice will flow.

The plastic flow of ice also contains an element of viscous behavior in which the flow velocity is proportional to the amount of shear stress. An equation called *Glen's Flow Law* approximates the visco-plastic flow of ice. This equation can be stated as

$$\dot{\gamma} = A\tau^n$$

Eq. 16-1

where $\dot{\gamma}$ is the strain rate, or velocity, of the ice, τ is the shear stress, and A and n are empirical constants. The coefficient A is temperature dependent, and the value of n ranges between 1.9 and 4.5. Like the flow of water in streams, the flow of ice is opposed by friction between the bed and the glacial ice. Therefore, velocity is greatest at the ice surface and decreases toward the bed (Figure 16-3). An important difference between glacier and stream flow, however, is that the flow of glacial ice is laminar rather than turbulent.

One complication of the flow process is caused by the brittle rupture of the upper portion of the glacier (Figure 16-2). Because of the lower pressure near the surface of the glacier, a network of vertical cracks called *crevasses* is often present (Figure 16-4), particularly where the glacier passes over a steeply sloping bedrock surface.

A second major component of glacier movement, which may take place along with visco-plastic flow, is *basal sliding*. The tendency of a glacier to slide as a block above its bed is dependent upon the thermal condition at the base of the glacier. Because the thermal conductivity of ice is low, a glacier several

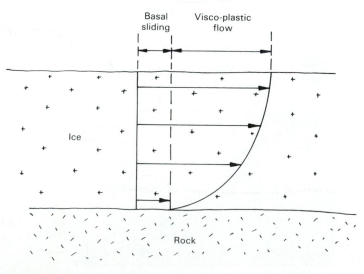

Figure 16-3 Velocity profile of a glacier showing the two components of motion.

Figure 16-4   Crevasses on the surface of a glacier. (M. R. Meier; photo courtesy of U.S. Geological Survey.)

kilometers thick acts as an insulating blanket to trap geothermal heat that is constantly flowing upward toward land surface. This heat may be sufficient to melt a small amount of ice at the base of the glacier. The melting point of ice is lowered as pressure increases, so ice will melt at a lower temperature at the base of a glacier than at the surface. Thus glaciers may be divided into two major types: *cold-based*, which contain no water at their beds, and *wet-based,* which do contain basal water. Cold-based glaciers are below the pressure melting point of ice throughout their vertical profile and are therefore frozen to their beds. For this reason, cold-based glaciers move by visco-plastic flow alone. Wet-based glaciers, on the other hand, are in effect decoupled from their beds by the presence of water between the ice and the bed. These glaciers may therefore slide upon their beds. Frictional heat produced by sliding provides an additional basal heat source. Typical estimates of basal sliding range between 10 and 25% of the total velocity of the glacier. Glaciers whose movement is intermediate between wet based and cold based are common. These include glaciers that are frozen to their beds in some areas and unfrozen in others. A common situation is to find a glacier unfrozen beneath its interior and frozen near its terminus.

An important aspect of glacial flow is the difference between accelerating and decelerating flow. Accelerating flow is common above the equilibrium line, where ice is thickening, and decelerating flow occurs below the equilibrium line, where the ice is either thinning or advancing upslope (Figure 16-5). A consequence of changes in flow velocity is that flow lines—imaginary lines oriented in the direction of movement at a particular point in the glacier—are not always parallel. Flow lines converge in areas of accelerating flow and diverge in areas of decelerating flow. At glacier margins, where the ice is thin and flow is decelerating, flow lines are directed upward toward the ice surface. This may cause thrusting along shear planes within the ice of active ice over stagnant or more slowly moving ice at the margin. Decelerating flow also provides a mechanism for bringing rock and sediment eroded from the bed upward into the ice and concentrating it at the glacier margin.

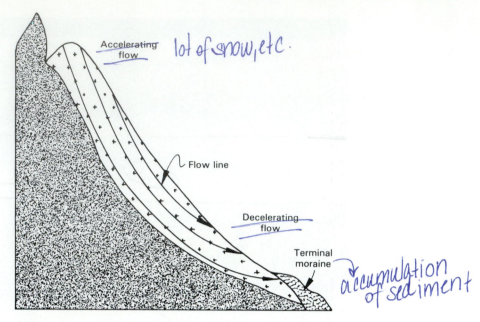

Accelerating flow *lot of snow, etc.*

Flow line

Decelerating flow

Terminal moraine *accumulation of sediment*

Figure 16-5 Flow lines converge in the accumulation area (accelerating flow) and diverge in the ablation zone (decelerating flow). Where flow lines curve upward near terminus, sediment is brought to glacier surface and concentrated by ablation.

Erosion —TRANSPORTED

AS GLACIER MOVES, IT PICKS ↑ SEDIMENT (EROSION)
-abrasion: scouring causing striations
-plucking: bedrock is picked up & inc.
-ice-thrusting: shearing of sediment

The amount of erosion accomplished by a glacier flowing over an area of land can vary from insignificant to deep scouring of the landscape. Variables that control glacial erosion are complex. They include the thermal condition of the glacier bed, the topography and lithology of the bed material, and many others. The processes of erosion, however, are relatively few in number.

The sliding movement of a wet-based glacier over a bedrock surface is unmistakably preserved by the polished and grooved bed evident after glacial retreat. Elongated grooves and other types of marks on a glaciated rock surface are an indication of glacial *abrasion.* The erosional marks vary from striations (Figure 16-6) on a single outcrop to grooves a meter or more in width and depth that can be traced for kilometers on the land surface. Abrasion is accomplished by rock fragments that are dragged along the bed rather than scour by the ice itself. This crushing and grinding action reduces the particle size of transported rock fragments. The marks produced by erosion are an excellent indicator of ice-movement direction. Abrasion is not effective in cold-based glaciers.

Incorporation of large blocks of bedrock into the ice is possible by a process known as *plucking.* Plucking is most prevalent in wet-based glaciers and occurs as the ice moves over bedrock obstacles that protrude above the surface of the bed (Figure 16-7). The process is controlled by the pressure distribution around the obstacle. On the up-glacier side, high pressures develop as the ice contacts the protrusion. If the pressure-melting temperature of the ice is reached, ice melts and flows around the obstacle to the region of lower pressure on the opposite side. Here the water refreezes around and within the joints and cracks of the rock. This weakens the rock mass, and larger fragments can be broken from the down-glacier rock faces and transported within the ice. Rock outcrops subjected to plucking during glaciation frequently have a characteristic shape. In profile, this shape consists of an abraded, gently sloping up-glacier—*stoss*—

Figure 16-6 Striations on bedrock produced by glacial erosion.
(N. K. Huber; photo courtesy of U.S. Geological Survey.)

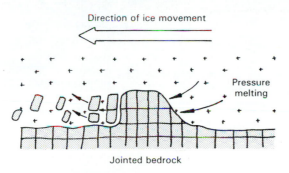

Direction of ice movement

Pressure
melting

Jointed bedrock

Figure 16-7 Plucking takes place when ice melts on the
upstream side of bed obstacles and refreezes within joints and
fractures on the downstream side. Large blocks can then be
incorporated into the flowing ice.

side, which terminates in a steeply sloping down-glacier, or *lee*, margin formed
by plucking (Figure 16-8).

In the marginal area of a glacier, large-scale erosion of rock or sediment
may occur by *ice-thrusting*, particularly when this part of the glacier is frozen
to its bed. (Figure 16-9). Shearing of the rock or sediment takes place below
the bed, mobilizing slabs of material up to tens of meters thick and kilometers
in length. These blocks then are thrust upward into the ice along the shear
planes that develop within decelerating flow zones. The ice-thrust blocks are
transported and deposited as isolated hills or ridges marking the former ice
margins (Case History 16-1).

Deposition

Drift is an umbrella term for all deposits associated with glaciers. These materials include sediment deposited directly by the ice itself as well as sediment
deposited on, within, beneath, or in front of the ice by streams or in lakes.
Glacial drift must be subdivided, however, in order to account for the various
depositional processes and characteristics of the resulting sediment (Table 16-
1).

Figure 16-8 Roches moutonnées, glacial landforms produced by abrasion and plucking. (G. K. Gilbert; photo courtesy of U.S. Geological Survey.)

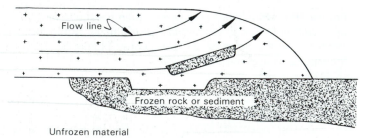

Flow line

Frozen rock or sediment

Unfrozen material

Figure 16-9 Ice thrusting takes place near the glacier terminus when the sediment or rock is frozen. Large slabs or blocks are carried up into the ice along shear planes oriented parallel to the diverging flow lines.

Poorly sorted, nonstratified sediment deposited directly in contact with glacial ice is called *till* (Figure 16-10). The lack of sorting and stratification in till, which contains particles ranging in size from clay to huge boulders, usually makes it easy to differentiate till from other sedimentary deposits. Till is the most common type of glacial drift; for example, it underlies most of the northern midwestern United States.

Till can be classified according to its depositional process; the differences in physical properties among the various types of till make their recognition a matter of engineering significance. In general, till can be deposited at the base or from the surface of a glacier (Figure 16-11). *Basal,* or *lodgement,* till

TABLE 16-1

Types and Characteristics of Glacial Drift *[handwritten: — ALL TYPES OF GLACIAL SEDIMENT]*

[handwritten left margin: POORLY SORTED]
[handwritten: till = "diamicton"]

Till: Nonsorted, nonstratified sediment deposited by glacial ice at the base, from the interior, or from the top of a glacier.

Basal (lodgement) till: Dense, compact till deposited at the base of a wet-based glacier. *[handwritten: STRONGER! LODGED...]*

[handwritten: JOINTS MAY WEAKEN & INC. PERM.] Ablation till: Till deposited as debris accumulates on top of a glacier and is gradually lowered as the ice melts from beneath. Tends to be less dense than basal till and more variable in grain size and sorting. *[handwritten: (MELTING)]* *[handwritten right: NO FORCE LIKE BASAL]*

Ice-contact stratified drift: Drift sorted and stratified by meltwater in contact with glacial ice. Bedding disrupted by subsequent melting. *[handwritten: @ terminus]*

Outwash: Coarse-grained sediment deposited by meltwater in front of a glacier in outwash plains or valley trains. *[handwritten: BETTER SORTED THAN TILL = GOOD AQUIFER!]*

[handwritten left: TILL ON BOTTOM] Lacustrine sediment: Deposits of glacial lakes, mostly fine grained and often varved. Coarse-grained deposits occur in fans and deltas associated with the lakes.

Eolian sediment: Wind-blown sediment derived from other forms of drift. Very well sorted; grain size ranges from sand to silt (loess). *[handwritten: B/C ONLY SILT-SIZE]*

Glacial-marine sediment: Mostly fine-grained, clay-rich sediment deposited by glaciers that advance to the coastline. Fine and coarse material dropped from calving ice and floating ice bergs.

[handwritten left margin: GLACIERS THAT TERMINATE IN OCEAN = MARITIME GLACIERS (form ICEBERGS)]

Figure 16-10 Exposure of glacial till. Notice poor sorting and lack of stratification. (D. R. Crandell; photo courtesy of U.S. Geological Survey.)

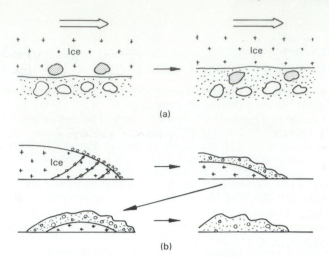

Figure 16-11 Deposition of (a) lodgement till and (b) ablation till. In (a), particles transported near the base of the sliding ice are separated from the ice and added to the bed material below because of frictional resistance to transport and pressure melting. Ablation till (b) originates from sediment accumulated on the glacier surface. It becomes thicker and is deposited as the ice melts from below.

is the name given to till deposited at the base of a glacier. Wet-based, sliding glaciers frequently contain a layer of debris-rich ice just above the bed. Deposition of single particles or particle aggregates from this zone can occur if the frictional resistance to movement between the particle and the bed is greater than the shear stress exerted by the glacier. The resulting lodgement till is a very densely compacted material because of the great weight of the overlying glacier.

Ablation till originates as debris is brought to the glacier surface by thrusting or flow along the upward-trending flow lines in the ablation area of the glacier (Figure 16-5). Debris is concentrated on the surface as the ice surface melts slowly downward. In the melting process, the debris is continually reworked by meltwater and mass movement (Figure 16-11). Debris flows are common as the saturated, clay-rich debris will flow down even gentle slopes. Deposition is not complete until all the underlying ice is melted. This depositional process results in a loosely compacted till that may be more sorted than basal till. These two types of till therefore have major differences in particle-size distribution, density, strength, and compressibility.

During the summer, abundant meltwater is produced at the surface of a glacier, particularly during the retreat of continental ice sheets in past glaciations. This water moves toward the ice margin either on top of, within, or beneath the glacier. Tunnels within or beneath the glacier frequently convey meltwater water to the terminus. Lakes on top of the ice may also be present from time to time. The sediment deposited by these streams or in ponded water is known as *ice-contact stratified drift*. This material is highly variable in grain size, sorting, thickness, and areal extent. In surface exposures, ice-contact stratified drift often can be recognized by folding, faulting, and other deformational structures produced during the melting of the glacier after deposition of the sediment. Isolated bodies, lenses, or beds of ice-contact stratified drift may be buried or interbedded with till in a glacial sequence. Location of these deposits is relevant to excavation, foundation design, and the successful siting of water wells.

Once the meltwater from the glacier reaches the terminus, it may either be ponded in *proglacial lakes* or flow along or away from the ice in meltwater streams. The sediment deposited in these environments is known as glacial-lacustrine sediment and outwash, respectively. When continental ice sheets retreated, large volumes of water were trapped in proglacial lakes. The largest of these in North America was glacial Lake Agassiz (Figure 16-12). Glacial-lacustrine sediment is mostly fine-grained, bedded silt and clay. Low strength and high compressibility are characteristic of these sediments. Glacial-lacustrine sediment often can be recognized by its *varved* appearance. Varves are thin, alternating laminations of clay and silt or sand (Figure 16-13). The dark-colored clay beds are thought to be deposited during the winter when the lakes are ice covered, and the coarser, light-colored silt or sand beds are deposited during the summer when glacial meltwater carries more sediment into the lake. Each varve pair, or couplet, therefore implies one year's sedimentation in the lake.

Figure 16-12 Map of glacial Lake Agassiz at its maximum extent. The dark areas are modern remnants of Lake Agassiz. (From J. P. Bluemle, 1977, The face of North Dakota, North Dakota Geological Survey Educational Series 11.)

Figure 16-13 Varved glacial lake deposits (light- and dark-colored banded unit). The large boulder at center was dropped from an iceberg floating on the lake.

Figure 16-14 Cross-bedded, coarse-grained outwash exposed in a gravel pit.

Glacial outwash (Figure 16-14) is alluvial sediment deposited by melt-water streams. Outwash is coarse grained and may be quite thick in valleys that carried glacial meltwater. Outwash is a good source of aggregate material for road construction, concrete, and other purposes. Outwash deposits are also highly permeable and may contain large quantities of groundwater.

Eolian sediment and *glacial-marine* sediment are less common but may be extremely important in some areas. Eolian sediment is derived from outwash and other types of glacial drift and then transported and deposited by winds. It is very well sorted and varies in grain size from sand to silt. Deposits of eolian silt are called *loess.* Glacial-marine sediment is deposited where glaciers terminate at the ocean. Large blocks of ice calve off from the glacier and float away. As they melt, debris sinks to the sea floor and is deposited as glacial-marine sediment. This material is clay rich and stratified but also contains the large particles that were incorporated into the ice.

VALLEY GLACIERS → INTERESTING LANDFORMS

Glaciers can be classified into two types: *valley glaciers,* which are confined to narrow valleys, and larger masses of ice that are not controlled by local topography. Except for several remaining ice caps in polar regions, most modern glaciers are valley glaciers in mountainous regions. During the Pleistocene, valley glaciers extended to much lower elevations and were much more common in the midlatitudes than at present. Today, extensive valley glaciers develop only in high latitudes in areas of abundant precipitation. These glaciers form networks analogous to river systems, with tributary glaciers joining larger glaciers that occupy major valleys (Figure 16-15).

A region that has contained valley glaciers can be recognized by characteristic landforms of both erosional and depositional origin. The most distinctive erosional landform is the shape of the glaciated valley itself. Glacial erosion modifies the cross-sectional profile of a valley from the V-shape common to valleys produced by stream erosion to a U-shape, characteristic of glaciated valleys (Figure 16-16). The U-shaped profile of glaciated valleys results in valley sides that are steeper than their counterparts in V-shaped stream valleys. Glaciated valleys are said to be oversteepened because the steep valley

[handwritten margin notes: —VALLEY —CONTINENTAL ICE-SHEETS (ICE-AGE) "LORENPLANCE" AROUND HERE]

Figure 16-15 Valley glaciers in the French Alps extending downslope from larger ice masses.

Figure 16-16 The U-shape of this valley in Yosemite National Park indicates that it was once occupied by a valley glacier. (Photo courtesy of Dexter Perkins, III.)

sides are prone to rockfall, rockslide, and other forms of mass wasting. These conditions may be hazardous for highway construction or other development in glaciated valleys.

In a glaciated mountainous area, valley glaciers extend, or once extended, downslope in all directions from the central region of highest elevation. In addition to *U*-shaped valleys, other erosional landforms are characteristic of these areas. The "bowl-shaped" depression at the base of a mountain peak where glaciers form and begin to flow down valley is called a *cirque* (Figure 16-17). Small lakes, or *tarns*, are common in the scoured floors of cirques that formerly contained glacial ice. The narrow ridge that separates two glaciated valleys often is quite steep and sharp. It is called an *arête,* a French word meaning "sharp edge." The glacial ice in a cirque erodes in a headward direction similar to stream systems and modifies the topography of the highest peaks in the process. A gap produced in a ridge by the headward erosion of glaciers in two

MORAINE

TARN—REMNANT WATER

arête— two cirques meet

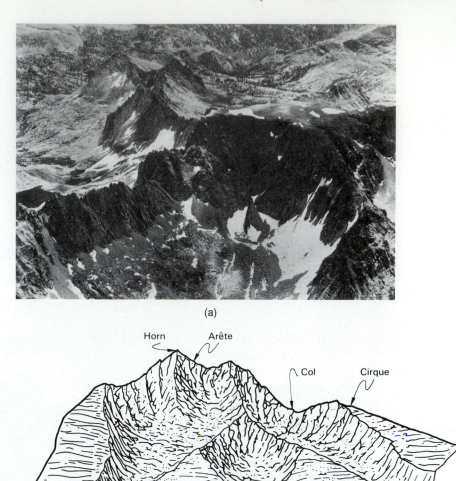

(a)

(b)

Figure 16-17 (a) Cirques and arêtes in the Sierra Nevada. (F. E. Matthes; photo courtesy of U.S. Geological Survey.) (b) Diagram of valley-glacier landforms. (From W. M. Davis, 1911, reproduced by permission from *Annals of the Association of American Geographers* 1:57.)

cirques from opposite sides of the ridge is a *col.* Where more than two glaciers erode headward at the base of a mountain peak, an isolated, sharply pointed pinnacle known as a *horn* is formed (Figure 16-17). The Matterhorn in the Alps of Switzerland is a horn.

The rock and sediment eroded from the cirques and valley walls by plucking and abrasion is transported down valley toward the glacier terminus and is deposited as drift in several types of landforms. A ridge of drift deposited by the glacier at its point of maximum advance is called a *terminal moraine* (Figure 16-18). Terminal moraines are usually composed of till and ice-contact stratified drift and may be deposited by subglacial or ablation processes. Frequently, meltwater emerges from the terminus in tunnels or from the glacier surface and deposits outwash in front of the glacier. These deposits are formed

Figure 16-18 A terminal moraine deposited by a valley glacier emerging from the U-shaped valley in the background. Faint channel markings can be seen on the gently sloping outwash plain in front of the moraine.

by braided fluvial systems because of the extremely high sediment load and may occupy the entire width of the valley bottom (Figure 12-20). In this case, the outwash deposit is known as a *valley train*. Valley-train sediments are composed of coarse sand and gravel and decrease rapidly in grain size with distance from the glacier (Figure 14-25).

As a valley glacier retreats, the terminal moraine may impound a proglacial lake (Figure 16-19). In addition, during a long period of net retreat, the glacier terminus may stabilize or even readvance short distances. The result may be a series of concentric *recessional* moraines in a glaciated valley.

A narrow ridge of drift is also deposited along the lateral edge of valley glaciers against the valley wall. The elevation of these *lateral moraines* is an indication of the maximum elevation of a particular glacier in the valley. In a series of active, coalescing valley glaciers, the lateral moraines from glaciers

Figure 16-19 A small proglacial lake impounded between a retreating valley glacier and its terminal moraine.

423

in tributary valleys can be traced as longitudinal debris bands in the glacier in the main valley. Upon the glacier's retreat, these bands of debris remain as ridges extending down the center of the valley. They are then classified as *medial moraines* because of their position on the valley floor.

CONTINENTAL GLACIERS

The distribution of deposits of Pleistocene glaciers in the Northern Hemisphere (Figure 16-1) indicates that huge continental glaciers that flowed south into the northern interior lowlands were much more widespread than valley glaciers. The portion of the United States covered by these immense ice sheets is shown in Figure 16-20. The ice sheet that covered most of Canada and the northern United States east of the Rocky Mountains is known as the Laurentide Ice Sheet. This glacier complex was as much as 4 km thick at its center in the Hudson Bay region.

The processes of erosion and deposition associated with continental ice sheets are similar to valley glaciers, although they vary tremendously throughout the extent of the glaciated area. Landforms and deposits present in any particular area are the result of the characteristics of the topography, bedrock material, and the glacier-bed thermal regime. For example, broad areas of the Canadian Shield, underlain by hard Precambrian igneous and metamorphic rocks, were subjected to only shallow erosion. Other regions, such as the basins of the Great Lakes, were deeply scoured into bedrock, and deposits of glacial drift tens of meters to more than 100 m thick blanket some parts of the plains of the United States and Canada.

Deposits of both lodgement till and ablation till are common. Thick accumulations of ablation till are common in terminal moraines and in other areas of decelerating flow where large amounts of sediment were sheared upward into the ice and concentrated at the glacier surface. The topography of

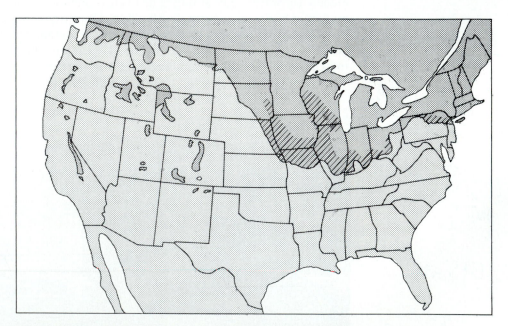

Figure 16-20 Extent of Pleistocene glaciation in the United States (darker area). (From R. F. Flint, 1945, *Glacial Map of North America*, Geological Society of America Special Paper 60.)

LANDSCAPE TOPOGRAPHY (IRREGULAR)

ablation-till deposits is called *hummocky* (Figure 16-21) because it consists of alternating mounds (hummocks) and depressions filled with lakes or swamps. Local relief can be as much as tens of meters in areas of thick ablation till.

Glacial meltwater deposits are concentrated in *outwash plains* (Figure 12-20), where braided fluvial systems transported sediment away from ice margins. Broad plains of outwash sediment often contain abundant resources of shallow groundwater. Unfortunately, these unconfined aquifers are highly susceptible to contamination.

Several types of continental glacial landforms deserve special mention. *Drumlins* are streamlined hills produced by wet-based, sliding glaciers (Figure 16-22). These hills, which occur in groups or fields (Figure 16-23), are the subject of more research and debate than perhaps any other glacial landform. Characteristics include an asymmetrical shape elongated in the direction of glacier flow (often called a streamlined shape) and a composition of till some-

Figure 16-21   High-relief hummocky topography resulting from deposition of thick ablation till.

Figure 16-22   A drumlin in Wayne County, New York. (G. K. Gilbert; photo courtesy of U.S. Geological Survey.)

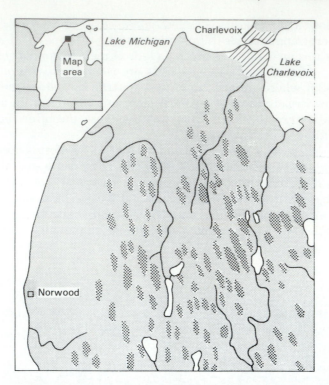

Figure 16-23 Map of drumlins in northwestern Michigan. Ice moved toward the south-southeast. (From F. Leverett and F. B. Taylor, 1915, U.S. Geological Survey Monograph 53.)

times deposited over a core of bedrock. The controversy surrounding the origin of drumlins may have arisen because there are multiple origins for these landforms. Drumlins may be formed by erosion, deposition, or a combination of both.

Another common glacial landform is the *esker* (Figure 16-24). An esker is a sinuous ridge composed of poorly sorted sand and gravel. These deposits

Figure 16-24 The Dahlen esker located in eastern North Dakota. (Photo courtesy of J. R. Reid.)

are formed by meltwater streams flowing in tunnels within or beneath a glacier. Because the stream channel is confined by ice walls and roof, the deposits remain as a raised mound after the ice melts. The sinuous, meandering shape of eskers is similar to subaerial stream channels.

ENGINEERING IN GLACIATED REGIONS

Glaciated regions present several types of problems relevant to construction and other types of land use. In general, these include differences in engineering properties among types of deposits, complex changes and variations in subsurface stratigraphy, and variations in depth to bedrock.

Of major importance are differences in engineering properties among tills, glacial-lacustrine, glacial-marine, and outwash deposits. Glacial-lacustrine deposits are clayey, compressible materials with low shear strength. The role of these materials in the collapse of the Transcona, Manitoba, grain elevator was described in Chapter 6. The low bearing capacity of glacial-lake sediments may require piles or other deep foundation types, for heavy structures. Piles are usually driven until they encounter a dense till or bedrock. Other engineering problems associated with glacial-lake deposits include settlement, shallow water tables, shrink-swell behavior, and instability of natural or constructed slopes.

Glacial-marine deposits are similar in texture to glacial-lacustrine sediments. The special slope-stability problems associated with these soils were discussed in Chapters 6 and 13. The Nicolet, Quebec, slide (Figure 13-11) is an excellent example of the construction hazards that must be considered.

The geotechnical properties and behavior of till are dependent upon its mode of deposition. Basal tills become dense and overconsolidated under the load of the overlying glacier. As a result, they tend to have favorable strength and compressibility in comparison with ablation tills. On the negative side, lodgement tills may be so dense and hard that excavation may require ripping or blasting.

An important characteristic of lodgement, and to some extent ablation tills, is that they often are observed to be jointed or fissured (Figure 16-25). In the case of lodgement tills, fissures may have been caused by unloading and rebound during and after the retreat of the glacier. Fissures have several important effects upon the engineering properties of the soil mass. First, fissures weaken the material. This can be shown by comparing the results of strength tests on samples of two sizes from the same construction site (Figure 16-26). The strength values measured from the larger samples are lower because of the greater number of fissures included in the sample. Fissures also influence the permeability of glacial deposits. Laboratory permeability tests conducted on small intact samples consistently yield lower values than field tests because of the absence of fissures. These discontinuities also provide preferred paths of movement for water through the deposit. A consequence of this condition is that tills may not be so favorable for waste-disposal sites as they may appear based on lab permeability tests.

The presence of isolated large boulders can sometimes prove troublesome in construction projects involving till. Difficulties associated with excavation, test drilling, and pile driving have been experienced. Boulders are often concentrated in *boulder pavements* between till units in sequences containing multiple tills.

Outwash deposits generally are composed of cohesionless materials of variable sorting and high permeability. They frequently constitute excellent

SORTING → VOID SPACES

Figure 16-25 Exposure of a joint in till. The surfaces of the joint are covered by manganese oxide.

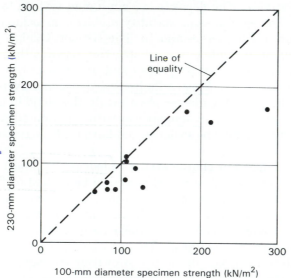

Figure 16-26 Effect of fissures on strength of till. Samples of two different sizes from identical materials were tested. Points plot below the line of equality because the smaller samples are stronger than the larger samples, which have more fissures. (From W. F. Anderson, "Foundation engineering in glaciated terrain." In *Glacial Geology: An Introduction for Engineers and Scientists,* edited by N. Eyles, copyright © 1983 by Pergamon Press, Ltd., Oxford.)

unconfined aquifers and favorable soils for irrigation if they occur at the surface. Outwash deposits are of great value as construction materials unless they contain excessive amounts of poor-quality material, such as shale. Therefore, states and provinces expend considerable effort in exploration for and evaluation of these deposits.

Sand and gravel deposits have moderate to high strength and low compressibility. Therefore, they provide high bearing capacities for foundations. Problems arise with high water inflow in excavations below the water table. Dewatering techniques (Chapter 12) may then be required. Glaciofluvial materials make very poor waste-disposal settings because of their high permeability and their potential utilization as sources of groundwater supplies.

THE PLEISTOCENE EPOCH

"ICE AGE"

The timing and causes of glaciations in the Pleistocene Epoch have been subject to much debate since the idea of a former great Ice Age was proposed by Swiss naturalist Louis Agassiz in the mid-nineteenth century. The beginning of the Pleistocene, the time at which late Cenozoic glaciations began, has been pushed back in time by evidence of older and older glaciations; it now stands between 1.5 and 3.0 million years, depending upon which particular chronology one accepts.

The most compelling evidence concerning the timing of glacial advances and retreats has been provided by studies of deep-sea sediment cores done in the past several decades. Deep-ocean floors are the only places on the earth that preserve a continuous sediment record through the Pleistocene. The type of evidence used to decipher Pleistocene climates is the oxygen isotope content of microfossils in the sediments. This method is based on the assumption that the $^{18}O/^{16}O$ ratio present in the shells secreted by shallow-water organisms is in equilibrium with the isotope ratio of seawater at the time of secretion. It is also known that the oxygen isotope ratio is temperature dependent. Thus ocean-bottom sediments contain a continuous record of temperature changes recorded in the oxygen isotope ratios of microfossils. The changes in mean surface seawater temperature between glacial and nonglacial periods are usually within several degrees Celsius. Fluctuations in oxygen isotope ratios thus indicate fluctuations in Pleistocene seawater temperature. These studies suggest that the Pleistocene Epoch was a period of time that fluctuated repeatedly between glacial and interglacial climates (Figure 16-27). Glacial periods tended to last 80,000 to 90,000 years, interspersed with warmer interglacial intervals 10,000 to 15,000 years in length. The Holocene Epoch, the last 10,000 years of geologic time in which we now live, may be simply another interglacial period to be followed in several thousand years by another glaciation.

The number of glaciations revealed by oxygen isotope studies is much larger than the traditional model developed by studies of glacial deposits on land. Figure 16-27 indicates that at least eight major glacial periods occurred during the last 700,000 years. In contrast, studies dating back to the nineteenth century have recognized four glacial periods. In the United States, these stages have been named the Nebraskan, Kansan, Illinoian, and Wisconsinan, listed in chronological order from oldest to youngest. The names are derived from the areas in which the deposits were first discovered or are most abundant. It is now realized that the concept of four glaciations is much too simple. The deposits of additional glaciations either could be present and inaccurately dated or could have been removed by erosion.

The debate as to why the earth's climate suddenly began to fluctuate wildly in late Cenozoic time is not fully resolved. Two major trends must be explained: a slow, gradual decrease in the earth's temperature during Cenozoic time and the climatic oscillations during the Pleistocene. The gradual decrease in temperature (Figure 16-28) may be related to the elevation and distribution of continents. Plate movements have resulted in a grouping of continents closer

Figure 16-27 Oxygen isotope variations during the last 800,000 years as measured in deep-sea sediments. Age in thousands of years on right, depth of core in centimeters on left. Periods or stages numbered in center. Curve moves to right of graph during interglacial periods and to left during glacial periods. (From N. Eyles, W. R. Dearman, and T. D. Douglas, "Glacial land systems in Britain and North America." In *Glacial Geology: An Introduction for Engineers and Scientists,* edited by N. Eyles, copyright © 1983 by Pergamon Press, Ltd., Oxford. Data from N. J. Shackleton and N. D. Opdyke, 1973, *Quaternary Research* 3.)

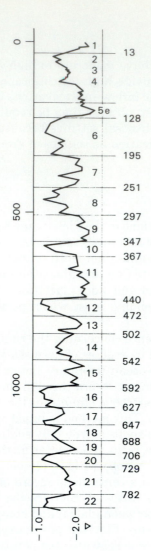

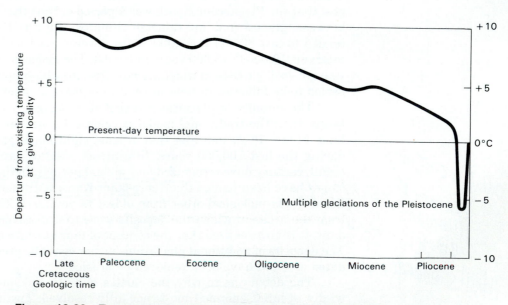

Figure 16-28 Temperature trends in Cenozoic time. The gradual decline prior to the Pleistocene may be due to plate interactions and movements. (From S. Judson, M. E. Kauffman, and L. D. Leet, *Physical Geology,* 7th ed., copyright © 1987 by Prentice-Hall, Inc., Englewood Cliffs, N.J.)

to the polar regions. In addition, these land masses are somewhat higher in elevation than at times in the past. These conditions would affect the transfer of heat by oceanic currents from the equator to the poles, thus cooling the polar regions, and also position the continents favorably for the formation and expansion of continental glaciers.

The fluctuations in climate during the Pleistocene can most likely be explained by periodic changes in the earth's rotation and revolution around the sun. These planetary movements affect climate by slight changes in tilt of the earth's axis or distance from the sun. A very close match has been obtained by comparing the periods of several types of orbital variations with the climatic changes indicated by deep-sea sediments. Thus it is reasonable to explain the glaciations of the Pleistocene Epoch by periodic changes in the earth's rotation superimposed on a gradually cooling climate caused by plate movements.

PERMAFROST

The development of petroleum resources in the North American arctic regions and the transport of oil southward through pipelines have required new concepts in construction in cold regions. The likelihood of continued resource development in these regions requires a better understanding of this unique environment.

Construction problems in arctic regions result from the presence of permanently frozen ground, or *permafrost.* The area of permafrost on the earth is extensive. Approximately 20% of the earth's land area is underlain by permafrost, including 50% of Canada and the U.S.S.R., and 85% of Alaska.

Extent and Subsurface Conditions

Permafrost areas can be classified as either continuous or discontinuous (Figure 16-29). Discontinuous permafrost can be identified by adjacent tracts of frozen and unfrozen ground. Continuous permafrost extends to great depths at high latitudes and gradually thins to the south (Figure 16-30).

Immediately below the land surface is the *suprapermafrost zone,* which is several meters thick and may contain pockets of material called *taliks* that never freeze, as well as ground that thaws during the summer and refreezes in the winter (Figure 16-31). This upper part of the suprapermafrost layer is called the *active zone.* At the base of the suprapermafrost layer, is the *permafrost table.* This is the upper boundary of the zone that remains frozen year-round, although taliks also may be present below the permafrost table.

Freeze and Thaw

Processes that shape the landscape in cold climates are called *periglacial* processes. These processes, which are dominated by freeze and thaw activity, occur in any cold climate whether or not a glacier is nearby. Therefore, areas of high elevation in the midlatitudes experience periglacial activity. During Pleistocene time, the extent of periglacial climate was much farther south than at present. Although the effects of past periglacial phenomena are relevant to engineering projects in the midlatitudes, we will focus our discussion on the regions of current permafrost. In addition to the direct effects of freeze and thaw, special types of mass wasting occur under periglacial conditions. These are discussed in the following section.

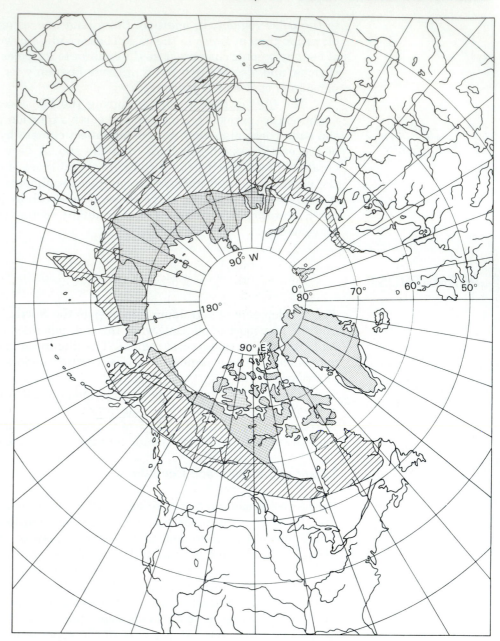

Figure 16-29 Extent of permafrost in the Northern Hemisphere. Stippled area represents continuous permafrost, and cross hatching indicates discontinuous permafrost. (From O. J. Ferrians et al., 1969, U.S. Geological Survey Professional Paper 678.)

The effects of ground freezing are extremely variable. Lithology of the soil is an important factor. Under appropriate conditions, isolated *ice lenses* form in the soil and begin to grow by drawing moisture by capillary movement toward the expanding ice mass. Soils of predominantly silt-size particle composition are most susceptible to ice-lens formation. Clay-rich soils retard the movement of soil water toward the ice lens, and coarse-grained soils contain pores that are too large for capillary movement. *Frost heave,* or uplift of the land surface, is most damaging in silty, frost-susceptible soils (Figure 16-32). Frost heave is a problem in any area where temperatures dip below freezing on a regular basis during the winter.

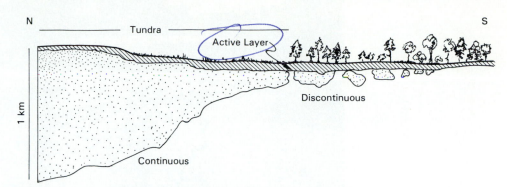

Figure 16-30 Distribution of permafrost along a northwest transect in northwest Canada. Numbers above the diagram represent degrees of latitude. (From N. Eyles and M. A. Paul, "Landforms and sediments resulting from former periglacial climates." In *Glacial Geology: An Introduction for Engineers and Scientists,* edited by N. Eyles, copyright © 1983 by Pergamon Press, Ltd., Oxford.)

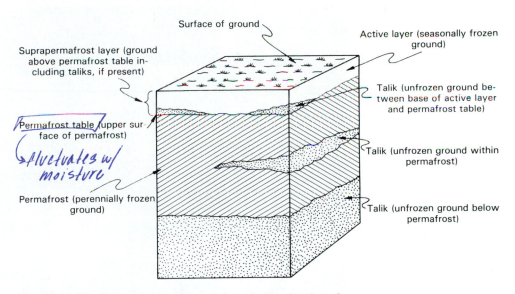

Figure 16-31 Subdivision of permafrost below land surface. (From O. J. Ferrians et al., 1969, U.S. Geological Survey Professional Paper 678.)

As the ground freezes in the cold season, vertical cracks may form by thermal contraction of the soil. These cracks, which extend below the bottom of the active zone, often form a polygonal pattern on the land surface. In the summer, meltwater from the active zone percolates downward into the crack and freezes. Over a period of years, large ice wedges can form by repeated frost cracking in the same location (Figure 16-33). The polygons at the surface are known as *ice-wedge polygons* (Figure 16-34). They are a good indication of permafrost conditions. Other characteristic permafrost landforms include *pingos* and *patterned ground*. Pingos are large circular or conical mounds, tens of meters high, that form by the growth of an ice core. Patterned ground de-

Figure 16-32 Cracking of a cement slab by frost heave. Notice bulging of the slab in the vicinity of the crack.

Figure 16-33 Ice wedge in soft sediment. (T. L. Péwé; photo courtesy of U.S. Geological Survey.)

velops by the sorting of surficial particles into coarse and fine fractions (Figure 16-35). The resulting shapes include polygons, circles, and stripes.

Mass Movement

A characteristic form of mass movement in permafrost regions is known as *solifluction*. During the summer, meltwater in the active zone is prevented from draining downward by the impermeable frozen soil below. Therefore, the soil within the active zone becomes saturated and flows as lobate shallow masses on very gentle slopes. The velocities of these solifluction lobes are

water that melts can't move down

"gelifluction"

Figure 16-34 Aerial view of ice wedge polygons in Alaska. (T. L. Péwé; photo courtesy of U.S. Geological Survey.)

Figure 16-35 Patterned ground formed by sorting of surficial material. Coarse particles outline the polygons.

variable, from imperceptibly slow to tens of centimeters per day. Internally, solifluction masses contain poorly sorted debris with a wide range of particle sizes.

Alpine regions contain moving masses of debris called *rock glaciers* (Figure 16-36), which form at high elevations in cirques or on slopes leading to mountain peaks. They consist of angular blocks of rubble that are produced by frost action on rock outcrops and accumulate on slopes by rockfall, snow avalanches, or other processes. Rock glaciers resemble true glaciers in form and movement. In addition to rock debris, they also contain ice, either as a core beneath the debris or as interstitial matrix between rock fragments.

Figure 16-36 An active rock glacier moving downslope. (F. H. Moffit; photo courtesy of U.S. Geological Survey.)

Engineering Problems in Permafrost Regions

Engineering problems in permafrost terrains result from unwanted thawing of frozen ground and frost heaving during seasonal freezing of the active zone. Problems are most severe in low-lying areas underlain by fine-grained soils. Upon thawing, these soils become water-logged and lose bearing capacity because the frozen soil beneath the active zone prevents drainage. A lesson learned from many mistakes is that permafrost exists in a fragile state of equilibrium. Vegetation insulates the frozen ground beneath, and if it is removed, thawing and depression of the permafrost table likely will result. Roads and railroads are constructed by building an insulating pad of non-frost-susceptible soil on the ground after vegetation has been removed. Failure to design roads correctly leads to expensive failures (Figure 16-37).

Buildings are often constructed upon insulated piles driven below the permafrost table. The bottom of the floor is elevated about a meter above land surface to allow for the circulation of air and dissipation of heat.

The trans-Alaska pipeline was perhaps the ultimate engineering project in an arctic region. Differential thaw settlement or frost heaving had to be minimized in order to prevent rupture of the pipe. A variety of soil and permafrost conditions were crossed by the pipeline. In an attempt to solve these

WHEN GROUND THAWS — STRENGTH & BEARING CAPACITY ↓

Figure 16-37   Disruption of railroad resulting from permafrost thaw after removal of vegetation. (O. J. Ferrians, Jr.; photo courtesy of U.S. Geological Survey.)

problems, a variety of designs was utilized. In areas of favorable soils, the line was buried. Where the soils were conducive to freeze and thaw movements, the pipeline was suspended above ground on pile foundations. In the future the lessons learned from the construction of the trans-Alaska pipeline no doubt will have to be applied to other engineering projects in the arctic regions of the world.

CASE STUDY 16-1
SUBSURFACE COMPLEXITY IN GLACIAL TERRAIN

Complex glacial stratigraphy beneath deceptively simple glacial landforms can cause frustrating and expensive foundation problems. Boston's historic Beacon Hill traditionally was interpreted as a drumlin. Under this assumption, soils composed mainly of till with favorable conditions for foundations and excavations were predicted. The site investigations for several large modern buildings, however, revealed a much more complicated subsurface stratigraphy (Kaye 1976). One of these structures was a garage constructed near the south end of the cross section shown in Figure 16-38. Till was expected to be present for the entire 13-m depth of the foundation excavation. Instead, subsurface folds containing gravel were encountered. Groundwater inflow from these gravels into the excavation required the installation of an expensive dewatering system and caused months

of construction delay. The construction of a large office building on the north side of the hill also was affected by inadequate subsurface information. The building was designed to rest upon piles driven to dense till beneath a surficial bed of clay. When test piles were driven, the surficial clay was found to extend to a much greater depth than anticipated because of folding and faulting of the section. The piles were driven 15 m beyond the predicted depth of the till unit with very little resistance to penetration.

Data gathered from these and other projects indicated that Beacon Hill is not a drumlin composed mostly of till. The hill is stratigraphically and structurally complex. Beneath a surficial layer of till, north-dipping slabs of sand, gravel, clay, and till are stacked against each other like books standing tilted to one side on

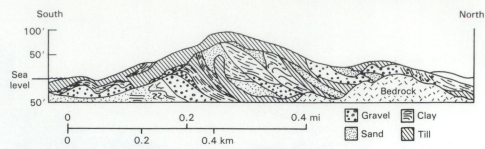

Figure 16-38 Cross section of Beacon Hill showing the complex structure and stratigraphy. Ice advanced from the north and transported frozen slabs of bed sediment, which were emplaced by ice thrusting at the glacier terminus near Beacon Hill. (From C. A. Kaye, "Beacon hill end moraine, Boston: A new explanation of an important urban feature." In Geological Society of America Special Paper 174, edited by D. A. Coates, 1976.)

a bookshelf. The depositional units are separated by thrust faults, and beds are intensely folded in some parts of the hill.

A reasonable explanation for the origin of Beacon Hill is that it consists of a series of individual thrust sheets emplaced by a glacier frozen to its bed near its terminus. Under this hypothesis, slices of frozen bed sediment were sheared into the ice and transported as intact slabs to the ice margin. There the blocks were thrust upward against other slabs of subglacial bed material. Therefore the landform is a type of terminal moraine rather than a drumlin. Construction experience in this area exemplifies the necessity for detailed subsurface site investigations in glaciated terrain.

SUMMARY AND CONCLUSIONS

An understanding of glacial processes and landforms is important because of the large area covered by glaciers during the Pleistocene Epoch. The expansion and contraction of a glacier is determined by its mass balance—the relationship between accumulation and ablation. The position of the terminus is controlled by this relationship even though the ice may be flowing downslope continuously. Flow occurs by visco-plastic deformation and basal sliding. Both types of movement occur in wet-based glaciers, whereas only visco-plastic flow is characteristic of cold-based glaciers.

Glacial erosion includes abrasion and plucking by sliding glaciers. Glaciers frozen to their beds can incorporate huge slabs of bed material into the ice by ice thrusting. Eroded rock and sediment are then transported and deposited at a later time. Deposition may take place beneath the ice to form lodgement till or above the ice to form ablation till. Other types of glacial drift are deposited by glacial meltwater streams as outwash or in proglacial lakes as lacustrine sediment.

Valley glaciers produce a characteristic set of erosional landforms in mountainous terrain. Chief among these is the U-shaped valley with over-steepened valley walls. Eroded sediment is then deposited in terminal, lateral, and medial moraines. Modern glaciers cannot compare with the huge continental ice sheets that buried the landscape repeatedly during Pleistocene time. The landforms and deposits of these glaciers are the substrate upon which all human activities take place in northern latitudes.

Each type of glacial deposit has unique engineering characteristics. Glacial outwash deposits may contain valuable resources of aggregate material and groundwater. Glacial lacustrine sediment and tills present varied construction conditions depending upon their method of deposition, thickness, stratigraphy, and many other factors. The contrast in engineering properties between lodgement and ablation till is particularly significant.

The factors that led to repeated glaciations extending into the midlatitudes during Pleistocene time are not fully known; however, it is thought likely that periodic perturbations in the earth's motion can explain the climatic cycles of cooling and warming. Long-term climatic changes were probably caused by lateral movement and elevation of the continents by plate interactions.

Permafrost causes severe engineering problems in cold regions over more than 20% of the earth's land surface. Most of these problems are related to instability of the active layer of the soil. Differential settlement and mass movement are common during thaw periods, and frost heave takes place during freezing of the active layer. These conditions require special design and construction procedures for almost all engineering projects.

REFERENCES AND SUGGESTIONS FOR FURTHER READING

ANTEVS, E. 1929. *Maps of the Pleistocene Glaciations.* Geological Society of America Bulletin 40:636.

BLUEMLE, J. P. 1977. *The Face of North Dakota.* North Dakota Geological Survey Educational Series 11.

DAVIS, W. M. 1911. The Colorado Front Range. *Annals of the Association of American Geographers,* v. 1.

EYLES, N. 1983. *Glacial Geology: An Introduction for Scientists and Engineers.* Elmsford, N.Y.: Pergamon Press.

FERRIANS, O. J., R. KACHADOORIAN, and G. W. GREENE. 1969. *Permafrost and Related Engineering Problems in Alaska.* U.S. Geological Survey Professional Paper 678.

FLINT, R. F. 1945. *Glacial Map of North America.* Geological Society of America Special Paper 60.

KAYE, C. A. 1976. Beacon Hill end moraine, Boston: New explanation of an important urban feature. In *Urban Geomorphology,* edited by D. A. Coates. Geological Society of America Special Paper 174, pp. 1–20.

JUDSON, S., M. E. KAUFFMAN, and L. D. LEET. 1987. *Physical Geology,* 7th ed. Englewood Cliffs, N.J.: Prentice-Hall, Inc.

LEVERETT, F., and F. B. TAYLOR. 1915. The Pleistocene of Indiana and Michigan and the history of the Great Lakes. U.S. Geological Survey Monograph 53.

PROBLEMS

1. Summarize the mechanics of glacier movement.
2. What effect does the condition of the base of the glacier have upon movement?
3. Contrast the different types of glacial erosion.
4. How is till deposited?
5. What types of sediment other than till are associated with glaciers?
6. What topographic features indicate whether or not a mountainous area was glaciated during Pleistocene time?

7. What construction problems can be expected in glaciated terrain?
8. What is the best explanation for Pleistocene glaciations?
9. What landforms suggest the presence of permafrost?
10. What is the most difficult problem associated with construction in arctic regions?

INDEX

A

Aa lava, 56
Ablation, glacial, 411
Accelerograph, 158–59
Advection, 288
Alaska earthquake:
  landsliding, 306
Alluvial aquifer, 282–83
Alluvial fan, 355
Amphibole group, 38 (*see also* Silicates)
Andesite, 53, 55, 62
Angle of internal friction, 129
Anticline, 209
Antidune, 342
Aphanitic texture, 50
Appalachian Mountains, 7
Aquifer:
  confined, 277–79
    flowing artesian, 279
  geologic setting, 282–86
    alluvial, 282–83
    coastal plain, 283
    fractured rock, 285
    glacial, 283
    tectonic valley, 282
  perched, 279
  unconfined, 277–79
Aquitard, 277
Artesian aquifer, 279 (*see* Aquifer, confined)
Aseismic creep, 164–65
Asthenosphere, 22
Atoll, 395–96
Atom:
  atomic number, 27
  electron, 26
  ion, 27
  isotope, 27 (*see also* Isotopes, radioactive)
  nucleus, 26
  neutron, 27
  oxidation number, 28
  proton, 27
Attenuation, of contaminants, 290
Atterberg limits, 140–41, 258
Augite, 68
Average linear velocity, of groundwater, 288

B

Baldwin Hills reservoir, 164–65
Barrier island, 391, 398, 405–6
Basalt, 53 (*see also* Volcanism)
Base flow, of streams, 272
Base level, 359
Batholith, 65, 106
Bauxite, 252
Bay of Fundy, tides, 385
Beach, 392–94
Beacon Hill, glacial geology of, 437–38
Bearing capacity, soil, 143–44
Bedding, 85–86
Bedforms, 342–43
Bingham substance, 126 (*see also* Flow)
Biotite, 43, 109, 115
  cleavage, 33
Bond, chemical:
  covalent, 28
  ionic, 28
Bonneville, Lake, 83
Bouguer anomaly, 196
Bowen's Reaction Series, 68–69
Braided stream, 282, 352–55
Breakwater, 403
Breccia:
  sedimentary, 80
  volcanic, 57
Brittle substance, 127–28
Buried valley aquifer, 283

C

Calcite, 46
Caldera, 63
Caliche, 252
Carbon 14, 12
Carbonate mineral group, 46
Cascade Mountains, 55, 62
Catastrophism, theory of, 2

Cation exchange, 42 (*see also* Groundwater, contamination)
Cementation, 79
Channelization, 363
Chelation, 252
Chezy formula, 343
Chinese earthquakes, 180
Chromium, in groundwater, 297
Cinder cone, 63 (*see also* Volcanism)
Classification tests, soils, 137
Clastic texture, 79
Clay:
    role in soil swelling, 255–57
Clay minerals:
    diffuse double layer, 147
    structure, 38–43
Cleavage:
    metamorphic rock, 108
    mineral, 32–34
Coastal engineering structures:
    breakwater, 403
    groin, 403–4
    jetty, 402–3
    seawall, 403
Coastal hazards:
    tropical cyclone, 398
    tsunami, 399
Coastal-plain aquifer, 283
Coasts, types of, 394–96
Cohesion, 130
Cohesionless soils, 137
Cohesive soils, 137
Colluvium, 305–6
Color:
    mineral, 35
    sedimentary rocks, 90
Colorado Plateau, 88
Columbia River Plateau, 60
Columnar jointing, 61
Compaction, 78
Composite volcano, 61
Compressibility, soil, 144–45
Compression index, 144–45
Compressive stress, 123
Cone of depression, 280
Confining pressure, 204
Conglomerate, 80
Consistency, soil, 139
Consolidation, 145–46
    test, 144
Constancy of Interfacial Angles, Law of, 31
Contact, geologic, 9
Continental drift, 200, 220–22
Continental glaciers, 425–27
Continental margins, 379–81
Continents, 23
Convection, in mantle, 190–91
Copper, 46
Correlation, 9
Crater, 59
Crater Lake, 64
Creep, 310
Crevasse:
    fluvial, 357
    glacial, 412–13
Cross bedding, 86
Cross-cutting Relationships, Principle of, 9
Cross sections, geologic, 209
Crystal, mineral, 31–33
Crystallization, magma, 68–69
Crystallization axes, 31

Crystal systems, 31
Curie point, 199

D

Dam construction, 294
    relation of geology, 225–27
Darcy's Law, 268–70
Darwin, Charles, 11 (*see also* Atoll)
Declination, magnetic, 199
Deere, D. U., 131
Deformation, rock, 204–6
    effect of confining pressure, 204
    effect of temperature, 204–5
    effect of time, 205
Deltas:
    deposition of, 356–58
    Mississippi, 358
Density, soil:
    in-place, 139
    relative, 139
Depositional environment, 76
    table, 79
Detrital rocks, 80
Dewatering, 295
Diamond, 28–29, 46
Diatreme, 65
Differentiation of earth, 19–21, 190
Dilatency model, 178
Diorite, 53
Direct shear test, 129
Discharge:
    groundwater, 272–75
    stream, 342
Dispersed fabric, 139
Divergent margin, 55 (*see also* Plate tectonics)
Dolomite, 46, 81
Drainage basin, 334–40
Drift, glacial, 415–20
Drumlin, 425
Ductile substance, 127–28
Dune, 87, 342

E

Earthquake prediction:
    animal behavior, 180
    dilatency model, 179
    precursors, 178–79
Earthquakes:
    causes, 155–58
    dilatency model, 178
    engineering, 174–77
    hazards, 164–68
    location, 168–70
    magnitude and intensity, 155–58
    precursors, 178–79
    prediction, 178–79
Effective stress:
    Principle of, 280
    role in slope movements, 319
Elastic limit, 127
Elastic material, 124–27

Elastic rebound theory, 155
Engineering:
    earthquake, 174–77
    in igneous rocks, 69–70
    in metamorphic rocks, 112
    in sedimentary rocks, 94
Engineering classification:
    of rock, 130–34
    of soil, 141–43
Ephemeral stream, 272
Erosion:
    in geologic cycle, 7
    in sedimentary environments, 76
    by water:
        control by contour farming, 242
        control by strip cropping, 242–43
        control by terracing, 242–43
        overland flow, 239
        raindrop erosion, 239
        rill erosion, 239
        sedimentation, 243–44
        sheet erosion, 239
        Universal Soil Loss Equation, 23–41
    by wind, 245–46
Erosion control, 242
Esker, 426
Evaporites, 84
Evapotranspiration, 275, 338
Evolution, Theory of, 11
Exfoliation, 233, 306
Expansive soils, 255–58

F

Fabric, soil, 139
Failure, rock, 127
Falls, 303
Faults:
    nomenclature, 216–17
    types, 215–19
Fault scarp, 165–66
Felsic rocks, 53
Feldspars:
    internal structure, 43
Ferromagnesian minerals, 53
Field Act, 182
Fissure eruption, 59
Fjord, 395–96
Flocculated fabric, 139
Flood basalts, 60
Flooding:
    control, 367–68
        at Rapid City, S.D., 370
        at Winnipeg, Manitoba, 371
    frequency, 366–67
    magnitude, 364–66
Flood plain, 283, 347, 350–64, 367–68
Flow:
    creep, 310–11
    debris flows, 311–12
Flow net, groundwater, 273–75
Focus, earthquake, 156
Folds:
    parts of, 210–12
    types, 209–10, 213–14
Foliation, metamorphic, 107
Fossils, 91
    assemblage, 10
    in relative age dating, 10

Fractional crystallization, 69
Fracture, mineral, 33–34
Frost action, 231
Frost heave, 432
Froude number, 341

G

Gabbro, 53
Geologic column, 11
Geologic cross section, 209
Geologic cycle, 6
Geologic hazards, 15–16
    frequency, 16
    magnitude, 16
    recurrence interval, 15
Geologic maps, 208–9
Geologic time:
    absolute, 11–14
    relative, 9–11
Geology and engineering, 14–19
    construction, 14–16
    natural resources, 16–18
    water resources, 18–19
Geophysics:
    gravity, 194–98
    heat flow, 190–94
    magnetism, 198–202
Geothermal energy, 194
Geothermal gradient, 102
Glacial aquifer, 283
Glacier:
    causes of glaciation, 429–31
    continental glaciers, 424–27
    deposition by, 415–20
    engineering in glaciated areas, 427–28
    erosion by, 414–15
    flow of, 412–14
    mass balance, 411–12
    valley glaciers, 420–24
Glassy texture, 52
Glen's Flow Law, 412
Gneiss, 109, 115
Gold, 46
Goldich's stability series, 236–37
Gradation, soils, 138
Graded bedding, 86 (see also Turbidity
    current)
Graded stream, 358–59
Grain flow, 315
Grain-size distribution, 138
Granite, 57, 66
Granitization, 67
Granodiorite, 53, 66
Graphite, 28–29, 46
Gravity:
    Bouguer correction, 196
    gravity anomalies, 196–97
    gravity meter, 196
    isostasy, 197–98
    isostatic rebound, 198
    measurement of, 195–96
Great Salt Lake, 83
Gros Ventre slide, 149
Groundmass, 52
Ground shaking, earthquake, 167
Groundwater:
    aquifers, 277–86
        cone of depression, 280

Groundwater (*continued*)
  confined, 277–79
    flowing artesian, 279
    Principle of Effective Stress, 280
    production of water from, 279–82
    unconfined, 277–79
  capillary fringe, 272
  and construction, 294–95
  contamination, 286–94
  Darcy's Law, 268–70
  discharge, 272–75
  flow net, 272–73
  flow of, 268–77
  natural quality, 286–87
  recharge, 272–75
  springs, 277
  waste disposal, 291–94
  water quality standards, 286–88
  water table, 270
Guatemalan earthquake, 177–78
Gypsum, 46, 81

H

Half-life, 11
Halide group, 45
Hardness, mineral, 34–35
  Mohs hardness scale, 35
Hardness, water, 286
Harold D. Roberts Tunnel, 115
Hawaiian Islands, 55
Heat flow:
  differentiation of earth, 190
  geothermal energy, 194
  thermal conductivity, 190
Hebgen Lake earthquake, 165
Helical flow, 350
Hematite:
  streak, 35
High Plains aquifer, 295–96
Hjulstrom diagram, 345
Hornblende, 38, 57, 69
Hutton, James, 3, 76
Hydration, 236
Hydraulic conductivity, 123, 270, 277, 292
Hydraulic gradient, 269
Hydraulic head, 268
Hydrocompaction, 259–60
Hydrodynamic dispersion, 288
Hydrolysis, 236
Hydrothermal solution, 103

I

Ice lenses, 432 (*see also* Permafrost)
Idaho batholith, 66
Igneous rocks:
  color and composition, 52–55
  in geologic cycle, 6
  texture, 50–52
Illite, 43
Inclination, magnetic, 199
Index properties, soil, 136–40
Infiltration capacity, 275

Intensity, earthquake, 158–59
Intrinsic permeability, 122–23
Intrusive processes:
  crystallization of magma, 68–69
  pluton types, 65–67
Isostasy, 197–98
Isostatic rebound, 295
Isotopes, radioactive, 11
  table, 12

J

Jetty, 402–3
Joints, 214–15

K

Kaolinite, 40
Kyanite, 104

L

Laccolith, 65
Lahar, 58
Lateral spread, 306–7
Laterite, 252
Lava:
  aa, 56
  pahoehoe, 56
Lava tube, 56
Leachate, 288
Leaning Tower of Pisa, 145–46
Limestone, 81 (*see also* Subsidence)
Liquefaction, of soils, 261
Lithification, 6, 78
Lithosphere, 22
Load, stream, 346–47
Long Island, N.Y., groundwater
    contamination, 296–97
Longshore bar, 393–94
Longshore currents, 388–89
Longshore drift, 389
Love Canal, 287
Lower Van Norman Dam, 167, 182
Luster, 35–36

M

Magma, 6, 120
Magma chamber, 52
Magma, crystallization of, 68
Magnetism:
  Continental Drift theory, 200
  Curie point, 199
  declination, 197
  inclination, 199

paleomagnetism, 199–200
 and sea-floor spreading, 200–201
Magnitude, earthquake, 159–60
Mafic rocks, 53
Manning equation, 343
Mantle plumes, 55–56
Marble, 109, 112
Martinique, 57
Mass movement:
 causes, 316–23
 complex movements, 312–16
  rock-slide-debris avalanche, 313–15
 falls and topples, 303
 flows, 309–12
 lateral spreads, 306–9
 slides, 303–6
 slope stability, 323–28
Mass spectrometer, 12
Meandering stream, 282, 348–52
Mechanical analysis, soils, 137
Metamorphic aureole, 103
Metamorphic grade, 103
Metamorphic rock, 6, 112
Metamorphism:
 cleavage, 108
 contact, 103
 dynamic, 106
 foliation, 107
 metamorphic rocks, 112
 regional, 103–6
Meteorites, age of, 12
Micas, 38–43
Midoceanic ridges, 55, 59, 376
Migmatite, 106, 115
Mineral groups:
 carbonates, 46
 halides, 45
 native elements, 46
 oxides, 45
 sulfates, 46
 sulfides, 45
Minerals:
 cleavage, 32–34
 crystal, 31–33
 crystal systems, 31
 internal structure, 26–30
 mineral groups, 36–46
 physical properties, 32–36
  table, 39
 unit cell, 26–30
  x-ray diffraction, 29
Modified Mercalli Scale, 158–59
Modulus of elasticity, 125, 131
Modulus ratio, 131
Mohr-Coulomb equation, 130
Monitoring well, 292
Montmorillonite group, 42–43
Mount Pelée, 57
Mount St. Helens, 55, 62
 Case Study, 70–72
Mud cracks, 90
Mudstone, 80
Muscovite, 43, 69
 cleavage, 33

N

Native elements, mineral group, 46
Neptunism, Theory of, 2

Nevados Huascaran, rock-slide-debris
 avalanche, 314–15
Niagara Falls, 340
Nicolet, Québec, lateral spread, 309
Normal fault, 216
Nuée ardente, 57

O

Obsidian, 52, 54
Ocean basins, 23, 376–79
Oceanic currents, 376
Olivine, 30, 37, 53, 68
Original Continuity, Principle of, 2
Original Horizontality, Principle of, 2
Orthoclase, 43, 53
Outwash, 419–20
Overbreak, 116
Overconsolidated clay, 317
Overland flow, 239, 275
Oxidation, 236
Oxides, 45

P

Pahoehoe lava, 56
Paleomagnetism, 199–200
Palmdale bulge, 178–79
Partial melting, 62
Particle shape, 139
Payline, 116
Pedalfer, 254
Pedocal, 255
Perennial stream, 272
Period, seismic waves, 175
Permafrost:
 engineering problems, 436
 frost heave, 322
 ice lenses, 432
 rock glacier, 435
 solifluction, 434
Permeability (table), 122
Petrified wood, 91
Petroleum, 120–21
Phaneritic texture, 51
Phenocryst, 52
Phreatic zone, 271–72
Phyllite, 109
Plagioclase, 43, 47, 53, 68
Plasticity, soil, 141–43
Plastic material, 124–27
Plate tectonics:
 asthenosphere, 22
 development of theory, 223–24
 lithosphere, 22
 plates, 22
Pleistocene Epoch, 429–31
Pluton, 65
Point bar, 350
Porosity, 120–22
Porphyritic texture, 52
Precambrian, 11
Precursors, earthquake, 178–79
Principal stress, 124
Principle of Effective Stress, 281

Pyrite, 45
Pyroclastic material, 56
Pyroxene, 38, 53

Q

Quartz, 43–44, 53, 69
Quartzite, 109, 115
Quick clay, 306

R

Radioactive waste disposal, 94–99, 294
Radon gas, 179
Rational equation, 366
Recharge, groundwater, 272–75
Recurrence interval, 366
Reynold's number, 341
Rhyolite, 53, 56
Richter scale, 159
Rill erosion, 239
Rip current, 388
Ripple marks, 90
Ripples, 87, 342
Rivers:
    deposition, 347–58
        alluvial fans, 355
        braided streams, 352–55
        deltas, 356–58
        meandering streams, 348–52
    development of drainage networks, 338–
        40
    equilibrium, 358–63
        graded stream, 358–60
        terraces, 360–61
    flooding, 364–68
    hydrologic budget, 337–38
    stream hydraulics, 340–44
    stream order, 336–37
    stream sediment, 344–47
Rock glacier, 435
Rock mass:
    classification, 136
    properties of, 135–36
    Rock Quality Designation, 136
    slope stability of, 318
Rock Quality Designation, 136
Rocky Mountains, 7

S

Safety factor, 316
St. Francis Dam, 112–13
St. Venant model, 126
Salt dome, 96
Salt weathering, 232–33
San Andreas fault, 167, 170, 179, 217–18
San Fernando earthquake, 181–85
San Francisco earthquake, 154, 170–71
Sanitary landfill, 288, 292
Sawtooth batholith, 66
Schist, 109, 115
Sea-floor spreading, 200–201, 222–24
Sea-level changes, 396–98
Sedimentary rocks:
    characteristics of, 79–93

classification of (table), 80
in geologic cycle, 6
origin of, 76–79
Sedimentary structures, 85
Sedimentation, 243–44
    and reservoir capacity, 246–47
Seismic attenuation, 170
Seismic waves, 160–64
    P waves, 160
    surface waves, 160
    S waves, 160
Seismograph, 157
Sensitivity, soil, 139
Settlement, soil, 144–45
Shale, 80
Shear strength, 128–30
    soil, 142–43
Shear stress, 123–29
Sheet silicates, 38 (see also Micas, Clay
    minerals)
Shelterbelts, 246
Shield volcano, 61
Shiprock, 65
Shorelines:
    beaches, 392–94
    longshore currents, 388–89
    morphology, 389–92
    sea-level changes, 395–98
    tides, 385–88
    types, 394–96
    waves, 381–85
Silica content, 53
Silicates, 37–45
Silica tetrahedron, 37
Silicic rocks, 53
Sillimanite, 104
Silver, 46
Sinkhole, 81, 264–65
Slate, 108
Slaty cleavage, 108
Slides, 303–6
Slip, along fault, 155
Slope stability:
    analysis, 325–26
    preventative and remedial measures, 326–
        28
    recognition of unstable slopes, 323–24
Slump, 304
Smectite group, 42–43
Smith, William, 10, 76
Snake River plateau, 60
Soil:
    bearing capacity, 143
    classification, 254–56
    clays and clay properties, 147–48
    consistency, 139
    consolidation, 144–45
    engineering classification, 141–43
    engineering properties, 136–47
    fabric, 139
    hazards, 255–61
        expansive soil, 255–59
        hydrocompaction, 259–61
        liquefaction, 261
    horizons, 250–51
    index properties, 136–40
    profile, 250–51
    sensitivity, 139
    settlement, 144
    shear strength, 142–43
    soil-forming factors, 252–53
Soil-moisture active zone, 257
Solar system:
    origin of, 19

solar nebula, 19
   terrestrial planets, 19
Solid-solution series, 30, 37–38
Solifluction, 435
Sorting, sedimentary rocks, 85
Specific gravity, mineral, 36
Spheroidal weathering, 233
Spit, 391
Springs, 277
Steno, Nicolaus, 2
Stock, 65
Strain, 124–27
Stratigraphy, 91–93
Streak, mineral, 35
Stream order, 336–37
Stream patterns, 334–35
Strength, 128–30
Stress:
   compressive, 123
   shear, 123–24
   tensile, 123
Strike and dip, 207–8
Structures, geologic:
   folds, 209–14
   fractures, 214–20
   measurement of, 207–9
Subduction zone, 224
Submarine canyons, 380
Subsidence, 261–65
Sulfate mineral group, 46
Sulfide mineral group, 45
Sulfur, 46
Superposition, Principle of, 2
Syncline, 210
System, rock, 4

T

Tectonic valley aquifer, 282
Tensile stress, 123
Terraces, stream, 360–61
Texture:
   igneous:
      aphanitic, 50
      glassy, 52
      phaneritic, 51
      porphyritic, 52
      vesicular, 52
      table, 51
   sedimentary, 79–85
Thalweg, 348
Thermal conductivity, 190
Thrust fault, 216
Tides, 385–88
Till, 416–17
Time-distance relations, 162–164
Topples, 303
Transcona grain elevator, 148
Transform fault, 224
Transportation, sediments, 76
Travertine, 46, 82
Triaxial test, 204
Tropical cyclone, 398
Tsunami, 166, 399
Tuff, 47
Tunneling, 295
Turbidity current, 88, 380–81

U

Unconfined compression test, 128
Unconfined compressive strength, 130
Ultramafic rocks, 57
Unconformity, 6
Unified Soil Classification System, 141–43
Uniformitarianism, Principle of, 4, 76
Universal Soil Loss Equation, 239–41
Uranium isotopes, 11

V

Vaiont Dam, and slide, 305
Valley glaciers, 420–24
Vesicles, 52
Vesicular texture, 52
Viscous material, 124–27
Viscosity, 125
Void ratio, 120–22
Volcanic ash, 56
Volcanism:
   eruptive products, 55–56
   and plate tectonics, 55–56
   remnant landforms, 63–65
   types of eruptions, 59–63

W

Water, role in slope movements, 318
Water quality standards, 286–87
   table, 288
Water table, 270
Waves, 381–85
Weathering:
   chemical:
      hydration, 236
      hydrolysis, 236
      oxidation, 236
      role of CO_2, 235
   and engineering, 70, 238
   mechanical:
      frost action, 231–32
      moisture changes, 233
      salt weathering, 232–33
      temperature changes, 233
   and sedimentary rocks, 76
   spheroidal, 233
   stability, 236–38
Wegener, Alfred, 220
Welded tuffs, 57
Wentworth scale (table), 81
Wind erosion, 245–46

X

X-ray diffraction, 29

Y

Yellowstone National Park, 46, 82

UNIFIED SOIL CLASSIFICATION SYSTEM

Major Divisions			Group Symbols[a]	Typical Names	Field Identification Procedures (excluding particles larger than 75 mm and basing fractions on estimated weights)
1	2		3	4	5
Coarse-grained soils — More than half of material is *larger* than No. 200[b] (75 μm) sieve size.	Gravels — More than half of gravel fraction is larger than No. 4 sieve size (4.75 mm)	Clean gravels (little or no fines)	GW	Well-graded gravels, gravel sand mixtures, little or no fines.	Wide range in grain sizes and substantial amounts of all intermediate particle sizes.
			GP	Poorly graded gravels, gravel-sand mixtures, little or no fines.	Predominantly one size or a range of sizes with some intermediate sizes missing.
		Gravels with fines (appreciable amount of fines)	GM	Silty gravels, gravel-sand-silt mixtures.	Nonplastic fines or fines with low plasticity (for identification procedures see ML below).
			GC	Clayey gravels, gravel-sand-clay mixtures.	Plastic fines (for identification procedures see CL below).
	Sands — More than half of coarse fraction is smaller than No. 4 sieve size (4.75 mm) (for visual classification, 5 mm may be used as equivalent to the No. 4 sieve size)	Clean sands (little or no fines)	SW	Well-graded sands, gravelly sands, little or no fines.	Wide range in grain sizes and substantial amounts of all intermediate particle sizes.
			SP	Poorly graded sands, gravelly sands, little or no fines.	Predominantly one size or a range of sizes with some intermediate sizes missing.
		Sands with fines (appreciable amount of fines)	SM	Silty sands, sand-silt mixtures.	Nonplastic fines or fines with low plasticity (for identification procedures see ML below).
			SC	Clayey sands, sand-clay mixtures.	Plastic fines (for identification procedures see CL below).

The No. 200 sieve size is about the smallest particle visible to the naked eye.

Major Divisions			Group Symbols	Typical Names	Identification Procedures on Fraction Smaller than No. 40 Sieve Size		
					Dry strength (crushing characteristics)	Dilatancy (reaction to shaking)	Toughness (consistency near PL)
Fine-grained soils — More than half of material is *smaller* than No. 200 (75 μm) sieve size.	Silts and clays — Liquid limit less than 50		ML	Inorganic silts and very fine sands, rock flour, silty or clayey fine sands or clayey silts with slight plasticity.	None to slight	Quick to slow	None
			CL	Inorganic clays of low to medium plasticity, gravelly clays, sandy clays, silty clays, lean clays.	Medium to high	None to very slow	Medium
			OL	Organic silts and organic silty clays of low plasticity.	Slight to medium	Slow	Slight
	Silts and clays — Liquid limit greater than 50		MH	Inorganic silts, micaceous or diatomaceous fine sandy or silty soils, elastic silts.	Slight to medium	Slow to none	Slight to medium
			CH	Inorganic clays of high plasticity, fat clays.	High to very high	None	High
			OH	Organic clays of medium to high plasticity, organic silts.	Medium to high	None to very slow	Slight to medium
Highly organic soils			Pt	Peat and other highly organic soils	Readily identified by color, odor, spongy feel, and frequently by fibrous texture.		

After U.S. Army Engineer Waterways Experiment Station (1960), "The Unified Soil Classification System," *Technical Memorandum* No. 3-357, Appendix A, Characteristics of Soil Groups Pertaining to Embankments and Foundations, 1953, and Appendix B, Characteristics of Soil Groups Pertaining to Roads and Airfields, 1957; and A. K. Howard (1977), "Laboratory Classification of Soils—Unified Soil Classification System," *Earth Sciences Training Manual* No. 4, U.S. Bureau of Reclamation, Denver, 56 pp.

[a]Boundary classifications: soils possessing characteristics of two groups are designated by combinations of group symbols. For example: GW-GC well-graded gravel sand mixture with clay binder.

[b]All sieve sizes on this chart are U.S. Standard.